M.J. Bertin
A. Decomps-Guilloux
M. Grandet-Hugot
M. Pathiaux-Delefosse
J.P. Schreiber

Pisot and Salem Numbers

1992 Springer Basel AG

Authors' addresses

Dr. M.J. Bertin
Université Pierre et Marie Curie
Mathématiques
4 place Jussieu
F-75252 Paris Cedex 05

Dr. A. Decomps-Guilloux
Université Pierre et Marie Curie
Mathématiques
4 place Jussieu
F-75252 Paris Cedex 05

Prof. M. Grandet-Hugot
Université de Caen
Mathématiques
Esplanade de la Paix
F-14032 Caen Cedex

Dr. M. Pathiaux-Delefosse
Université Pierre et Marie Curie
Mathématiques
4 place Jussieu
F-75252 Paris Cedex 05

Prof. J.P. Schreiber
Université d'Orléans,
Château de la Source
B.P. 6749
F-45067 Orléans Cedex

A CIP catalogue record for this book is available from the Library of Congress, Washington D.C., USA

Deutsche Bibliothek Cataloging-in-Publication Data

Pisot and Salem numbers / M. J. Bertin... - Basel; Boston: Berlin; Birkhäuser, 1992

NE: Bertin, Marie-José

Originally published by Birkhäuser Verlag Basel

Printed from the authors' camera-ready manuscripts on acid-free paper in Germany
ISBN 978-3-0348-9706-8 ISBN 978-3-0348-8632-1 (eBook)
DOI 10.1007/978-3-0348-8632-1

CONTENTS

Chapter 4
Generalities concerning distribution modulo 1 of real sequences

Chapter 5
Pisot numbers, Salem numbers and distribution modulo 1

Chapter 6
Limit points of Pisot and Salem sets

Chapter 7
Small Pisot numbers.

Chapter 8
Some properties and applications of Pisot numbers

Chapter 9
Algebraic number sets

Chapter 10
Rational functions over rings of adeles

Chapter 11
Generalizations of Pisot and Salem numbers to adeles

Chapter 12
Pisot elements in a field of formal power series

Chapter 13
Pisot sequences, Boyd sequences and linear recurrence

Chapter 14
Generalizations of Pisot and Boyd sequences

Chapter15
The Salem–Zygmund theorem

PREFACE

The publication of Charles Pisot's thesis in 1938 brought to the attention of the mathematical community those marvelous numbers now known as the Pisot numbers (or the Pisot–Vijayaraghavan numbers). Although these numbers had been discovered earlier by A. Thue and then by G. H. Hardy, it was Pisot's result in that paper of 1938 that provided the link to harmonic analysis, as discovered by Raphaël Salem and described in a series of papers in the 1940s. In one of these papers, Salem introduced the related class of numbers, now universally known as the Salem numbers.

These two sets of algebraic numbers are distinguished by some striking arithmetic properties that account for their appearance in many diverse areas of mathematics: harmonic analysis, ergodic theory, dynamical systems and algebraic groups.

Until now, the best known and most accessible introduction to these numbers has been the beautiful little monograph of Salem, *Algebraic Numbers and Fourier Analysis*, first published in 1963. Since the publication of Salem's book, however, there has been much progress in the study of these numbers. Pisot had long expressed the desire to publish an up-to-date account of this work, but his death in 1984 left this task unfulfilled.

Fortunately, a group of his students, the authors of this book, have taken up the challenge of completing Pisot's work. This book organizes and makes accessible to a wide audience much material previously found only in the research literature. It is a fitting tribute to those mathematicians Charles Pisot and Raphaël Salem whose work laid the foundation of so much of the work described here, and it should provide the starting point for those who, it is to be hoped, will penetrate even further into the mysteries still surrounding these numbers.

David W. Boyd
Vancouver, Canada

INTRODUCTION

The present volume, dedicated to the memory of Charles Pisot (1910–1984), is intended to be an expression of the warm esteem of his students.

At the end of a meeting of the *Séminaire de Théorie des Nombres*, Charles Pisot once evoked the long path he had followed since his thesis, in which he introduced that *remarkable set of algebraic integers*, usually denoted by S, and which he alone refused to call Pisot numbers. It seemed to him that the time had come to make a synthesis of all the results that had been obtained over more than forty years. In view of its size, such an undertaking necessarily required a collective effort, so that Pisot wished the work to be the result of our group. We set to work, detailed the plan of the book, and had already begun to write the text when Charles Pisot left us, but his ambition remained.

Pisot's method of proving the algebraicity of a real number is that of *generating functions*: one associates to every real a certain power series which is then proved to represent a rational function; hence the importance of criteria of rationality.

The study of the set S leads to that of certain families of rational functions (Chapters 1, 2 and 3); it also requires a knowledge of the *distribution modulo 1* of certain sequences (Chapter 4).

Chapters 5, 6, 7 and 8 are wholly devoted to the study of the sets S (Pisot numbers) and T (Salem numbers): properties concerning the distribution modulo 1, accumulation points, small elements, applications; in view of their importance, an entire chapter (Chapter 15) is devoted to applications in harmonic analysis. Most of the results discussed were obtained between 1938 and 1985, the older ones being due to Pisot and Salem.

Chapters 9 and 10 begin the study of generalizations, which is continued in the following chapters: generalizations to adeles (Chapter 11) and to formal power series (Chapter 12).

In his thesis Pisot also introduced various sequences of rational integers related to the sets S and T. These sequences (called Pisot sequences), together with their generalization (Boyd sequences), are discussed in Chapter 13. Chapter 14 discusses their extensions to adeles and formal power series.

CHAPTER 1

RATIONAL SERIES

In this chapter we denote by A an integral domain with quotient field K, by $A[X]$ (resp. $K[X]$) the ring of polynomials in one variable with coefficients in A (resp. K) and by $A[[X]]$ (resp. $K[[X]]$) the ring of formal power series with coefficients in A (resp. K). A for instance will be the ring $\mathbf{Z}$ and K the field $\mathbf{Q}$.

A formal power series F, F belonging to $K[[X]]$, is called a rational series or a rational fraction if and only if there exist two polynomials P and Q in $K[X]$ such that Q is invertible in $K[[X]]$ with $F = PQ^{-1} = P/Q$.

We recall that a formal power series in $K[[X]]$ is invertible if and only if the coefficient of X^0 is non-zero. Let F be an element of $K[[X]]$; we will denote by $F(0)$ the coefficient of X^0 of F.

The aim of this chapter is to determine criteria of rationality, i.e., necessary and sufficient conditions on the coefficients of the series F for F to be a rational series. These criteria will be used in the following chapters.

If the ring A has no special property, the criterion will be algebraic, but if A can be extended to a complete field for a particular absolute value (for instance $\mathbf{Z}$ to $\mathbf{C}$ or $\mathbf{C}_p$), we obtain more general criteria, which can be interpreted in terms of analytic functions.

We end the chapter with the generalized Fatou's lemma, which says that in certain rings, if F is a rational series in $A[[X]]$, then there exist P and Q in $A[X]$ such that $F = P/Q$, $Q(0) = 1$.

1.1 Algebraic criteria of rationality

Proposition 1.1. *A formal series $F = \sum_{n\in\mathbf{N}} a_n X^n$, an element of $K[[X]]$, is a rational series if and only if there exist two integers s and n_0 and $s+1$ elements of K denoted $q_0, q_1, \ldots, q_s$ with $q_0 \neq 0$, such that for all $n \geq n_0$ the coefficients a_n satisfy the relation*

$$q_0 a_n + q_1 a_{n-1} + \cdots + q_s a_{n-s} = 0. \qquad (*)$$

Proof. If F is a rational series, it can be written as $F = \sum_{n\in\mathbf{N}} a_n X^n = P/Q$ with $Q = \sum_{0\leq i\leq s} q_i X^i$, $P \in K[X]$, $Q \in K[X]$, $d^\circ P = r$.

From the formal series identity $QF = P$, we can deduce the relation $(*)$ by equating coefficients for $n \geq n_0 = \sup(r+1, s)$. Conversely if we set $Q = \sum_{0\leq i\leq s} q_i X^i$, the relations $(*)$ show that $QF = P$ with $d^\circ P \leq n_0 - 1$, and that F is a rational series.

Definition 1.1. Let $F = \sum_{n\in\mathbf{N}} a_n X^n$ be an element of $K[[X]]$. We call Kronecker determinants of F the determinants $D_n(F)$ defined by

$$D_n(F) = \begin{vmatrix} a_0 & a_1 & \cdots & a_n \\ a_1 & a_2 & \cdots & a_{n+1} \\ \vdots & \vdots & \ddots & \vdots \\ a_n & a_{n+1} & \cdots & a_{2n} \end{vmatrix}.$$

Theorem 1.1.1. *A formal series F of $K[[X]]$ is a rational series if and only if the Kronecker determinants $D_n(F)$ are zero starting from a certain index.*

Proof. Necessary condition: Let $F = \sum_{n\in\mathbf{N}} a_n X^n$ be a rational series of $K[[X]]$. According to Proposition 1.1, there exist an integer n_0 and $s+1$ elements of K: $q_0, q_1, \ldots, q_s$ such that

$$q_0 a_n + q_1 a_{n-1} + \cdots + q_s a_{n-s} = 0 \qquad \text{for } n \geq n_0.$$

Hence for $n \geq \sup(n_0, s)$ the last column of $D_n(F)$ is a linear combination of the previous s columns because $q_0 \neq 0$ and $D_n(F)$ is zero for $n \geq \sup(n_0, s)$.

Sufficient condition: Assume that there exists $n_0 \in \mathbf{N}$ such that $D_n(F) = 0$, $\forall n \geq n_0$. If $n_0 = 0$ then $F \equiv 0$; otherwise let s be an integer in $[1, \ldots, n_0]$ such that $D_{s-1}(F) \neq 0$ and $D_n(F) = 0$, $\forall n \geq s$. By the choice of s, the last column of $D_s(F)$ is a linear combination of the previous ones; hence there exist s elements of K $q_1, \ldots, q_s$ such that $a_n + q_1 a_{n-1} + \cdots + q_s a_{n-s} = 0$ for $s \leq n \leq 2s$.

Denote by v_n the element of K defined by $v_n = a_n + q_1 a_{n-1} + \cdots + q_s a_{n-s}$; we verify by induction on n that v_n is zero for $n \geq 2s$.

Assume that $v_n = 0$ for $s \leq n \leq m-1$ with $m - 1 \geq 2s$; then $D_{m-s}(F) = 0$, i.e.,

$$D_{m-s}(F) = \begin{vmatrix} a_0 & \cdots & a_{s-1} & a_s & \cdots & a_{m-s} \\ \vdots & \ddots & \vdots & \cdots & \ddots & \\ a_{s-1} & \cdots & a_{2s-2} & a_{2s-1} & \cdots & a_{m-1} \\ \vdots & \ddots & \vdots & \vdots & \ddots & \vdots \\ a_{m-s} & \cdots & a_{m-1} & a_m & \cdots & a_{2m-2s} \end{vmatrix} = 0.$$

Denote by $A_0, A_1, \ldots, A_{m-s}$ the columns of this determinant. For $s \leq j \leq m-s$, replace the column A_j by the column $A_j + q_1 A_{j-1} + \cdots + q_s A_{j-s}$. The relations $v_j = 0$ for $s \leq j \leq m-1$ induce the equalities

$$D_{m-s}(F) = \begin{vmatrix} a_0 & \cdots & a_{s-1} & 0 & \cdots & 0 \\ \vdots & \ddots & \vdots & \vdots & \ddots & \vdots \\ a_{s-1} & \cdots & a_{2s-2} & 0 & \cdots & 0 \\ a_s & \cdots & a_{2s-1} & 0 & \cdots & v_m \\ \vdots & \ddots & \vdots & \vdots & \ddots & \vdots \\ a_{m-s} & \cdots & a_{m-1} & v_m & \cdots & v_{2m-2s} \end{vmatrix} = \mp D_{s-1} v_m^{m+1-2s}.$$

Hence $v_m = 0$ and $v_m = 0$ for $n \geq$s. According to Proposition 1.1, F is a rational series. ∎

The following criterion concerns the special case $A = \mathbf{Z}$.

Notation. If r belongs to $\mathbf{N}$, $\mathcal{L}_r$ denotes the set of strictly increasing sequences of $r+1$ elements of $\mathbf{N}$. If L_r belongs to $\mathcal{L}_r$, L_r is denoted $L_r = (l_0, l_1, \ldots, l_r)$; we then have $l_i \in \mathbf{N}$, $\forall i \in [0, \cdots, r]$, and $0 \leq l_0 < l_1 < \cdots < l_r$.

The sequence $(0, 1, \ldots, r)$ is denoted I_r.

Let $F = \sum_{n \in \mathbf{N}} a_n X^n$ and $T = \sum_{n \in \mathbf{N}} t_n X^n$ be two elements of $K[[X]]$. We set, for $(m, n) \in \mathbf{N}^2$, $x_{m,n} = \sum_{0 \leq i \leq m} \sum_{o \leq j \leq n} t_i t_j a_{m+n-(i+j)}$.

Then the matrix $A(T, L_r, F)$ is defined by the equality

$$A(T, L_r, F) = \begin{pmatrix} x_{l_0,0} & x_{l_0,1} & \cdots & x_{l_0,r} \\ x_{l_1,0} & x_{l_1,1} & \cdots & x_{l_1,r} \\ \vdots & \vdots & \ddots & \vdots \\ x_{l_r,0} & x_{l_r,1} & \cdots & x_{l_r,r} \end{pmatrix}.$$

We denote the matrix $A(1, L_r, F)$ by $H(L_r, F)$.

We then get $\det H(I_r, F) = D_r(F)$ (Definition 1.1).

Remark. If we define the matrix T_r by the equality

$$T_r = \begin{pmatrix} t_0 & 0 & \cdots & 0 \\ t_1 & t_0 & \cdots & 0 \\ \vdots & \vdots & \ddots & \vdots \\ t_r & t_{r-1} & \cdots & t_0 \end{pmatrix}$$

an easy calculation shows that

$$A(T, I_r, F) = T_r\, H(I_r, F)\, T_r^t.$$

Hence $\det A(T, I_r, F) = t_0^{2(r+1)}\, D_r(F)$.

Theorem 1.1.2. *Let $F = \sum_{n\in\mathbf{N}} a_n X^n$ be an element of $\mathbf{Z}[[X]]$ such that $a_0 \neq 0$. We assume that there exist a series $T = \sum_{n\in\mathbf{N}} t_n X^n \in \mathbf{C}[[X]]$ such that $t_0 = 1$, and an integer $r > 0$ such that*

$$|\det A(T, L_r, F)| < 1 \qquad \forall L_r \in \mathcal{L}_r.$$

Then F is a rational series.

Proof. We fix F and T. For simplicity we set

$$A(T, L_r, F) = A(L_r), \quad D_r(F) = D_r, \quad H(L_r, F) = H(L_r).$$

According to the hypothesis $|\det A(I_r)| < 1$. Hence $|\det H(I_r)| < 1$, since $t_0 = 1$. As $\det H(I_r)$ is an integer, $\det H(I_r) = D_r = 0$.

Let s be an integer $\leq r$ such that $D_n = 0$ for $n \in [s, \ldots, r]$ and $D_{s-1} \neq 0$. We have $s \geq 1$ because $a_0 \neq 0$. The equality $D_s = 0$ and the inequality $D_{s-1} \neq 0$ imply that there exist s rationals $q_1, q_2, \ldots, q_s$ such that

$$v_n = a_n + q_1 a_{n-1} + \cdots + q_s a_{n-s} = 0 \qquad \forall n \in [s, \ldots, 2s].$$

We wish to show that $v_n = 0$ for any $n \geq s$. We proceed by induction on the integer n, and distinguish two cases: $2s < n \leq r + s$, and $n > s + r$.

a) $2s < n \leq r + s$. The proof is the same as that of Kronecker's criterion, because $D_s, D_{s+1}, \ldots, D_r$ are zero.

b) $r + s < n$. We assume again that $v_n = 0$ for $n \in [s, s+1, \ldots, m-1]$ when $m - 1 \geq r + s$, but we no longer have the hypothesis that $D_{m-s} = 0$ as in a). We therefore introduce a matrix K_m of order $r + 1$ defined by $K_m = H(0, 1, \ldots, s-1, m-r, \ldots, m-s)$; we will show first that $\det K_m = 0$ and then that v_m is zero.

We denote by W_i the element of $\mathbf{Z}^{r+1}$

$$W_i = (a_i, a_{i+1}, \dots, a_{i+r}).$$

The rows $W_s, W_{s+1}, \dots, W_{m-r-1}$ that do not appear in the matrix are linear combinations of $W_0, W_1, \dots, W_{s-1}$. In the matrix K_m we replace successively W_i by $W_i' = \sum_{l=0}^{i} t_l W_{i-l}$ for $i = m-s, \dots, m-r, \dots, 0$.

The matrix K_m' thus obtained has as rows

$$W_i' = \left(\sum_{l=0}^{l=i} t_l a_{i-l}, \sum_{l=0}^{l=i} t_l a_{i+1-l}, \dots, \sum_{l=0}^{l=i} t_l a_{i+r-l} \right),$$

and $\det K_m' = \det K_m$ since $t_0 = 1$.

Let $V_0', V_1', \dots, V_r'$ be the columns of the matrix K_m'; we replace V_j' by

$$V_j'' = \sum_{k=0}^{k=j} t_k V_{j-k}'$$

successively for $j = r, r-1, \dots, 0$. We then obtain the matrix $A(0, 1, \dots, s-1, m-r, \dots, m-s)$. Hence $|\det K_m| < 1$; which implies $\det K_m = 0$ since $\det K_m$ belongs to $\mathbf{Z}$.

We end the proof as in a); we replace successively the columns A_j of K_m by $A_j' = \sum_{l=0}^{l=s} q_l A_{j-l}$ for $s \le j \le r$; we get a matrix whose determinant is equal to $(-1)^k v_m^{r+1-s} D_{s-1}$. Hence $v_m = 0$. ∎

1.2 Criteria of rationality in C

Notation. Let f be an analytic function in the neighborhood of zero. We denote by $S(f)$ the Taylor expansion of f about zero.

Let H^p where $p \in \mathbf{R}^{*+}$ (resp. H^0) denote the set of functions f analytic on $D(0,1)$ such that

$$\lim_{r \longrightarrow 1} \frac{1}{2\pi} \int_{-\pi}^{+\pi} |f(re^{i\theta})|^p d\theta < +\infty$$

$$\text{(resp.} \quad \lim_{r \longrightarrow 1} \frac{1}{2\pi} \int_{-\pi}^{+\pi} \log^+ |f(re^{i\theta})| d\theta < +\infty).$$

In particular for $p = 2$ and if $S(f) = \sum_{n \in \mathbf{N}} a_n z^n$, f belongs to H^2 if and only if $\sum_{n \in \mathbf{N}} |a_n|^2 < +\infty$.

We recall the following propositions:

Proposition 1.2.1. *Let f be a meromorphic function on $D(0,1)$. Then f has a bounded characteristic if and only if f is the quotient of two analytic and bounded functions on $D(0,1)$.*

Proposition 1.2.2. *Let f be a function of H^p with $p \in \mathbf{R}^+$. Then f can be written in the form $f = f_1/f_2$, where f_1 and f_2 are analytic and bounded functions on $D(0,1)$.*

We deduce from the above propositions the following statement:

Proposition 1.2.3. *If f is a meromorphic function on $D(0,1)$, then f has a bounded characteristic if and only if f is the quotient of two functions in H^2 or more generally in any two spaces H^p.*

Lemma 1.2.1. *If f is a meromorphic function with bounded characteristic on $D(0,1)$ having no pole at the origin, then*

$$\lim_{n \longrightarrow +\infty} (D_n(S(f)))^{1/n} = 0 \ .$$

Proof. According to Proposition 1.2.3, we can assume that $f = s/t$ with s and t in H^2, $t_0 = 1$.

Let $S = \sum_{n \in \mathbf{N}} s_n z^n$, $T = \sum_{n \in \mathbf{N}} t_n z^n$, and $F = S/T = \sum_{n \in \mathbf{N}} a_n z^n$ be the Taylor expansions of s, t, f at zero.

We consider the following product of two determinants of order $2n+1$:

$$D_n(F) = \begin{vmatrix} t_0 & 0 & \cdots & 0 \\ t_1 & 0 & \cdots & 0 \\ \vdots & \vdots & \ddots & \vdots \\ t_{2n} & \cdots & \cdots & t_0 \end{vmatrix} \begin{vmatrix} 1 & \cdots & 0 & 0 & 0 & \cdots & 0 & a_0 \\ 0 & \cdots & 0 & 0 & 0 & \cdots & a_0 & a_1 \\ \vdots & & \vdots & \vdots & \vdots & & \vdots & \vdots \\ 0 & \cdots & 1 & 0 & a_0 & \cdots & a_{n-2} & a_{n-1} \\ 0 & \cdots & 0 & a_0 & a_1 & \cdots & a_{n-1} & a_n \\ \vdots & & \vdots & \vdots & \vdots & & \vdots & \vdots \\ 0 & \cdots & 0 & a_n & a_{n+1} & \cdots & a_{2n-1} & a_{2n} \end{vmatrix} .$$

Then

$$D_n(F) = \begin{vmatrix} t_0 & 0 & \cdots & 0 & 0 & 0 & \cdots & 0 & s_0 \\ t_1 & t_0 & \cdots & 0 & 0 & 0 & \cdots & s_0 & s_1 \\ \vdots & \vdots & & \vdots & \vdots & \vdots & & \vdots & \vdots \\ t_{n-1} & t_{n-2} & \cdots & t_0 & 0 & s_0 & \cdots & s_{n-2} & s_{n-1} \\ \vdots & \vdots & & \vdots & \vdots & \vdots & & \vdots & \vdots \\ t_{k+n} & t_{k+n-1} & \cdots & t_{k+1} & s_k & s_{k+1} & & s_{n-1+k} & s_{n+k} \\ \vdots & \vdots & & \vdots & \vdots & \vdots & & \vdots & \vdots \\ t_{2n} & t_{2n-1} & \cdots & t_{n+1} & s_n & s_{n+1} & \cdots & s_{2n-1} & s_{2n} \end{vmatrix}.$$

Let $y(k) = \sum_{i=k}^{+\infty} |t_i|^2 + |s_i|^2$. By applying the Hadamard inequality we get, for $k \leq n+1$:

$$|D_n(F)|^2 \leq y(0)^{n+k} y(k)^{n+1-k}, \quad \text{and} \quad |D_n(F)|^{1/n} \leq y(0)^{\frac{1}{2}+\frac{k}{2n}} y(k)^{\frac{1}{2}+\frac{1-k}{2n}}.$$

Since k is fixed, we obtain

$$\limsup_{n \longrightarrow +\infty} |D_n(F)|^{1/n} \leq (y(0)y(k))^{1/2}.$$

As $\lim_{k \longrightarrow +\infty} y_k = 0$, we have $\lim_{n \longrightarrow +\infty} |D_n(F)|^{1/n} = 0$. ■

Theorem 1.2.1. *Let f be a meromorphic function on $D(0,1)$ with bounded characteristic and no pole at zero, such that the Taylor expansion of f about zero has integer coefficients. Then f is a rational function.*

Proof. By hypothesis, $D_n(S(f))$ is an element of $\mathbf{Z}$ for any $n \in \mathbf{N}$. Now according to Lemma 1.2.1, $D_n(S(f))$ has absolute value smaller than 1 starting from a certain index n_0 and $D_n(S(f)) = 0$ for $n \geq n_0$. The Kronecker criterion allows us to assert that $S(f)$ is a rational series and f a rational function.

To see the applications of this theorem, we now give some examples of meromorphic functions on $D(0,1)$ with bounded characteristic.

Proposition 1.2.4. *Let f be a meromorphic functions on $D(0,1)$, we suppose that there exist $\alpha \in \mathbf{C}$, $\eta \in [0,1[$, $\delta \in \mathbf{R}^{+*}$ such that*

$$|f(z) - \alpha| \geq \delta \qquad \forall z \in \left\{ z \in \mathbf{Z},\, \eta \leq |z| < 1 \right\}.$$

Then the characteristic of f is bounded.

Proof. We consider the function g defined by $g(z) = f(z) - \alpha$. Then g is meromorphic on $D(0,1)$ and is non-zero on the set $\{ z \in \mathbf{C},\ \eta \leq |z| < 1 \}$. So the function g has only a finite number of zeros on $D(0,1)$. Therefore there exists a polynomial P such that $h = P/g$ is analytic on $D(0,1)$. Now according to the maximum modulus principle, we have $|h(z)| \leq \frac{1}{\delta} \sup_{|z|=1} |P(z)| \quad \forall z \in D(0,1)$ and h is bounded on $D(0,1)$. Hence h belongs to H^2. As $f = \dfrac{\alpha h + P}{h}$, we deduce that the characteristic of f is bounded. ∎

Proposition 1.2.5. *Let f be a meromorphic function on $D(0,1)$. We suppose that there exist $k \in \mathbf{R}$ and $\eta \in [0,1[$ such that the inequality $\operatorname{Re} f(z) \leq k$ holds $\forall z \in \{ z \in \mathbf{C}, \eta \leq |z| < 1 \}$. Then the characteristic of f is bounded.*

Proof. We have $|f(z) - (k+1)| \geq |k + 1 - \operatorname{Re} f(z)| = k + 1 - \operatorname{Re} f(z) \geq 1$ for z belonging to the considered set. The preceding proposition then allows us to conclude f is bounded. ∎

We now give a criterion of rationality, which will be used in the following chapters.

Theorem 1.2.2. *Let f be a function meromorphic on $D(0,1)$ with no pole at zero such that the Taylor series of f at zero has integer coefficients. We suppose that $f = s/t$, s and t being analytic on $D(0,1)$ with $s(z) = \sum_{n=0}^{+\infty} s_n z^n$, $t(z) = \sum_{n=0}^{+\infty} t_n z^n$ for $|z| < 1$, and that there exist two real numbers α and β such that:*

i) $$(\alpha,\beta) \in \{ (\alpha,\beta) \in \mathbf{R}^2, \alpha > 0, \beta > 0 \} \cup \{ (\alpha,\beta) \in \mathbf{R}^2, \alpha = 0, \beta > 1 \} \cup \{ (\alpha,\beta) \in \mathbf{R}^2, \alpha > 1, \beta = 0 \}$$

ii) $$\sum_{m=n}^{m=2n-1} |s_m|^2 = o(n^{-\alpha}), \quad \sum_{m=n}^{m=2n-1} |t_m|^2 = o(n^{-\beta}).$$

Then f is a rational function.

The proof will use the following lemmas.

Lemma 1.2.3. *Let $(y_n)_{n\in\mathbf{N}}$ be a sequence of positive real numbers and $s \in \mathbf{N}$. We set $\delta_n = \sup_{m\geq n} \sum_{i=m}^{2m-1} y_i$ for $n \in \mathbf{N}$ and $\delta_0 = \sup(y_0, \delta_1)$.*

If $(i_0, i_1, \ldots, i_s)$ *is a strictly increasing sequence of natural integers, then we have the following inequality:*

$$\sum_{h=0}^{h=s} \sum_{m=0}^{m=s} y_{i_h+m} \leq 8 \sum_{j=0}^{j=s} \delta_j .$$

Proof. For $s = 0$ the inequality is trivial. Assume then that $s \neq 0$. In the proof of the lemma we will need sums of the form $\sum_{m=n}^{2n-1}$; hence it will be useful to introduce a symbol for the greatest power of 2 smaller than a given integer. We denote by Γ the set:

$$\Gamma = \left\{ n = 2^p,\, n \in \mathbf{N},\, p \in \mathbf{N} \right\} .$$

We fix h in $[1, 2, \ldots, s]$, and denote by j_h, l_h, g respectively the unique elements of Γ such that

$$j_h \leq 1 + \frac{s}{i_h} < 2j_h, \quad l_h \leq h < 2l_h, \quad g \leq s < 2g.$$

We then have the following inequalities:

$$0 \leq \sum_{m=0}^{m=s} y_{i_h+m} \leq \delta_{i_h} + \delta_{2i_h} + \delta_{4i_h} + \cdots + \delta_{j_h i_h} .$$

As $i_h \geq h \geq l_h$ and the sequence δ_n is decreasing, we obtain

$$0 \leq \sum_{m=0}^{m=s} y_{i_h+m} \leq \delta_h + \delta_{2h} + \cdots + \delta_{j_h h} \leq \delta_{l_h} + \delta_{2l_h} + \cdots + \delta_{j_h l_h} .$$

As $j_h l_h \leq l_h + hs/i_h \leq 2s$, $j_h l_h \in \Gamma$, we have $j_h l_h \leq 2g$ and we obtain, if $h \geq 1$:

$$0 \leq \sum_{m=0}^{m=s} y_{i_h+m} \leq \delta_{l_h} + \delta_{2l_h} + \cdots + \delta_{2g} .$$

Similarly we obtain, if $h = 0$ and $i_0 \geq 1$:

$$0 \leq \sum_{m=0}^{m=s} y_{i_0+m} \leq \delta_1 + \delta_2 + \cdots + \delta_{2g} ,$$

and, if $h = 0$, $i_0 = 0$:

$$0 \leq \sum_{m=0}^{m=s} y_{i_0+m} \leq \delta_0 + \delta_1 + \delta_2 + \delta_4 + \cdots + \delta_g.$$

We set

$$Y_s = \sum_{h=0}^{h=s} \sum_{m=0}^{m=s} y_{i_h+m}.$$

Hence we obtain

$$Y_s \leq \delta_0+\delta_1+\delta_2+\delta_4+\cdots+\delta_g+(\delta_1+\delta_2+\cdots+\delta_{2g})+2(\delta_2+\cdots+\delta_{2g})+\cdots+g(\delta_g+\delta_{2g}).$$

As the sequence (δ_n) is decreasing we obtain

$$Y_s \leq 2(\delta_0 + \delta_1 + \cdots) + 2(\delta_1 + \cdots + \delta_g) + \cdots + g(\delta_{g/2} + \delta_g).$$

We use the inequality

$$2^{q+1}\delta_{2^q} \leq 4(\delta_{2^{q-1}+1} + \cdots + \delta_{2^q})$$

to get

$$\sum_{h=0}^{h=s} \sum_{m=0}^{m=s} y_{i_h+m} \leq 8 \sum_{j=0}^{j=s} \delta_j.$$

■

Lemma 1.2.4. *If γ is a positive number and $(v_n)_{n\in\mathbf{N}}$ is a sequence of complex numbers such that* $\sum_{m=n}^{2n-1} |v_m|^2 = o(n^{-\gamma})$, *we set* $\rho_i = \sup_{n\geq i} \sum_{h=n}^{2n-1} |v_h|^2$, *and then:*

i) $\sum_{i=0}^{i=r} |v_i|^2 = O(1)$ *if* $\gamma > 0$

ii) $\left(\sum_{i=0}^{i=r} |v_i|\right)^2 = \begin{cases} o(r^{1-\gamma}) & \text{if } \gamma < 1 \\ O(1) & \text{if } \gamma > 1 \end{cases}$

iii) $\sum_{i=0}^{i=r} \rho_i = \begin{cases} o(r^{1-\gamma}) & \text{if } \gamma < 1 \\ O(1) & \text{if } \gamma > 1 \end{cases}.$

Proof. **i)** With r fixed in $\mathbf{Z}$, let l be the integer such that $2^l \leq r < 2^{l+1}$. Then we have

$$\sum_{i=0}^{i=r} |v_i|^2 \leq |v_0|^2 + \sum_{n=0}^{l} \sum_{i=2^n}^{2^{n+1}-1} |v_i|^2. \text{ Hence } \sum_{i=0}^{i=r} |v_i|^2 = O(1) \text{ if } \gamma > 0.$$

ii) The Cauchy–Schwarz inequality implies

$$\sum_{i=n}^{2n-1} |v_i| \leq n^{1/2} (\sum_{i=n}^{2n-1} |v_i|^2)^{1/2}. \text{ Hence } \sum_{i=n}^{2n-1} |v_n| = o(n^{(1-\gamma)/2}).$$

Let l be fixed as in i). We then have the inequalities

$$\sum_{i=0}^{i=r} |v_i| \leq |v_0| + \sum_{n=0}^{l} \sum_{i=2^n}^{2^{n+1}-1} |v_i|. \text{ Hence}$$

$$\sum_{i=0}^{i=r} |v_i| = \begin{cases} o(r^{l(1-\gamma)/2}) & \text{if } \gamma < 1 \\ O(1) & \text{if } \gamma > 1 \end{cases}.$$

iii) We have $\rho_n = o(n^{-\gamma})$; hence

$$\sum_{i=0}^{i=n} \rho_i = \begin{cases} o(r^{1-\gamma}) \text{ if } \gamma < 1 \\ O(1) \text{ if } \gamma > 1 \end{cases}.$$

■

Lemma 1.2.5. *Let $X = (x_{i,j})$ be a square matrix of order $n+1$ with complex components such that*

$$\sum_{j=0}^{j=n} \sum_{i=0}^{i=n} |x_{i,j}|^2 < n+1.$$

Then $|\det X| < 1$.

Proof. By Hadamard's inequality and the arithmetic-geometric mean inequality we obtain

$$|\det X|^2 \leq \prod_{i=0}^{i=n} \sum_{j=0}^{j=n} |x_{i,j}|^2 \leq \Big(\frac{1}{n+1} \sum_{i=0}^{i=n} \sum_{j=0}^{j=n} |x_{i,j}|^2 \Big)^{n+1} < 1.$$

■

Lemma 1.2.6. *Let $F = \sum_{n \in \mathbf{N}} a_n X^n$ be an element of $K[[X]]$ that can be written as $F = S/T$, with $S = \sum_{n \in \mathbf{N}} s_n X^n \in K[[X]]$ and $T = \sum_{n \in \mathbf{N}} t_n X^n \in K[[X]]$, $t_0 = 1$. Then*

$D_n(F) = \det(x_{h,k})$ *with* $x_{h,k} = u_{h,k} + v_{h,k}$ *and*

$$u_{h,k} = \sum_{j=0}^{j=k} t_j s_{h+k-j}, \qquad v_{h,k} = \sum_{j=0}^{k-1} s_j t_{h+k-j}, \qquad v_{h,0} = 0.$$

Proof. We consider the following product of matrices:

$$\begin{pmatrix} t_0 & 0 & \cdots & 0 \\ t_1 & t_0 & \cdots & 0 \\ \vdots & \vdots & \ddots & \vdots \\ t_n & t_{n-1} & \cdots & t_0 \end{pmatrix} \begin{pmatrix} a_0 & a_1 & \cdots & a_n \\ a_1 & a_2 & \cdots & a_{n+1} \\ \vdots & \vdots & \ddots & \vdots \\ a_n & a_{n+1} & \cdots & a_{2n} \end{pmatrix} \begin{pmatrix} t_0 & t_1 & \cdots & t_n \\ 0 & t_0 & \cdots & t_{n-1} \\ \vdots & \vdots & \ddots & \vdots \\ 0 & 0 & \cdots & t_0 \end{pmatrix}.$$

An easy calculation shows that the general term of the product is

$$x_{h,k} = \sum_{i=0}^{i=h} \sum_{j=0}^{j=k} t_i t_j a_{h+k-(i+j)}.$$

Hence

$$x_{h,k} = \sum_{j=0}^{j=k} t_j \sum_{i=0}^{h+k-j} t_i a_{h+k-(i+j)} - \sum_{j=0}^{k-1} \sum_{i=h+1}^{h+k-j} t_i t_j a_{h+k-(i+j)}$$

and

$$x_{h,k} = \sum_{j=0}^{j=k} t_j s_{h+k-j} - \sum_{i=h+1}^{h+k} t_i \sum_{j=0}^{h+k-i} t_j a_{h+k-(i+j)}.$$

Then

$$x_{h,k} = u_{h,k} - \sum_{i=h+1}^{h+k} t_i s_{h+k-i} = u_{h,k} - \sum_{j=0}^{k-1} s_j t_{h+k-j}$$

and $x_{h,k} = u_{h,k} + v_{h,k}$; since $t_0 = 1$ we have $D_n(F) = \det(x_{h,k})$. ∎

Proof of Theorem 1.2.2:

i) Assume that $\alpha > 0$ and $\beta > 0$. According to Lemma 1.2.4 the two series $\sum_{n\in\mathbf{N}} |s_n|^2$, $\sum_{n\in\mathbf{N}} |t_n|^2$ are convergent, and by Theorem 1.2.1, f is a rational function.

ii) Assume that

$$(\alpha, \beta) \in \{ (\alpha, \beta) \in \mathbf{R}^2, \alpha = 0, \beta > 1 \} \bigcup \{ (\alpha, \beta) \in \mathbf{R}^2, \alpha > 1, \beta = 0 \}.$$

According to Lemma 1.1.6 we have $D_r(F) = \det(x_{m,n})$ with

$$x_{m,n} = u_{m,n} + v_{m,n}, \quad u_{m,n} = \sum_{i=0}^{n} t_i s_{m+n-i}, \quad \text{and} \quad v_{m,n} = -\sum_{i=0}^{n-1} s_i t_{m+n-i}$$

We now use Lemma 1.2.5. We first evaluate the sums

$$U = \sum_{m=0}^{r}\sum_{n=0}^{r} |u_{m,n}|^2 \text{ and } V = \sum_{m=0}^{r}\sum_{n=0}^{r} |v_{m,n}|^2.$$

a) Majoration of U.

We define the polynomial $P_{m,r}$ by

$$P_{m,r} = \sum_{k=0}^{k=r} t_k z^k \sum_{j=m}^{m+r} s_j z^j.$$

Then if n is an integer smaller than r, $u_{m,n}$ is the coefficient of z^{m+n} in the expansion of $P_{m,r}$. Parseval's formula now implies the following inequality

$$\sum_{n=0}^{n=r} |u_{m,n}|^2 \leq \frac{1}{2\pi}\int_0^{2\pi} \left| \sum_{k=0}^{r} t_k e^{ik\theta} \times \sum_{j=m}^{m+r} s_j e^{ij\theta} \right|^2 d\theta$$

Hence

$$\sum_{n=0}^{n=r} |u_{m,n}|^2 \leq \left(\sum_{k=0}^{r} |t_k|\right)^2 \frac{1}{2\pi}\int_0^{2\pi} \left| \sum_{j=m}^{m+r} s_j e^{ij\theta} \right|^2 d\theta;$$

moreover Parseval's formula allows us to write

$$\sum_{n=0}^{r} |u_{m,n}|^2 \leq \left(\sum_{k=0}^{r} |t_k|\right)^2 \sum_{j=m}^{m+r} |s_j|^2 = \left(\sum_{k=0}^{r} |t_k|\right)^2 \sum_{j=0}^{r} |s_{j+m}|^2.$$

Hence

$$\sum_{m=0}^{r}\sum_{n=0}^{r} |u_{m,n}|^2 \leq \left(\sum_{k=0}^{r} |t_k|\right)^2 \sum_{m=0}^{r}\sum_{j=0}^{r} |s_{j+m}|^2$$

According to Lemma 1.2.3 we now get

$$\sum_{m=0}^{r}\sum_{n=0}^{r} |u_{m,n}|^2 \leq 4\left(\sum_{k=0}^{r} |t_k|\right)^2 \times \sum_{i=0}^{r} \delta_i$$

with

$$\delta_i = \sup_{j\geq i} \sum_{h=j}^{2j-1} |s_h|^2 \text{ if } i \neq 1, \quad \delta_0 = \sup(|s_0|^2, \delta_1)$$

b) Majoration of $\sum_{m=0}^{r} \sum_{n=0}^{r} |v_{m,n}|^2$.

We define similarly the polynomial $Q_{m,r}$ by the equality

$$Q_{m,r} = \sum_{k=0}^{r} s_k z^k \sum_{j=m+1}^{m+r} t_j z^j.$$

Then if n is an integer smaller than r, $v_{m,n}$ is the coefficient of z^{m+n} in the expansion of $Q_{m,r}$ and we similarly have the inequalities

$$\sum_{n=0}^{r} |v_{m,n}|^2 \leq \left(\sum_{k=0}^{r} |s_k|\right)^2 \sum_{j=m}^{m+r} |t_j|^2$$

and

$$\sum_{m=0}^{r} \sum_{n=0}^{r} |v_{m,n}|^2 \leq 8 \left(\sum_{k=0}^{r} |s_k|\right)^2 \sum_{i=0}^{r} \mu_i$$

with $\mu_i = \sup_{j\geq i} \sum_{h=j}^{2j-1} |t_h|^2$ if $i \neq 0$, $\mu_0 = \sup(|t_0|^2, \mu_1)$.

c) Conclusion of proof:

As the Cauchy–Schwarz inequality implies $|x_{m,n}|^2 \leq 2(|u_{m,n}|^2 + |v_{m,n}|^2)$, we have

$$\sum_{m=0}^{r} \sum_{n=0}^{r} |x_{m,n}|^2 \leq 16 \left[\left(\sum_{k=0}^{r} |t_k|\right)^2 \left(\sum_{i=0}^{r} \delta_i\right) + \left(\sum_{k=0}^{r} |s_k|\right)^2 \left(\sum_{i=0}^{r} \mu_i\right) \right].$$

Since the second member is symmetric in s and t, we can suppose for instance that $\alpha = 0$, $\beta > 1$.

Then Lemma 1.2.4 implies the following equalities:

$$\left(\sum_{k=0}^{r} |t_k|\right)^2 = O(1), \text{ since } \beta > 1, \quad \sum_{i=0}^{r} \delta_i = o(r), \text{ since } \alpha = 0,$$

$$\left(\sum_{k=0}^{r} |s_k|\right)^2 = o(r), \text{ since } \alpha = 0, \quad \sum_{i=0}^{r} \mu_i = O(1), \text{ since } \beta > 1, \text{ and}$$

$$\sum_{m=0}^{r} \sum_{n=0}^{r} |x_{m,n}|^2 = o(r).$$

So there exists $r_0 \in \mathbf{N}$ such that for all $r \geq r_0$ we have the inequality:

$$\sum_{m=0}^{r} \sum_{n=0}^{r} |x_{m,n}|^2 < r + 1.$$

According to Lemma 1.2.5 we have $|D_r(F)| < 1, \quad \forall r \geq r_0$. As $D_r(F)$ is an element of $\mathbf{Z}$, we have $D_r(F) = 0, \quad \forall r \geq r_0$.

By Kronecker's criterion, F is a rational fraction and f a rational function. ∎

We deduce from Theorem 1.2.2, by taking $\alpha = 0$ and β greater than 1, the following corollary.

Corollary. *Let f be a meromorphic function on $D(0,1)$ with no pole at the origin, such that the Taylor series of f at zero has integer coefficients. We suppose that $f = s/t$, where t is a polynomial with complex coefficients, s is analytic on $D(0,1)$ with $s(z) = \sum_{n \in \mathbf{N}} s_n z^n$ if $|z| < 1$, and $\sum_{m=n}^{2n-1} |s_m|^2 = o(1)$. Then f is a rational function.*

1.3 Generalized Fatou's lemma

Once the given function f has been shown to be rational with the methods used in the preceding sections, it is often useful for applications to determine more precisely the denominator of f. We will need the following theorem.

Theorem 1.3. *Let A be a Dedekind ring and F a rational series in $A[[X]]$. Then there exist two polynomials P and Q in $A[X]$ such that $F = P/Q$, where P and Q are relatively prime and $Q(0) = 1$.*

Proof. By hypothesis, $F = P/Q$ with $P \in A[X], Q \in A[X], Q(0) \neq 0$. We may suppose that P and Q are relatively prime. So there exist U and V in A[X] and $d \in A$ such that $UP + VQ = dX^{r-1}$ with $r = d^\circ Q$. We then have

$$UF + V = \frac{dX^{r-1}}{Q} \quad \text{and} \quad \frac{dX^{r-1}}{Q} \in A[X].$$

Let
$$\sum_{i\geq r-1} a'_i X^i = \frac{dX^{r-1}}{Q}, \qquad Q = \sum_{0\leq i\leq r} q_i X^i.$$

We denote by $\mathcal{B}$ the greatest common divisor of the ideals $q_0 A, q_1 A, \ldots, q_r A$ and by $\mathcal{A}$ the greatest common divisor of the sequence of ideals $(a'_i A)_{i\geq r-1}$.

Let $\mathcal{P}$ be a prime ideal such $\mathcal{PB}$ divides $q_0 A$ and k the smallest integer $\geq r-1$ such that $a'_k A$ is not divisible by $\mathcal{AP}$. We have

$$\begin{aligned} q_1 a'_k &= -(q_0 a'_{k+1} + q_2 a'_{k-1} + \cdots + q_r a'_{k-r+1}) \\ &\vdots \\ q_r a'_k &= -(q_0 a'_{k+r} + q_1 a'_{k+r-1} + \cdots + q_{r-1} a'_{k+1}). \end{aligned}$$

From the first equation we deduce that $\mathcal{PAB}$ divides $q_1 a'_k A$; as $q_1 A = \mathcal{BB}_1$, $a'_k A = \mathcal{AA}_k$ and $\mathcal{P}$ does not divide $\mathcal{A}_k$, $\mathcal{P}$ divides $\mathcal{BB}_1$ and $q_1 A$ is divisible by $\mathcal{PB}$.

We prove, by using the other equations, that $q_1 A, q_2 A, \ldots, q_r A$ are divisible by $\mathcal{PB}$, which contradicts the definition of $\mathcal{B}$. Hence $\mathcal{B} = q_0 A$. Therefore there exists $q'_i \in A$ such that $q_i = q_0 q'_i$ for $1 \leq i \leq r$.

Let $Q_1 = \sum_{0\leq i\leq r} q'_i X^i$ with $q'_0 = 1$. Then $Q = q_0 Q_1$, $Q_1 \in A[X]$, $Q_1(0) = 1$.

From the equation $q_0 Q'_1 F = P$ we deduce that the coefficients of P are the product of an element of A by q_0. So P can be written as $P = q_0 P_1$ with $P_1 \in A[X]$, and we get

$F = \dfrac{q_0 P_1}{q_0 Q_1} = \dfrac{P_1}{Q_1}$, where P_1 and Q_1 have the required properties. ∎

Notes

Criterion 1.1.1 is due to Kronecker [9]. Instead of the classical Hankel criterion that can be found, for instance, in the book of Amice [1], we preferred presenting the criterion 1.1.2 proved by Cantor [5], which will be used in Chapter 5.

There have been several versions of Theorem 1.2.2, each generalizing the preceding one, and proved in chronological order by Borel [3], Salem [11], Chamfy [6], and Cantor [4].

The definition of a function with bounded characteristic and the equivalence written in Proposition 1.2.1 can be found in the book of Tsuji [12]. Proposition 1.2.2 is Theorem 17.6 in the book of Hervé [8]. We cannot express the fact that a function f belongs to H^p through conditions on the coefficients of the Taylor series of f at zero except for the case $p = 2$, which is insufficient for the applications treated in Chapter 5; hence the interest of Theorem 1.2.2. The first version of this theorem, with a more restricted condition on the pair $(\alpha, \beta) \in \mathbf{R}^2$, is due to Cantor [5]; the authors have extended the set of (α,β) for which the proposition is true.

Theorem 1.3 is a generalized version of Fatou's lemma [7] for the case $A = \mathbf{Z}$. The proof is an adaptation of the one given by Pisot [10] for the case where A is a ring of integers in an algebraic number field.

More generally Benzaghou [2] proved that Fatou's lemma remains true in Krull rings. We recall that principal ideal rings, Dedekind rings and factorial rings are all Krull rings. This proof was omitted because it uses p-adic analysis.

References

[1] Y. AMICE, *Les nombres p-adiques*, Presses Universitaires de France.

[2] B. BENZAGHOU, Anneaux de Fatou, *Séminaire Delange-Pisot-Poitou: Théorie des nombres, 9e année*, (1968/69), n°9, 8p.

[3] E. BOREL, Sur une application d'un théorème de M. Hadamard, *Bull. Sci. Math.*, 18 (1894), 22-25.

[4] D. G. CANTOR, Power series with integral coefficients, *Bull. Amer. Math. Soc.*, Vol. 69, (1963), 362-366.

[5] D. G. CANTOR, On power series with only finitely many coefficients (mod 1): Solution of a problem of Pisot and Salem, *Acta Arith.*, 34, (1977), 43-55.

[6] C. CHAMFY, Fonctions méromorphes dans le cercle unité et leurs séries de Taylor, *Ann. Inst. Fourier*, 8, (1958), 211-261.

[7] P. FATOU, Séries trigonométriques et séries de Taylor, *Acta Math.*, Uppsala, Vol. 30, (1906), 335-400.

[8] M. HERVE, *Les fonctions analytiques*, Presses Universitaires de France.

[9] L. KRONECKER, Zur Theorie der Elimination einer Variablen aus zwei algebraischen Gleichungen, *Monatsber*, Berlin, (1881), 535-600.

[10] C. PISOT, La répartition modulo 1 et les nombres algébriques, *Ann. Di. Sc. Norm. Sup. Pisa*, Ser 2, 7, (1938), 205-248.

[11] R. SALEM, Power series with integral coefficients, *Duke Math. J.*, Vol. 12, (1945), 153-172.

[12] M. TSUJI, *Potential theory in modern function theory*, Chelsea.

CHAPTER 2

COMPACT FAMILIES OF RATIONAL FUNCTIONS

The main aim of this book is to determine closed families of algebraic numbers. We can for instance associate to an algebraic number θ the rational function $z \in \mathbf{C} \longrightarrow \dfrac{P(z)}{P^*(z)}$, P being the minimal polynomial of θ and P^* the reciprocal polynomial of P. We therefore need to study families of rational functions with coefficients in $\mathbf{Z}$.

The aim of this chapter is to exhibit compact sets of rational functions for the uniform convergence topology on the compacts of $D(0,1)$. This will allow us in the following chapters to find closed families of algebraic numbers. We will mainly use the results of Chapter 1.

If a function $f\colon \mathbf{C} \longrightarrow \mathbf{C}$ is analytic in the neighborhood of zero, we will denote by $S(f) \in \mathbf{C}[[z]]$ the Taylor expansion of f at zero.

We recall that if $G = \sum_{n\in\mathbf{N}} a_n z^n \in \mathbf{C}[[z]]$, then $\operatorname{ord} G$ is defined by

$$\operatorname{ord} G = \inf\left\{ n \in \mathbf{N}, a_n \neq 0 \right\}.$$

Moreover if $(F_n)_{n\in\mathbf{N}}$ is a sequence of formal series of $\mathbf{C}[[z]]$, we will say that the sequence $(F_n)_{n\in\mathbf{N}}$ converges to $F \in \mathbf{C}[[z]]$ if and only if

$$\lim_{n\longrightarrow+\infty} \operatorname{ord}(F - F_n) = +\infty.$$

2.1 Properties of formal series with rational coefficients

Lemma 2.1.1. *If $F = \sum_{n\in\mathbf{N}} a_n X^n$ is a formal series of $\mathbf{Q}[[X]]$ such that $F = S/T$ with $S \in \mathbf{Z}[[X]]$, $T \in \mathbf{Z}[[X]]$, $t_0 = q \in \mathbf{N}^*$, then:*

i) $a_n q^{n+1} \in \mathbf{Z} \qquad \forall n \in \mathbf{N}$

ii) $q^{2n+1} D_n(F) \in \mathbf{Z} \qquad \forall n \in \mathbf{N}$.

Proof. **i)** Let

$$S = \sum_{n \in \mathbf{N}} s_n X^n, \quad T = \sum_{n \in \mathbf{N}} t_n X^n \quad \text{and} \quad F_1 = \sum_{n \in \mathbf{N}} a_n q^{n+1} X^n;$$

then $F_1 = \dfrac{S_1}{T_1}$ with $S_1 = \sum_{n \in \mathbf{N}} q^n s_n X^n$ and $T_1 = \sum_{n \in \mathbf{N}} q^{n-1} t_n X^n$.

As $T_1(0) = 1$, $S_1 \in \mathbf{Z}[[X]]$, $T_1 \in \mathbf{Z}[[X]]$, F_1 is an element of $\mathbf{Z}[[X]]$ and $a_n q^{n+1}$ is an integer $\forall n \in \mathbf{N}$.

ii) By considering the determinant product used in Lemma 1.2.1, we obtain directly that $q^{2n+1} D_n(F)$ is an integer $\forall n \in \mathbf{N}$. ■

Notation. If $q \in \mathbf{N}^*$, we denote by Φ_q the set of rational fractions F with coefficients in $\mathbf{Q}$ that can be written in the form

$$F = A/B \text{ with } A \in \mathbf{Z}[X], \quad B \in \mathbf{Z}[X], \quad B(0) = q.$$

Theorem 2.1. *If $(F_n)_{n \in \mathbf{N}}$ is a sequence of elements of Φ_q that converges to a rational fraction F (with coefficients in $\mathbf{Q}$) then F belongs to Φ_q.*

Proof. By induction on q.

If $q = 1$, then $F_n \in \mathbf{Z}[[X]]$. Moreover the sequence $(F_n)_{n \in \mathbf{N}}$ converges to F, hence $F \in \mathbf{Z}[[X]]$ and Fatou's lemma allows us to assert that F belongs to Φ_1.

If q belongs to $\mathbf{N}^*$, assume that the property is true for all integers $q' < q$. Set

$$F_n = \frac{A_n}{B_n} = \sum_{i \in \mathbf{N}} u_{i,n} X^i, \quad F = \frac{A}{B} = \sum_{i \in \mathbf{N}} u_i X^i$$

with A and $B \in \mathbf{Z}[[X]]$, $u_i \in \mathbf{Q} \quad \forall i \in \mathbf{N}$.

If $u_i \in \mathbf{Z} \quad \forall i \in \mathbf{N}$, then according to Fatou's lemma, F belongs to Φ_1 and so to Φ_q. If this is not the case, there exists an integer r such that $(u_0, u_1, \dots, u_{r-1}) \in \mathbf{Z}^r$ and $u_r \notin \mathbf{Z}$.

As (F_n) converges to F, there exists an integer n_0 such that

$$u_{i,n} = u_i \qquad \forall n \geq n_0, \quad \forall i \in [0, 1, \dots, r].$$

The equality $F_n B_n = A_n$ shows, by comparing the coefficients of X^r in both members, that $q u_{r,n}$ is an integer $\forall n \geq n_0$, because $(u_{0,n}, u_{1,n}, \dots, u_{r-1,n}) \in \mathbf{Z}^r$ and $B_n(0) = q$.

Hence qu_r is an element of $\mathbf{Z}$. We can thus write

$$u_r = u'_r + \frac{q'}{q} \text{ with } u'_r \in \mathbf{Z} \text{ and } 0 < q' < q, \quad q' \in \mathbf{N}^*.$$

Set $P = \sum_{i=0}^{r-1} u_i X^i + u'_r X^r$. Then $F = \frac{A}{B} = P + X^r(\frac{q'}{q} + \cdots)$ and

$F_n = \frac{A_n}{B_n} = P + X^r(\frac{q'}{q} + \cdots) \quad \forall n \geq n_0.$

That is, $A - BP = X^r W$ with $W \in \mathbf{Z}[X]$, $W(0) = B(0)\frac{q'}{q}$ and similarly $A_n - B_n P_n = X^r W_n$ with $W_n \in \mathbf{Z}[X]$, $W_n(0) = q' \quad \forall n \geq n_0$.

Set $G = \frac{B}{W}$ and $G_n = \frac{B_n}{W_n}$ $\forall n \geq n_0$; then the sequence of formal series $(G_n)_{n \geq n_0}$ converges to G. Moreover $W_n(0) = q'$ and G_n belongs to $\Phi_{q'}$ $\forall n \geq n_0$; the recurrence hypothesis implies that G belongs to $\Phi_{q'}$ and that G can be written as $G = H/M$ with H and $M \in \mathbf{Z}[X]$, $M(0) = q'$. Since $G(0) = \frac{B(0)}{W(0)} = \frac{q}{q'} = \frac{H(0)}{q'}$, $H(0) = q$. As $F = P + \frac{X^r M}{H} = \frac{PH + X^r M}{H}$, F belongs to Φ_q. ∎

Lemma 2.1.2. *Let be $F = A/B$ with A and $B \in \mathbf{Z}[X]$, $B(0) \neq 0$. Then there exist A_1 and $B_1 \in \mathbf{Z}[X]$ relatively prime and such that $F = A_1/B_1$ and $B_1(0) = B(0)$.*

The proof of this lemma, which uses Gauss's lemma, is easy and left to the reader.

2.2 Compact families of rational functions

Definition 2.2. *If $q \in \mathbf{N}^*$, $k \in \mathbf{N}$, $\delta \in \mathbf{R}^{+*}$, we denote by $\mathcal{F}(q, k, \delta)$ the set of rational functions f that can be written in the form $f = \frac{A(z)}{Q(z)}$ where A and Q are polynomials with integer coefficients such that*

i) $Q(0) = q$, $A(0) \neq 0$

ii) Q *has no more than k zeros in $D(0,1)$ and is non-zero in* $D(0,\delta) \cup \{z \in \mathbf{C}, |z| = 1\}$

iii) $\left|\frac{A(z)}{Q(z)}\right| \leq 1$ *if* $|z| = 1$.

Theorem 2.2.1. *The family $\mathcal{F}(q,k,\delta)$ is compact for the uniform convergence topology on the compacts of $D(0,\delta)$.*

Proof. The proof is divided into three steps. If $(f_n)_{n\in\mathbf{N}}$ is a sequence of distinct elements of $\mathcal{F}(q,k,\delta)$, we prove that we can extract a sub-sequence (also denoted by f_n) such that

a) the sequence (f_n) converges uniformly on every compact of $D(0,\delta)$ to a function f analytic on $D(0,\delta)$ and meromorphic on $D(0,1)$;

b) f can be extended to a rational function in $\mathbf{C}$;

c) $f \in \mathcal{F}(q,k,\delta)$.

a) From the sequence (f_n) we can extract a sub-sequence (also denoted by f_n) such that

- the denominators of the functions f_n have each h zeros $(\alpha_{1,n}, \alpha_{2,n}, \ldots, a_{h,n})$ in the annulus $\{ z \in \mathbf{C}, \delta \le |z| < 1 \}$, where $h \le k$;
- $\lim\limits_{n \longrightarrow +\infty} \alpha_{j,n} = \alpha_j$ with $\delta \le |\alpha_j| \le 1 \; \forall j \in [1,2,\ldots,h]$.

Denote by φ_n and φ the functions defined by

$$\varphi_n(z) = \prod_{j=1}^{h} \frac{1-\overline{\alpha}_{j,n} z}{\alpha_{j,n} - z}, \qquad \varphi(z) = \prod_{|\alpha_j|<1} \frac{1-\overline{\alpha}_j z}{\alpha_j - z}.$$

Then the sequence (φ_n) converges uniformly on every compact of $D(0,\delta)$ to φ. Moreover the functions f_n/φ_n are analytic on $D(0,1)$ and bounded by 1 on $D(0,1)$. By Montel's Theorem, there exists a sub-sequence (also denoted byf_n/φ_n) and a function h such that

- the function h is analytic and bounded by 1 on $D(0,1)$;
- the sequence (f_n/φ_n) converges uniformly on all compacts of $D(0,1)$ to h.

Then the sequence (f_n) converges uniformly on every compact of $D(0,\delta)$ to the function $f = \varphi h$. Therefore f is analytic on $D(0,\delta)$ and meromorphic on $D(0,1)$.

b) Let $S(f_n) = \sum_{i\in\mathbf{N}} u_{i,n} z^i$ (resp. $S(f) = \sum_{i\in\mathbf{N}} u_i z^i$) the Taylor series of f_n (resp:f) in the neighborhood of zero. As (f_n) converges uniformly on all compacts of $D(0,\delta)$ to f, we have $\lim\limits_{n \longrightarrow +\infty} u_{i,n} = u_i \quad \forall i \in \mathbf{N}$.

By Lemma 2.1.1 we have $u_{i,n}q^{i+1} \in \mathbf{Z}$ $\forall i \in \mathbf{N}$, $\forall n \in \mathbf{N}$. We deduce that $q^{i+1}u_i$ is an integer and $u_i = u_{i,n}$ for $n \geq n_0(i)$.

So the sequence $S(f_n)$ converges to $S(f)$. In particular for fixed i we have $D_i(S(f_n)) = D_i(S(f)$ for $n \geq n_0(i)$.

Since according to Lemma 2.1.1 $q^{2i+1}D_i(S(f_n))$ is an integer, we conclude that $q^{2i+1}D_i(S(f))$ is an integer $\forall i \in \mathbf{N}$. As $f = \varphi h$, where h and $1/\varphi$ are analytic and bounded by 1 on $D(0,1)$, f has a bounded characteristic on $D(0,1)$. By Lemma 1.2.1 $\lim_{i \longrightarrow +\infty} |D_i(S(f))|^{1/i} = 0$.

So there exists $i_o \geq 1$ such that $|D_i(S(f))| \leq \dfrac{1}{(q+1)^{4i}}$ $\forall i \geq i_0$.

Hence $|q^{2i+1}D_i(S(f))| \leq \dfrac{q^{2i+1}}{(q+1)^{4i}} < 1$ $\forall i \geq i_o$ and $D_i(S(f)) = 0$ $\forall i \geq i_0$.

Kronecker's criterion implies that $S(f)$ is a rational fraction. As $S(f_n)$ belongs to Φ_q and $S(f_n)$ converges to $S(f)$, Theorem 2.1 implies that $S(f)$ belongs to Φ_q. Lemma 2.1.2 shows that $S(f) = A/Q$, A and Q being two polynomials of $\mathbf{Z}[z]$, relatively prime with $Q(0) = q$.

So the function f can be extended on $\mathbf{C}$ by the rational function $z \longrightarrow \dfrac{A(z)}{Q(z)}$.

c) i) $\dfrac{A(0)}{q} = f(0) = u_0 = u_{0,n}$ for $n \geq n_0$, so $A(0) \neq 0$.

iii) $f(z) = \dfrac{A(z)}{Q(z)} = h(z) \prod_{|\alpha_j|<1} \dfrac{1 - \overline{\alpha}_j z}{\alpha_j - z}$ $\forall z \in D(0,1)$

with h analytic and bounded by 1 on $D(0,1)$. Hence

$$\left|\frac{A(z)}{Q(z)}\right| \leq \prod_{|\alpha_j|<1} \left|\frac{1 - \overline{\alpha}_j z}{\alpha_j - z}\right| \qquad \forall z \in D(0,1).$$

When z tends to a point of absolute value 1, we have $\left|\dfrac{A(z)}{Q(z)}\right| \leq 1$ if $|z| = 1$.

ii) Since A and Q are relatively prime, by part iii) of the definition of $\mathcal{F}$, Q has no zero on $\{ z \in \mathbf{C}, |z| = 1 \}$. As $A/Q = \varphi h$, the zeros of Q in $D(0,1)$ belong to the set $\{ \alpha_1, \alpha_2, \ldots, \alpha_h \}$. They are therefore in the annulus $\{ z \in \mathbf{C}, \delta \leq |z| < 1 \}$. ∎

The following lemma will help us specify the limit-points of the set $\mathcal{F}(q,k,\delta)$.

Lemma 2.2.1. *If f and g are two analytic functions on $D(0,r)$ with $r>1$ such that*

i) $|f(z)| \leq |g(z)|$ *if* $|z|=1$

ii) $f(z)-g(z)=\gamma_n z^n+\cdots\cdots$ *where* $\gamma_n \neq 0$,

then g has at least n zeros in $D(0,1)$.

Proof. Let k be the number of zeros of g in $D(0,1)$.

a) Assume that $g(z)$ is non-zero if $|z|=1$; consider the function $h_\lambda = f-\lambda g$ with $\lambda>1$. By Rouché's theorem h_λ possesses k zeros in $D(0,1)$. Letting λ tend to 1 gives us that $h_1=f-g$ has at most k zeros in $D(0,1)$. Hence $n \leq k$.

b) Let $\alpha_1, \alpha_2, \ldots, \alpha_s$ be the zeros of g of modulus 1. Set:

$$P(z)=\prod_{i=1}^{s}(z-\alpha_i).$$

From inequality i) we deduce that $f_1=f/P$ and $g_1=g/P$ are analytic on $D(0,1)$. Moreover $f_1(z)-g_1(z)=\dfrac{\gamma_n}{P(0)}z^n+\cdots \quad \forall z \in D(0,1)$ and g_1 has k zeros in $D(0,1)$. By a) we have $n \leq k$. ■

Proposition 2.2.1. *If $f=A/Q$ is a limit point of the set $\mathcal{F}(q,k,\delta)$, then A is different from $\mp Q^*$, where Q^* is the reciprocal polynomial of Q.*

Proof. Let $f_n=A_n/Q_n$ be a sequence of distinct elements of $\mathcal{F}(q,k,\delta)$ that converges uniformly to f on every compact of $D(0,\delta)$. Then by the proof of Theorem 2.2.1, the sequence of rational functions A_n/Q_n converges to A/Q.

Assume that $A=\varepsilon Q^*$ with $\varepsilon=\mp 1$. Set $s=1+k+d^\circ Q$. There exists $n_0(s)$ such that

$$\operatorname{ord}\left(\frac{A_n}{Q_n}-\varepsilon\frac{Q^*}{Q}\right) \geq s \qquad \forall n \geq n_0(s),$$

i.e., $\operatorname{ord}(A_nQ-\varepsilon Q^*Q_n) \geq s \quad \forall n \geq n_0(s)$.

As f_n belongs to $\mathcal{F}(q,k,\delta)$, we have $|A_n(z)Q(z)| \leq |\varepsilon Q^*(z)Q_n(z)|$ if $|z|=1$.

Moreover by the choice of s, Q^*Q_n has at most $s-1$ zeros in $D(0,1)$. According to Lemma 2.2.1, we have $f_n=f \quad \forall n \geq n_0$, which contradicts the fact that all the f_n are distinct. ■

Notes

The first proof of Theorem 2.2.1 is due to Pisot [2]. It uses p-adic analysis which is essential to the proof. The proof given in this chapter is due to Dress [1].

References

[1] F. Dress, Familles de séries formelles et ensembles de nombres algébriques, *Ann. Scient. Ec. Norm. Sup.*, 4e Série, 1, (1968), 1-44.

[2] C. Pisot, Familles compactes de fractions rationnelles et ensembles fermés de nombres algébriques. *Ann. Scient. Ec. Norm. Sup.*, 3e Série, 81, (1964), 165-199.

CHAPTER 3

MEROMORPHIC FUNCTIONS ON D(0,1). GENERALIZED SCHUR ALGORITHM

At the beginning of this century, Schur showed by introducing an algorithm defined on $\mathbf{C}[[z]]$, that there exist necessary and sufficient conditions for an element of $\mathbf{C}[[z]]$ to be the Taylor series at zero of an analytic function bounded by 1 on $D(0,1)$.

In the first part of this chapter we expose this algorithm and moreover we give a criterion of "hyper-rationality" in $\mathbf{C}[[z]]$.

Schur's algorithm was deficient in certain cases and did not permit the characterization of other functions. In the second part of this chapter we introduce a more general algorithm than that of Schur. We then determine explicit necessary and sufficient conditions for a series in $\mathbf{C}[[z]]$ to be the Taylor series of a function f satisfying

- f is meromorphic on $D(0,1)$
- f has no pole at zero
- f has only a finite number of poles in $D(0,1)$
- $\limsup\limits_{\substack{z \longrightarrow e^{i\theta} \\ |z|<1}} |f(z)| \leq 1 \quad \forall \theta \in [0, 2\pi]$.

3.0 Notation

Let $F = \sum_{n \in \mathbf{N}} a_n z^n$ be an element of $\mathbf{C}[[z]]$. We denote by F_n and F_n^* the square matrices of order $n+1$ defined by

$$F_n = \begin{pmatrix} a_0 & 0 & \cdots & 0 \\ a_1 & a_0 & \cdots & 0 \\ \vdots & \vdots & \ddots & \vdots \\ a_n & a_{n-1} & & a_0 \end{pmatrix} \qquad F_n^* = \begin{pmatrix} \overline{a_0} & \overline{a_1} & \cdots & \overline{a_n} \\ 0 & \overline{a_0} & \cdots & \overline{a_{n-1}} \\ \vdots & \vdots & \ddots & \vdots \\ 0 & 0 & & \overline{a_0} \end{pmatrix}$$

and define $\delta_n(F)$ by

$$\delta_n(F) = \det \begin{pmatrix} I_{n+1} & F_n^* \\ F_n & I_{n+1} \end{pmatrix}.$$

These determinants were introduced by Schur. We have for instance

$$\delta_o(F) = 1 - |a_0|^2.$$

If $A = (a_{i,j})$ is a matrix with complex coefficients, we denote by $\overline{A}$ the matrix $(\overline{a_{i,j}})$; we denote by I_{n+1} the square identity matrix of order $n+1$, by J_{n+1} the square matrix whose coefficients are 1 on the non-principal diagonal and zero elsewhere. We denote by $O_{i,j}$ the matrix with i rows and j columns all of whose coefficients are zero.

If P belongs to $\mathbf{C}[z]$ with $P = p_0 + p_1 z + \cdots + p_n z^n$, $p_n \neq 0$, we denote by P^* (the reciprocal polynomial of P) the polynomial $P^* = \overline{p}_n + \overline{p}_{n-1} z + \cdots \overline{p}_0 z^n$.

3.1 Properties of Schur's determinants

Lemma 3.1.1. *If F and G belong to $\mathbf{C}[[z]]$ then:*

a) $(FG)_n = F_n G_n = G_n F_n$; *if* $G(0) \neq 0$ $(FG^{-1})_n = F_n G_n^{-1} = G_n^{-1} F$

b) $J_{n+1} F_n^* = \overline{F}_n J_{n+1}$ *and* $J_{n+1} \overline{F}_n = F_n^* J_{n+1}$

c) $\delta_n(F) = \det(I_{n+1} - F_n F_n^*) = \det(I_{n+1} - F_n^* F_n)$

d) *if* $F \in \mathbf{R}[[z]]$, $\delta_n(F) = \det(I_{n+1} + J_{n+1} F_n) \det(I_{n+1} - J_{n+1} F_n)$

e) *if* $F(0) \neq 0$, $\delta_n(\frac{1}{F}) = (-1)^{n+1} |F(0)|^{-2(n+1)} \delta_n(F)$.

Proof. a) and b) can be easily verified. c) is deduced from the following equalities:

$$\begin{pmatrix} I_{n+1} & O_{n+1,n+1} \\ -F_n & I_{n+1} \end{pmatrix} \begin{pmatrix} I_{n+1} & F_n^* \\ F_n & I_{n+1} \end{pmatrix} = \begin{pmatrix} I_{n+1} & F_n^* \\ O_{n+1,n+1} & I_{n+1} - F_n F_n^* \end{pmatrix}$$

$$\begin{pmatrix} I_{n+1} & F_n^* \\ F_n & I_{n+1} \end{pmatrix} \begin{pmatrix} I_{n+1} & O_{n+1,n+1} \\ -F_n & I_{n+1} \end{pmatrix} = \begin{pmatrix} I_{n+1} - F_n^* F_n & F_n^* \\ O_{n+1,n+1} & I_{n+1} \end{pmatrix}.$$

d) is deduced from b) and c) by noticing that if $F \in \mathbf{R}[[z]]$, then

$$(I_{n+1} + J_{n+1} F_n)(I_{n+1} - J_{n+1} F_n) = I_{n+1} - J_{n+1} F_n J_{n+1} F_n = I_{n+1} - F_n^* F_n.$$

e) is deduced from c). ■

Lemma 3.1.2. *If* $F = \sum_{n \in \mathbf{N}} a_n z^n$ *is an element of* $\mathbf{C}[[z]]$ *such that* $|a_0| \neq 1$ *we set*

$$F^1 = \frac{F - a_0}{z(1 - \bar{a}_0 F)}.$$

Then F^1 *belongs to* $\mathbf{C}[[z]]$ *(called Schur transform of* F*) and*

$$\delta_n(F) = \delta_0(F)^{n+1} \delta_{n-1}(F^1) \qquad \forall n \in \mathbf{N}^*.$$

Proof. We wish to reduce the hermitian matrix $I_{n+1} - F_n F_n^*$. We notice first that we have the equality:

$$(I_{n+1} - \bar{a}_0 F_n)(I_{n+1} - a_0 F_n^*) - (F_n - a_0 I_{n+1})(F_n^* - \bar{a}_0 I_{n+1}) = (1 - |a_0^2|)(I - F_n F_n^*).$$

Assume that $|a_0| \neq 1$; then the matrix $I_{n+1} - \bar{a}_0 F_n$ is invertible.

Let G_n be the matrix $(I_{n+1} - \bar{a}_0 F_n)^{-1}(F_n - a_0 I_{n+1})$. An easy calculation shows that

$$I_{n+1} - F_n F_n^* = \frac{1}{1 - a_0 \bar{a}_0} (I_{n+1} - \bar{a}_0 F_n)(I_{n+1} - G_n G_n^*)(I_{n+1} - a_0 F_n^*) \qquad (*)$$

According to the property a) of the previous lemma we have

$$G_n = \left(\frac{F - a_0}{1 - \bar{a}_0 F} \right)_n.$$

Denote $F_{n-1}^1 = \left(\dfrac{F - a_0}{z(1 - \bar{a}_0 F} \right)_{n-1}$; then $G_n = \begin{pmatrix} O_{1,n} & O_{1,1} \\ F_{n-1}^1 & O_{n,1} \end{pmatrix}$. So we have

$$G_n G_n^* = \begin{pmatrix} O_{1,n} & O_{1,1} \\ F_{n-1}^1 & O_{n,1} \end{pmatrix} \begin{pmatrix} O_{n,1} & F_{n-1}^{1*} \\ O_{1,1} & O_{1,n} \end{pmatrix} = \begin{pmatrix} O_{1,1} & O_{1,n} \\ O_{n,1} & F_{n-1}^1 F_{n-1}^{1*} \end{pmatrix}$$

and

$$I_{n+1} - G_n G_n^* = \begin{pmatrix} 1 & O_{1,n} \\ O_{n,1} & I_n - F_{n-1}^1 F_{n-1}^{1*} \end{pmatrix};$$

hence $\det(I_{n+1} - G_n G_n^*) = \delta_{n-1}(F^1)$.

Since by $(*)$, $\delta_n(F) = (1 - a_0 \bar{a}_0)^{n+1} \det(I_{n+1} - G_n G_n^*)$, we obtain

$$\delta_n(F) = (\delta_0(F))^{n+1} \, \delta_{n-1}(F^1).$$

■

Remark 3.1.1. If F belongs to $\mathrm{C}[[z]]$ and can be written in the form $F = z^s R$ with $s \in \mathbf{N}$, $R \in \mathbf{C}[[z]]$, then $\delta_n(F) = \delta_{n-s}(R) \quad \forall n \geq s$. (We just apply the preceding lemma s times with $a_0 = 0$.)

We now give a criterion for hyper-rationality in $\mathbf{C}[[z]]$.

Theorem 3.1.1. *If F belongs to $\mathbf{C}[[z]]$, then F has the form $F = z^r \dfrac{P^*}{P}$ with $r \in \mathbf{N}$, $P(0) \neq 0$, $P \in \mathbf{C}[z]$, P and P^* relatively prime, and $d^0(P) = n_0$, if and only if $\delta_n(F) = 0 \quad \forall n \geq n_0 + r$ and $\delta_{n_0+r-1}(F) \neq 0$.*

For the proof of this theorem, we need the following lemmas.

Lemma 3.1.3. *If F belongs to $\mathbf{C}[[z]]$ and verifies $F(0) \neq 0$, then the equality $\delta_n(F) = 0$ is equivalent to the following property: there exists a polynomial P of $\mathbf{C}[z]$ and $r \in \mathbf{N}$ such that*

i) $d^0(P) = n - 2r \geq 0$

ii) $\operatorname{ord}(FP - P^*) \geq n + 1 - r$

iii) $P(0) \neq 0$.

Proof. The equality $\delta_n(F) = 0$ is equivalent to writing that there exist two column matrices B and C with $n+1$ rows such that either B or C is not identically zero and such that

$$(1)\begin{cases} B + F_n^* C = O_{n+1,1} \\ C + F_n B = O_{n+1,1} \end{cases} \Longleftrightarrow \begin{cases} J_{n+1}\overline{B} + F_n J_{n+1}\overline{C} = O_{n+1,1} \\ C + F_n B = O_{n+1,1} \end{cases}$$

$$\Longleftrightarrow \begin{cases} -(J_{n+1}\overline{iB} - iC) + F_n(iB - J_{n+1}\overline{iC}) = O_{n+1,1} \\ -(J_{n+1}\overline{B} - C) + F_n(B - J_{n+1}\overline{C}) = O_{n+1,1} \end{cases}$$

As one of the two column matrices $iB - J_{n+1}\overline{iC}$, $B - J_{n+1}\overline{C}$ is different from zero, $\delta_n(F) = 0$ implies the existence of a column matrix Q, non-identically zero, such that $F_n Q - J_{n+1}\overline{Q} = O_{n+1,1}$.

Conversely, if $Q \not\equiv 0$ satisfies $F_n Q - J_{n+1}\overline{Q} = O_{n+1,1}$, then by taking $B = Q$ and $C = -J_{n+1}\overline{Q}$, B and C satisfy (1) and $\delta_n(F) = 0$.

So the equation $\delta_n(F) = 0$ is equivalent to writing that there exists a polynomial Q of $\mathbf{C}[z]$, $Q \not\equiv 0$, $d^\circ(Q) \leq n$ such that, if $Q = q_0 + q_1 z + \cdots + q_n z^n$:

$$a_0 q_0 + (a_1 q_0 + a_0 q_1)z + \cdots + (a_n q_0 + \cdots + a_0 q_n)z^n - (\overline{q}_n + \overline{q}_{n-1} z + \cdots + \overline{q}_0 z^n) = 0,$$

i.e., $\operatorname{ord}(FQ - (\overline{q}_n + \overline{q}_{n-1} z + \cdots + \overline{q}_0 z^n)) \geq n + 1$.

Let r be the index of the first non-zero coefficient of Q. Then $q_{n-i} = 0$ for $i \in [0, \ldots, r-1]$ and $\overline{q}_{n-r} \neq 0$ because $F(0) \neq 0$. Hence:

$\mathrm{ord}((q_r + \cdots + q_{n-r}z^{n-2r})F - (\overline{q}_{n-r} + \cdots + \overline{q}_r z^{n-2r})) \geq n+1-r$. So the polynomial $P = q_r + \cdots + q_{n-r}z^{n-2r}$ satisfies i), ii), iii). ∎

Lemma 3.1.4. *If F belongs to $\mathbf{C}[[z]]$, with $F(0) \neq 0$ and there exist $n \in \mathbf{N}$ and $k \in \mathbf{N}^*$ such that $\delta_{n-1}(F) \neq 0$ and $\delta_n(F) = \delta_{n+1}(F) = \cdots = \delta_{n+k-1}(F)$, then there exists a polynomial $P \in \mathbf{C}[z]$ such that*

i) $d^\circ(P) = n$, $P(0) \neq 0$

ii) P *and* P^* *are relatively prime (if* $n \neq 0$*)*

iii) $\mathrm{ord}(FP - P^*) \geq n + 1 + [k/2]$.

Proof. i) Since $\delta_n(F)$ is zero, Lemma 3.1.3 implies that there exists a polynomial P in $\mathbf{C}[z]$ such that $P(0) \neq 0$, $\mathrm{d}^\circ(P) = n - 2r$ with $r \in \mathbf{N}$, and $\mathrm{ord}(FP - P^*) \geq n + 1 - r$.

Assume that $n \neq 0$ and $r \geq 1$, and set $Q = (1+z)P$, then $d^\circ(Q) = n + 1 - 2r = n - 1 - 2(r-1)$ and $\mathrm{ord}(FQ - Q^*) \geq n - (r-1)$ with $r - 1 \in \mathbf{N}$. By Lemma 3.1.3, $\delta_{n-1} = 0$. So $r = 0$.

ii) Assume that $n \neq 0$, and set $R = P\mathrm{gcd}(P, P^*)$. As $P(0) \neq 0$, $R(0) \neq 0$. We can suppose that $R = R^*$ because if $P = RQ$, $P^* = RS$ then $P^* = R^*Q^*$, $P = R^*S^*$. Then R^* divides P and P^* and there exists $\lambda \in \mathbf{C}^*$ such that $R^* = \lambda R$. Hence $R = \overline{\lambda} R^* = \dfrac{R^*}{\lambda}$ and $\lambda\overline{\lambda} = 1$. By setting $\lambda = e^{i\theta}$ we have $e^{i\theta/2}R = (e^{i\theta/2}R)^*$.

Assume that $d^\circ R \geq 1$. Take $Q = \dfrac{P}{R}(1+z)^{d^\circ R - 1}$. Then $\mathrm{ord}(FQ - Q^*) \geq n + 1 \geq n$ and $d^\circ Q = n - 1$, $Q(0) \neq 0$ and by Lemma 3.1.3 $\delta_{n-1} = 0$. So $d^\circ R = 0$ and P and P^* are relatively prime.

iii) Verify that for $0 \leq s \leq k-1$ the inequality $\mathrm{ord}(FP - P^*) \geq n + 1 + [\frac{s}{2}]$ implies $\mathrm{ord}(FP - P^*) \geq n + 1 + [\frac{s+1}{2}]$.

If s is even this is true because $[\frac{s}{2}] = [\frac{s+1}{2}]$.

If s is odd set $s = 2p + 1$ with $s \leq k - 1$. So we suppose that

$$\mathrm{ord}(FP - P^*) \geq n + 1 + p. \tag{1}$$

By hypothesis $\delta_{n+2p+1} = 0$. By Lemma 3.1.3, there exist a polynomial $H \in \mathbf{C}[z]$ and $r \in \mathbf{N}$ such that

$$d^\circ H = n + 2p + 1 - 2r = (n-1) - 2(r-p-1)$$

$$\operatorname{ord}(FH - H^*) \geq n + 2p + 2 - r \tag{2}$$

$$H(0) \neq 0.$$

Since $\delta_{n-1} \neq 0$, necessarily $r - p - 1 < 0$; so $r \leq p$. Moreover the inequalities (1) and (2) imply $\operatorname{ord}(H^*P - P^*H) \geq \min(n+1+p, n+2p+2-r) = n+p+1$.

Set $T = H^*P - P^*H$. Assume that $T \not\equiv 0$ and set $l = \operatorname{ord}(T)$. Then:

$T = t_l z^l + \cdots - \bar{t}_l z^{2n+2p+1-2r-l}$ with $l \geq n+p+1$ and $l \leq 2n+2p+1-2r-l$. Hence $n+p+1 \leq l \leq n+p-r$ and $r \leq -1$.

Consequently $T \equiv 0$ and $H^*P = P^*H$. As P and P^* are relatively prime, $H^* = CP^*$, $H = CP$ with $C = C^*$, $C(0) \neq 0$, which together with (2) implies

$$\operatorname{ord}(FP - P^*) \geq n + 2p + 2 - r \geq n + p + 2 = n + 1 + [\frac{s+1}{2}].$$

Proof of Theorem 3.1.1.

By Remark 3.1.1, we can suppose that $F(0) \neq 0$.

a) Sufficient condition: Assume that $\delta_{n_0-1}(F) \neq 0$ and $\delta_n(F) = 0 \quad \forall n \geq n_0$. By Lemma 3.1.4 there exists a polynomial P that has the required properties.

b) Necessary condition: If $F = P^*/P$, and P has the required properties then $\operatorname{ord}(FP - P^*) = +\infty$. Set $Q_i = (1+z)^i P$ where $i \in \mathbf{N}$. Then $\operatorname{ord}(FQ_i - Q_i^*) = +\infty$ and by Lemma 3.1.3 $\delta_{n_0+i} = 0 \quad \forall i \in \mathbf{N}$. Moreover, if $\delta_{n_0-1} = 0$ (if $n_0 \neq 0$), the sufficient condition shows that $F = S^*/S$ with $d^\circ S \leq n_0 - 1$. Hence P and P^* cannot be relatively prime. ∎

3.2 Characterization of functions belonging to $\mathcal{M}$

Notation 3.2.1. The set of functions analytic and bounded by 1 on $D(0,1)$ is denoted by $\mathcal{M}$.

Theorem 3.2.1. *If F belongs to $\mathbf{C}[[z]]$, then F is the Taylor series in the neighborhood of zero of a function in $\mathcal{M}$ if and only if*

- *either $\delta_n(F) > 0 \quad \forall n \in \mathbf{N}$*
- *or there exists $n_0 \in \mathbf{N}$ such that $\delta_n > 0 \quad \forall n < n_0$ and $\delta_n = 0 \quad \forall n \geq n_0$.*

The proof of this theorem uses the Schur algorithm. More precisely, we need the following lemma.

Lemma 3.2.1. *Let F be an element of $\mathbf{C}[[z]]$ such that $\delta_0(F)$, $\delta_1(F)$,...,$\delta_n(F)$ are different from zero.*

a) Then the Schur transforms F^i are defined up to rank $n+1$

b) $\forall i \in [0,1,\ldots,n]$, there exist two polynomials A_n and B_n in $\mathbf{C}[z]$ such that:

i) $d^\circ A_i \leq i$, $d^\circ Q_i \leq i$, $Q_i(0) = 1$

$$F = \frac{A_i + z\,\widetilde{Q}_i F^{i+1}}{Q_i + z\,\widetilde{A}_i F^{i+1}} \quad \text{with} \quad \begin{cases} \widetilde{A}_i(z) = z^i \overline{A}_i(\frac{1}{z}) \\ \widetilde{Q}_i(z) = z^i \overline{Q}_i(\frac{1}{z}) \end{cases}$$

$$ii)\ Q_i\,\widetilde{Q}_i - A_i\,\widetilde{A}_i = \omega_i z^i \ \text{with} \begin{cases} \omega_i = \prod_{k=0}^{i} (1 - \gamma_k \overline{\gamma}_k) \\ \gamma_k = F^k(0) \end{cases}$$

$$iii)\ F - \frac{A_i}{Q_i} = \omega_i \gamma_{i+1} z^{i+1} + \cdots$$

c) $\delta_{n+k}(F) = (\delta_0(F))^{n+1+k} \left(\delta_0(F^1)\right)^{n+k} \cdots \left(\delta_0(F^{n-1})\right)^{k+1} \delta_k(F^n) \quad \forall k \in \mathbf{N}.$

Proof. a) First we prove by induction on i for $i \leq n+1$ that F^i is defined. Assume that F, F^1,..., F^i are defined where $i \leq n$. Then by Lemma 3.1.2, we have

$$\delta_k(F) = (\delta_0(F))^{k+1} \left(\delta_0(F^1)\right)^k \cdots \left(\delta_0(F^{i-1})\right)^{k-i+1} \delta_{k-i}(F^i) \qquad \forall k \geq i;$$

In particular for $k = i$, we have $\delta_o(F^i) \neq 0$ because $\delta_i(F) \neq 0$; so $F^{i+1}$is defined.

b) We verify property b) by induction on $i \leq n$:

For $i = 0$, we have $F^1 = \dfrac{F - a_0}{z(1 - \overline{a}_0 F)}$, $F = \dfrac{a_0 + zF^1}{1 + z\overline{a}_0 F^1}$; so

$A_0 = a_0 = \gamma_0$, $Q_0 = 1$, $F - \dfrac{A_0}{Q_0} = zF^1(1 - \overline{a}_0 F) = \gamma_1(1 - \gamma_0\overline{\gamma}_0)z + \cdots$ and

$$Q_0\,\widetilde{Q}_0 - A_0\,\widetilde{A}_0 = 1 - \gamma_0\overline{\gamma}_0 = \omega_0.$$

Assume that the properties b) are verified for the rank $i-1$; then from the equality $F^i = \dfrac{\gamma^i + zF^{i+1}}{1+z\overline{\gamma}_i F^{i+1}}$ and the recurrence hypothesis we obtain

$$F = \frac{(A_{i-1} + z\gamma_i\, \widetilde{Q}_{i-1}) + z(\overline{\gamma}_i A_{i-1} + z\, \widetilde{Q}_{i-1})F^{i+1}}{(Q_{i-1} + z\gamma_i\, \widetilde{A}_{i-1}) + z(\gamma_i Q_{i-1} + z\, \widetilde{A}_{i-1})F^{i+1}} \ .$$

So Q_i and A_i are defined by the equality

$$\begin{pmatrix} Q_i & A_i \\ \tilde{A}_i & \tilde{Q}_i \end{pmatrix} = \begin{pmatrix} 1 & \gamma_i z \\ \overline{\gamma_i} & z \end{pmatrix} \begin{pmatrix} Q_{i-1} & A_{i-1} \\ \tilde{A}_{i-1} & \tilde{Q}_{i-1} \end{pmatrix}.$$

So $Q_i(0) = Q_{i-1}(0) = 1$, $d^\circ Q_i \le i$ and $d^\circ A_i \le i$.

And by calculating the determinant of the matrices we obtain

$$Q_i\widetilde{Q}_i - A_i\widetilde{A}_i \ = z(1-\gamma_i\overline{\gamma}_i)(Q_{i-1}\widetilde{Q}_{i-1} - A_{i-1}\widetilde{A}_{i-1}\ ) = \omega_i z^i.$$

Moreover

$$F - \frac{A_i}{Q_i} = \frac{A_i + z\widetilde{Q}_i F^{i+1}}{Q_i + z\widetilde{A}_i F^{i+1}} - \frac{A_i}{Q_i} = \frac{z(Q_i\widetilde{Q}_i - A_i\widetilde{A}_i)}{Q_i(Q_i + z\widetilde{A}_i F^{i+1}} F^{i+1} = \omega_i\gamma_{i+1}z^{i+1} + \cdots.$$

c) We use Lemma 3.1.2.

Proof of Theorem 3.2.1.

Necessary condition: Consider $f \in \mathcal{M}$ and $F = \sum_{n\in\mathbf{N}} a_n z^n$, the Taylor series of f at zero.

Either $|a_0| = 1$ and then by the maximum principle we have $f(z) = a_0 \quad \forall z \in D(0,1)$ and $F = a_0$. Then the Theorem 3.1.1 shows that $\delta_n(F) = 0 \quad \forall n \in \mathbf{N}$.

Or $|a_0| < 1$ and then $\delta_0(F) > 0$. Moreover the function $f^1 \colon z \longrightarrow \dfrac{f(z) - a_0}{1 - \overline{a}_0 f(z)}$ belongs to $\mathcal{M}$ because the map $u \longrightarrow \dfrac{u - a_0}{1 - \overline{a}_0 u}$ is an automorphism of $D(0,1)$ if $|a_0| < 1$; furthermore the Taylor series of f^1 at zero is F^1.

Similarly either $|f^1(0)| = 1$ and then F^1 is equal to a constant of modulus 1 and $\delta_n(F^1) = 0 \quad \forall n \in \mathbf{N}$. By Lemma 3.1.2 $\delta_n(F) = 0 \quad \forall n \ge 1$, $\delta_0(F) > 0$. Or $|f^1(0)| < 1$ and then $\delta_0(F^1) > 0$. By Lemma 3.1.2 $\delta_1(F) > 0$ and $\delta_0(F) > 0$.

By reiterating the process, we obtain the result.

Sufficient condition: a) Assume that $\delta_n(F) > 0 \quad \forall n \in \mathbf{N}$. Then according to the Lemma 3.2.1, F^n is defined and $\delta_n(F^n) > 0 \quad \forall n \in \mathbf{N}$. Consider A_n and Q_n defined in Lemma 3.2.1. Then $Q_n\widetilde{Q}_n - A_n\widetilde{A}_n = \omega_n > 0$. Hence $|Q_n(z)|^2 - |A_n(z)|^2 > 0$ if $|z| = 1$.

Verify by induction on n that Q_n has no zero in $\overline{D(0,1)}$. This is true for $n = 0$ because $Q_0 = 1$. Assume that Q_{n-1} is non-zero on $\overline{D(0,1)}$. We know from the proof of Lemma 3.2.1 that $Q_n(z) = Q_{n-1}(z) + z\gamma_n A_{n-1}(z)$ with $|\gamma_n| < 1$.

As $|z\gamma_n A_{n-1}(z)| < |Q_{n-1}(z)|$ if $|z| = 1$, by Rouché's theorem Q_n has no zero in $\overline{D(0,1)}$. So the function $z \longrightarrow \dfrac{A_n(z)}{Q_n(z)}$ belongs to $\mathcal{M}$, $\forall n \in \mathbf{N}$. Cauchy's inequalities imply that all the coefficients of the series A_n/Q_n are bounded by 1 and by a(iii) of Lemma 3.2.1 the coefficients of F are bounded by 1. So the function $f : z \longrightarrow f(z) = \sum_{n=0}^{+\infty} a_n z^n$ is analytic on $D(0,1)$. Moreover by a(iii) of Lemma 3.2.1 we again have $|f(z) - \dfrac{A_n(z)}{Q_n(z)}| \leq \dfrac{2|z|^{n+1}}{1-|z|} \quad \forall z \in D(0,1)$.

Hence $|f(z)| \leq 1 + \dfrac{2|z|^{n+1}}{1-|z|}$ because A_n/Q_n belongs to $\mathcal{M}$.

By letting n tend to infinity for z fixed in $D(0,1)$ we obtain

$$|f(z)| \leq 1 \qquad \forall z \in D(0,1) \text{ and } f \in \mathcal{M}.$$

b) Assume that there exists $n_0 \in \mathbf{N}$ such that $\delta_i(F) > 0 \quad \forall i \leq n_0 - 1$ and $\delta_i(F) = 0 \quad \forall i \geq 0$. According to Lemma 3.2.1 F^i is defined up to rank n_0 and $\delta_i(F^{n_0}) = 0 \quad \forall i \in \mathbf{N}$. Theorem 3.1.1 shows that F^{n_0} is a constant ε of modulus 1. So F can be written in the form

$$F = \frac{A_{n_0-1} + z\varepsilon\, \widetilde{Q}_{n_0-1}}{Q_{n_0-1} + z\varepsilon\, \widetilde{A}_{n_0-1}}\,.$$

As in a), $Q_{n_0-1} + z\varepsilon A_{n_0-1}$ has no zero in $\overline{D(0,1)}$ and the function

$$f\colon z \longrightarrow \frac{A_{n_0-1}(z) + z\varepsilon\, \widetilde{Q}_{n_0-1}(z)}{Q_{n_0-1}(z) + z\bar{\varepsilon}\, \widetilde{A}_{n_0-1}(z)}$$

is analytic on $D(0,1)$. It is easy to verify that $|f(z)| < 1$ if $|z| < 1$, since $|A_{n_0-1}(z)| < |Q_{n_0-1}(z)|$ if $|z| < 1$. ■

To state the various corollaries of Theorem 3.2.1 we need the following notation.

Notation. Let f be a complex analytic function in the neighborhood of zero. We denote by F_n, F_n^*, $\delta_n(f) = \delta_n(F)$ the Schur matrices and determinants associated to the Taylor series F of f at zero. We say that f has indefinite rank if there is no N such that the determinants $\delta_n(f)$ do not all vanish for $n \geq N$.

Corollary 3.2.1 *If $f \in \mathcal{M}$ has indefinite rank, then $\delta_n(f) > 0 \quad \forall n \in \mathbf{N}$ and the sequence $(\omega_n)_{n\in\mathbf{N}}$ where $\omega_n = \dfrac{\delta_n(f)}{\delta_{n-1}(f)}$ is positive decreasing and converges.*

Proof. By Theorem 3.2.1 we have $\delta_n(f) > 0 \quad \forall n \in \mathbf{N}$. Moreover by Lemma 3.2.1 we have

$$\frac{\delta_n(f)}{\delta_{n-1}(f)} = \frac{(\delta_0(F))^{n+1}(\delta_0(F^1))^n \cdots \delta_0(F^n)}{(\delta_0(F))^n(\delta_0(F^1))^{n-1} \cdots \delta_0(F^{n-1})} = \delta_0(F)\delta_0(F^1)\cdots\delta_0(F^n) = \omega_n$$

where $\omega_n = \prod_{i=0}^{n}(1-\gamma_i\overline{\gamma}_i)$. As $|\gamma_i| < 1 \quad \forall i \in \mathbf{N}$, the sequence (ω_n) is decreasing positive; hence it converges. ∎

Corollary 3.2.2. *If f belongs to $\mathcal{M}$, then the hermitian matrices $I_{n+1} - F_nF_n^*$ are positive $\forall n \in \mathbf{N}$.*

Proof. The proof of Lemma 3.1.2 shows that if $\delta_0(f) > 0$, the matrix $I_{n+1} - F_nF_n^*$ is similar to the matrix

$$\begin{pmatrix} 1 & O_{1,n} \\ O_{n,1} & I_n - F_{n-1}^1F_{n-1}^{1*} \end{pmatrix}.$$

If $\delta_0(f) > 0, \ldots, \delta_n(f) > 0$, we obtain by induction that the matrix $I_{n+1} - F_nF_n^*$ is similar to the matrix

$$\begin{pmatrix} I_n & O_{n,1} \\ O_{1,n} & I_1 - F_0^nF_o^{n*} \end{pmatrix}.$$

As $I_1 - F_0^nF_0^{n*} = 1 - |\gamma_n|^2 = \delta_0(F^n) > 0$, the matrix $I_{n+1} - F_nF_n^*$ is similar to the matrix I_{n+1}. If f has indefinite rank then the hermitian matrices are positive $\forall n \in \mathbf{N}$.

On the other hand if f has finite rank n_0 then $F^{n_0} = \varepsilon$ with $|\varepsilon| = 1$.

Then for $n \leq n_0 - 1$ we obtain that $I_{n+1} - F_nF_n^*$ is similar to the matrix I_{n+1}.

For $n \geq n_0$, $I_{n+1} - F_n F_n^*$ is similar to the matrix

$$\begin{pmatrix} I_{n_0} & O_{n_0,n-n_0+1} \\ O_{n-n_0+1,n_0} & I_{n-n_0+1} - F_{n-n_0}^{n_0} F_{n-n_0}^{n_0 *} \end{pmatrix}.$$

As $F^{n_0} = \varepsilon$, we have $F_{n-n_0}^{n_0} F_{n-n_0}^{n_0 *} = I_{n-n_0+1}$. Hence $I_{n+1} - F_n F_n^*$ is similar to the matrix

$$\begin{pmatrix} I_{n_0} & O_{n_0,n-n_0+1} \\ O_{n-n_0+1,n_0} & O_{n-n_0+1,n-n_0+1} \end{pmatrix}$$

and $I_{n+1} - F_n F_n^*$ is positive $\forall n \in \mathbf{N}$. ■

Corollary 3.2.3. *If $f \in \mathcal{M}$ is such that the Taylor series F of f at zero belongs to $\mathbf{R}[[z]]$, then the matrices $I_{n+1} \pm J_{n+1} F_n$ are positive hermitian $\forall n \in \mathbf{N}$.*

Proof. By part b) of Lemma 3.1.1 the matrices are hermitian.

On $\mathbf{C}^{n+1}$ the scalar product is defined:

$$(X|Y) = \sum_{i=0}^{n} \overline{x}_i y_i \quad \text{where } X = (x_0, \dots, x_n) \in \mathbf{C}^{n+1},\ Y = (y_0, \dots, y_n) \in \mathbf{C}^{n+1}.$$

According to the preceding corollary we have: $(I_{n+1} - F_n F_n^* X|X) \geq 0 \quad \forall X \in \mathbf{C}^{n+1}$. So $(F_n F_n^* X|X) = (F_n^* X|F_n^* X) \leq (X|X) \quad \forall X \in \mathbf{C}^{n+1}$. Hence $(F_n^* J_{n+1} X|J_{n+1} X) \leq (J_{n+1} X|J_{n+1} X) = (X|X)$ because J_{n+1} is an isometry.

So $\|F_n^* J_{n+1}\| = \|J_{n+1} F_n\| \leq 1$. By the Cauchy–Schwarz inequality we have:

$|(J_{n+1} F_n X|X)| \leq \|J_{n+1} F_n X\| \|X\| \leq \|X\|^2$. So $\pm(J_{n+1} F_n X|X) \leq (X|X)$; and the matrices $I_{n+1} \pm J_{n+1} F_n$ are positive hermitian. ■

The aim of the following theorem is to express $\lim_{n \to +\infty} \dfrac{\delta_n(f)}{\delta_{n-1}(f)}$ for any function of $\mathcal{M}$ with infinite rank. We recall the following definitions and properties.

Definition 3.2.2. *Let g be a positive real function of $L^1(T)$, where T is the circumference of $D(0,1)$. We denote by $\mathcal{L}(g)$ the expression*

$$\mathcal{L}(g) = \exp \frac{1}{2\pi} \int_{-\pi}^{+\pi} \log g(e^{i\theta}) d\theta \qquad \textit{if } \log g \in L^1(T)$$

$$\mathcal{L}(g) = 0 \qquad \textit{if } \log g \notin L^1(T).$$

We recall the following properties:

i) if $g_1 \in L^1(T)$, $g_2 \in L^1(T)$, $g_1 g_2 \in L^1(T)$ are all real positive then

$$\mathcal{L}(g_1 g_2) = \mathcal{L}(g_1)\mathcal{L}(g_2)$$

ii) if $g \in L^1(T)$ and is real positive then

$$\mathcal{L}(g) = \inf_{P \in \mathbf{C}[z],\, P(0)=1} \frac{1}{2\pi} \int_0^{2\pi} |P(e^{i\theta})|^2 g(e^{i\theta}) d\theta$$

iii) if $\alpha \geq 1$ and f belongs to H^α then $\lim_{r \longrightarrow 1-} f(re^{i\theta})$ exists for almost all $\theta \in [0, 2\pi]$ and the function $\widehat{f}\colon e^{i\theta} \in T \longrightarrow \lim_{r \longrightarrow 1-} f(re^{i\theta})$ belongs to $L^\alpha(T)$

iv) Let f be a function of H^1, such that $f(0) \neq 0$. Then

$$\frac{1}{2\pi} \int_0^{2\pi} \log |\widehat{f}(e^{i\theta})| d\theta \geq \log |f(0)|.$$

In particular if f and $1/f$ belong to H^1 then $\mathcal{L}(|\widehat{f}|) = |f(0)|$.

Theorem 3.2.2. *Let f be a function of $\mathcal{M}$ of infinite rank and $\widehat{f} \in L^2(T)$ be defined by $\widehat{f}(e^{i\theta}) = \lim_{r \longrightarrow 1-} f(re^{i\theta})$ for almost all $e^{i\theta} \in T$. Then*

$$\mathcal{L}(1 - |\widehat{f}|^2) = \lim_{n \longrightarrow +\infty} \frac{\delta_n(f)}{\delta_{n-1}(f)}\,.$$

Proof. As f has indefinite rank, the sequence f_n of Schur transforms is defined for all $n \in \mathbf{N}$, and f_n belongs to $\mathcal{M}$, $\forall n \in \mathbf{N}$. Moreover by Lemma 3.2.1 and Theorem 3.2.1, f can be written in the form

$$f(z) = \frac{A_n(z) + z\, \widetilde{Q}_n(z) f_{n+1}(z)}{Q_n(z) + z\, \widetilde{A}_n(z) f_{n+1}(z)}$$

where Q_n has no zero in $\overline{D(0,1)}$ and $|Q_n(z)|^2 - |A_n(z)|^2 = \omega_n = \dfrac{\delta_n(f)}{\delta_{n-1}(f)} > 0$ if $|z| = 1$.

Similarly define $\widehat{f}_n \in L^2(T)$ by $\widehat{f}_n(e^{i\theta}) = \lim_{r \longrightarrow 1-} f(re^{i\theta})$ p.p.

An easy calculation shows that

$$1-|\widehat{f}(e^{i\theta})|^2 = \frac{\left(|Q_n(e^{i\theta})|^2 - |A_n(e^{i\theta})|^2\right)\left(1-|\widehat{f}_{n+1}(e^{i\theta})|^2\right)}{|Q_n(e^{i\theta}) + e^{i\theta}\ A(e^{i\theta})\widehat{f}_{n+1}(e^{i\theta})|^2}. \tag{1}$$

As the function $h: z \longrightarrow Q_n(z) + z\ A_n(z) f_{n+1}(z)$ is such that h and $1/h$ are analytic and bounded on $D(0,1)$, according to iv) we have

$$\mathcal{L}(|\widehat{h}|^2 = 1 \quad \text{because } Q_n(0) = 1.$$

Moreover, (1) implies that $\mathcal{L}(1-|\widehat{f}|^2) \leq \omega_n \mathcal{L}(|\widehat{h}|^2) = \omega_n$. So

$$\mathcal{L}(1-|\widehat{f}|^2) \leq \lim_{n\longrightarrow+\infty} \omega_n.$$

We now prove the inequality in the other direction.

By properties i) and iv) we have, since Q_n has no zero in $\overline{D(0,1)}$ and $Q_n(0) = 1$:

$$\mathcal{L}(1-\left|\frac{A_n}{Q_n}\right|^2) = \mathcal{L}(|Q_n|^2 - |A_n|^2) = \omega_n.$$

Consider a polynomial $P \in \mathbf{C}[z]$ such that $P(0) = 1$; by property ii) we have

$$\omega_n \leq \frac{1}{2\pi}\int_0^{2\pi} |P(e^{i\theta})|^2 d\theta - \frac{1}{2\pi}\int_0^{2\pi} |P(e^{i\theta})|^2 \left|\frac{A_n(e^{i\theta})}{Q_n(e^{i\theta})}\right|^2 d\theta.$$

As the function $z \longrightarrow \dfrac{P(z)A_n(z)}{Q_n(z)}$ belongs to H^2, it can be written in the form

$$\frac{P(z)A_n(z)}{Q_n(z)} = \sum_{i=0}^{+\infty} v_i z^i \qquad \text{with } \sum_{i=0}^{+\infty} |v_i|^2 < +\infty \quad \forall z \in D(0,1).$$

Hence

$$\frac{1}{2\pi}\int_0^{2\pi} |P(e^{i\theta})|^2 \left|\frac{A_n(e^{i\theta})}{Q_n(e^{i\theta})}\right|^2 d\theta = \sum_{i=0}^{+\infty} |v_i|^2$$

and

$$\lim_{n\longrightarrow\infty} \omega_n \leq \frac{1}{2\pi}\int_0^{2\pi} |P(e^{i\theta})|^2 d\theta - \sum_{i=0}^{+\infty} |v_i|^2 \qquad \forall n \in \mathbf{N}.$$

As the functions Pf and PA_n/Q_n have the same Taylor expansion in zero until the order n included, when n tends to infinity we have

$$\lim_{n\longrightarrow\infty} \omega_n \leq \frac{1}{2\pi}\int_0^{2\pi} |P(e^{i\theta})|^2 d\theta - \frac{1}{2\pi}\int_0^{2\pi} |P(e^{i\theta})|^2 |\widehat{f}(e^{i\theta})|^2 d\theta.$$

This inequality is true for all polynomials P in $\mathbf{C}[z]$ with $P(0)=1$, so by ii) we obtain

$$\lim_{n\longrightarrow+\infty} \omega_n \leq \mathcal{L}(1-|\widehat{f}|^2).$$

So

$$\lim_{n\longrightarrow+\infty} \omega_n = \lim_{n\longrightarrow+\infty} \frac{\delta_n(f)}{\delta_{n-1}(f)} = \mathcal{L}(1-|\widehat{f}|^2). \qquad \blacksquare$$

3.3 Generalized Schur algorithm

The following lemma resembles lemma 3.1.2 with $|a_0|=1$.

Lemma 3.3.1. *Let $F = a_0 + a_k z^k + \cdots$ be an element of $\mathbf{C}[[z]]$ such that $|a_0|=1$ and $|a_k|\neq 0$. Set*

$$C = \sum_{n\in\mathbf{N}} c_i z^i = \frac{a_0 z^k}{F-a_0} \in \mathbf{C}[[z]],$$

and

$$Q = c_0 + c_1 z + \cdots + c_{k-1}z^{k-1} - (\overline{c}_{k-1}z^{k+1} + \cdots + \overline{c}_0 z^{2k}).$$

Then

i) $\operatorname{ord}\left(F(Q-z^k) - a_0 Q\right) \geq 2k$, $\operatorname{ord}\left(\overline{a}_0 FQ - (Q+z^k)\right) \geq 2k$

ii) if $\operatorname{ord}\left(\overline{a}_0 FQ - (Q+z^k)\right) = 2k+s$ *with* $s \neq +\infty$ *and if d is the first non-zero coefficient of the series* $\overline{a}_0 FQ - (Q+z^k)$ *, then*

$$\delta_n(F) = (-1)^{k+s}|a_k|^{2(k+s)}|d|^{2(n+1-2k-s)}\delta_{n-2k-s}(H) \qquad \forall n \geq 2k+s$$

where

$$H = z^s \frac{F(Q-z^k) - a_0 Q}{\overline{a}_0 FQ - (Q+z^k)}$$

iii) if $\overline{a}_0 FQ - (Q+z^k) \equiv 0$ *then* $\delta_n(F) = (-1)^{n+1-k}|a_k|^{2(n+1-k)} \qquad \forall n \geq 2k.$

Proof. **i)** If we set $\dfrac{a_0 z^k}{F-a_0} = Q + z^k R$, then we have

$$R = c_k + (c_{k+1}+\overline{c}_{k-1})z + \cdots + (c_{2k}+\overline{c}_0)z^k + c_{2k+1}z^{2k+1} + \cdots \tag{1}$$

We easily establish that

$$F(Q-z^k) - a_0 Q = z^k(a_0 - F)(1+R), \; \overline{a}_0 FQ - (Q+z^k) = \overline{a}_0 z^k (F-a_0)R.$$

Remark. We have in particular $H = -z^s \dfrac{1+R}{\overline{a}_0 R}$ with $s = \operatorname{ord} R$.

ii) Define the square matrix of order $n+1$, M_{n+1-k} by

$$M_{n+1-k} = \begin{pmatrix} a_k & 0 & \cdots & 0 \\ a_{k+1} & a_k & \cdots & 0 \\ \vdots & \vdots & \ddots & \vdots \\ a_n & a_{n-1} & \cdots & a_k \end{pmatrix}.$$

Then $M_{n+1-k}^{-1} = \dfrac{1}{a_0} C_{n-k}$ where C_{n-k} is the Schur matrix of order $n+1-k$ associated with the series C. Set

$$C_{n-k} = \begin{pmatrix} C_{k,n+1-2k} & O_{k,k} \\ C_{n+1-2k} & D_{n+1-2k,k} \end{pmatrix}, \quad M = \begin{pmatrix} O_{k,n+1-k} & O_{k,k} \\ M_{n+1-k} & O_{n+1-k,k} \end{pmatrix}$$

where $C_{k,n+1-2k}$ (resp. $D_{n+1-2k,k}$) is a matrix with k (resp. $n+1-2k$) rows and $n+1-2k$ (resp. k) columns and C_{n+1-2k} is a square matrix of order $n+1-2k$ defined by

$$C_{n+1-2k} = \begin{pmatrix} c_k & c_{k-1} & \cdots & c_0 & \cdots & 0 \\ c_{k+1} & c_k & \cdots & c_1 & \cdots & 0 \\ \vdots & \vdots & \ddots & \vdots & \ddots & \vdots \\ c_{n-k} & c_{n-k-1} & \cdots & c_{n-2k} & \cdots & c_k \end{pmatrix}.$$

Then $F_n = a_0 I_{n+1} + M$ and $I_{n+1} - F_n F_n^* = -a_0 M^* - \overline{a}_0 M - MM^*$. Set

$$S = \begin{pmatrix} O_{n+1-k,k} & M_{n+1-k}^{-1} \\ I_k & O_{k,n+1-k} \end{pmatrix}.$$

Then we have

$$S(I_{n+1} - F_n F_n^*)S^* = -a_0 SMS^* - \overline{a}_0 SMS^* - SMM^*S^*.$$

We notice that $SM = M^*S^*$ with

$$SM = \begin{pmatrix} I_{n+1-k} & O_{n+1-k,k} \\ O_{k,n+1-k} & O_{k,k} \end{pmatrix} = \begin{pmatrix} I_k & O_{k,n+1-k} & O_{k,k} \\ O_{n+1-2k,k} & I_{n+1-2k} & O_{n+1-2k,k} \\ O_{k,k} & O_{k,n+1-2k} & O_{k,k} \end{pmatrix}.$$

Since we have

$$a_0 S = \begin{pmatrix} O_{k,k} & C_{k,n+1-2k} & O_{k,k} \\ O_{n+1-2k,k} & C_{n+1-2k} & D_{n+1-2k,k} \\ a_0 I_k & O_{k,n+1-2k} & O_{k,k} \end{pmatrix}$$

we easily obtain that the matrix $S(I - FF^*)S^*$ is equal to the matrix

$$-\begin{pmatrix} I_k & C_{k,n+1-2k} & \overline{a_0} I_k \\ C^*_{k,n+1-2k} & I_{n+1-2k} + C_{n+1-2k} + C^*_{n+1-2k} & O_{n+1-2k,k} \\ a_0 I_k & O_{k,n+1-2k} & O_{k,k} \end{pmatrix}.$$

Then let A and G be the matrices defined by

$$A = \begin{pmatrix} I_k & O_{k,n+1-2k} & O_{k,k} \\ O_{n+1-2k,k} & I_{n+1-2k} + C_{n+1-2k} + C^*_{n+1-2k} & O_{n+1-2k,k} \\ O_{k,k} & O_{k,n+1-2k} & -I_k \end{pmatrix}$$

$$G = \begin{pmatrix} I_k & C_{k,n+1-2k} & \overline{a_0} I_k \\ O_{n+1-2k,k} & I_{n+1-2k} & O_{n+1-2k,k} \\ O_{k,k} & C_{k,n+1-2k} & \overline{a_0} I_k \end{pmatrix}.$$

We easily verify that

$$S(I_{n+1} - F_n F_n^*)S^* = -G^* A G.$$

As $\det GG^* = 1$, $\det SS^* = (\dfrac{1}{|a_k|})^{2(n+1-k)}$, we obtain

$$\delta_n(F) = (-1)^{n+1-k} |a_k|^{2(n+1-k)} \det(I_{n+1-2k} + C_{n+1-2k} + C^*_{n+1-2k}) \tag{2}$$

Since the matrix $C_{n+1-2k} + C^*_{n+1-2k}$ is equal to the matrix

$$\begin{pmatrix} c_k + \overline{c_k} & c_{k-1} + \overline{c_{k+1}} & \cdots & c_0 + \overline{c_{2k}} & \cdots & \overline{c_{n-k}} \\ c_{k-1} + \overline{c_{k+1}} & c_k + \overline{c_k} & \cdots & c_1 + \overline{c_{2k-1}} & \cdots & \overline{c_{n-k-1}} \\ \vdots & \vdots & \ddots & \vdots & \ddots & \vdots \\ c_{n-k} & c_{n-k-1} & \cdots & \cdots & \cdots & c_k + \overline{c_k} \end{pmatrix}$$

we obtain

$$I_{n+1-2k} + C_{n+1-2k} + C^*_{n+1-2k} = I_{n+1-2k} + R_{n-2k} + R^*_{n-2k} \tag{3}$$

where R_{n-2k} is the Schur matrix of order $n+1-2k$ associated to the series R. By (2) and (3) we have

$$\delta_n(F) = (-1)^{n+1-k} |a_k|^{2(n+1-k)} \det(I_{n+1-2k} + R_{n-2k} + R^*_{n-2k}). \tag{4}$$

So if $R \equiv 0$, we obtain c).

We have the equation

$$I_{n+1} + R_n + R_n^* = (I_{n+1} + R_n)(I_{n+1} + R_n^*) - R_n R_n^* \qquad \forall n \in \mathbf{N}.$$

So we deduce for $R(0) \neq 0$ that

$$\det(I_{n+1-2k} + R_{n-2k} + R^*_{n-2k}) = \left(-|R(0)|^2\right)^{n+1-2k} \delta_{n-2k}\left(\frac{1+R}{R}\right). \quad (5)$$

As $\overline{a}_0 R(0) a_k = d$, (3) and (4) imply that

$$\delta_n(F) = (-1)^k |a_k|^{2k} |d|^{2(n+1-2k)} \delta_{n-2k}(H).$$

If $R(0) = 0$ and $R \not\equiv 0$, we set $R = z^s V$ with $V \in \mathbf{C}[[z]]$, $V(0) \neq 0$; similarly we have, since $1 + R(0) = 1$

$$\det(I_{n+1-2k} + R_{n-2k} + R^*_{n-2k}) = \delta_{n-2k}(\frac{R}{1+R}) = \delta_{n-2k}(\frac{z^s V}{1+R}). \quad (6)$$

By Remark 3.1.1 and e) of Lemma 3.1.1 we have

$$\delta_{n-2k}(\frac{z^s V}{1+R}) = \delta_{n-2k-s}(\frac{V}{1+R}) = \left((-1)|V(0)|^2\right)^{n+1-2k-s} \delta_{n-2k-s}(\frac{1+R}{V}).$$

As $\dfrac{1+R}{V} = \overline{a}_0 H$, we have $\delta_{n-2k-s}(\dfrac{1+R}{V}) = \delta_{n-2k-s}(H)$.

As $a_k \overline{a}_0 V(0) = d$, we obtain

$$\delta_{n-2k}(z^s \frac{V}{1+R}) = \left((-1)\frac{d^2}{|a_k|^2}\right)^{n+1-2k-s} \delta_{n-2k-s}(H). \quad (7)$$

From equations (3), (6), (7) we deduce

$$\delta_n(F) = (-1)^{k+s} |d|^{2(n+1-2k-s)} |a_k|^{2(k+s)} \delta_{n-2k-s}(H). \qquad \blacksquare$$

This last lemma reduces the hermitian matrix $I_{n+1} - F_n F_n^*$ when $|a_0| = 1$ and allows us to define generalized Schur transforms. More precisely, we introduce the following algorithm for $\mathbf{C}[[z]]$.

Definition 3.3.1 (Generalized Schur algorithm). *If F belongs to $\mathbf{C}[[z]]$, we define by induction the sequence $(F^n)_{n\in J}$ where J is a subset of $\mathbf{N}$ and F^n belongs to $\mathbf{C}[[z]]$, by setting $F = F^0$ and if F^i is defined and can be written in the form $F^i = \gamma_0 + \gamma_k z^k + \cdots$, then*

1) if $F^i \equiv \gamma_0$ with $|\gamma_0| = 1$ the sequence stops at F^i;

2) if $F^i \equiv \gamma_0$ with $|\gamma_0| > 1$ the sequence stops at F^i;

3) *if* $F^i = \gamma_0 + \gamma_k z^k + \cdots$ *with* $|\gamma_0| < 1$, *then* F^{i+1} *is defined as*

$$F^{i+1} = \frac{F^i - \gamma_0}{z(1 - \overline{\gamma}_0 F^i)};$$

4) *if* $F^i = \gamma_0 + \gamma_k z^k + \cdots$ *with* $|\gamma_0| > 1$, $\gamma_k \neq 0$, *then* F^{i+k} *is defined as*

$$F^{i+k} = z^k \left(\frac{1 - \overline{\gamma}_0 F^i}{F^i - \gamma_0} \right);$$

5) *if* $F^i = \gamma_0 + \gamma_k z^k + \cdots$ *with* $|\gamma_0| = 1$, $\gamma_k \neq 0$; *then consider the polynomial* Q *defined in the preceding lemma:*

   a) *if* $ord(\overline{\gamma}_0 F^i Q - (Q + z^k)) = 2k + s$ *with* $s \neq +\infty$, *then* F^{s+2k+i} *is defined by*

$$F^{s+2k+i} = z^s \frac{F^i(Q - z^k) - \gamma_0 Q}{\overline{\gamma}_0 F^i Q - (Q + z^k)}$$

   b) *if* $\overline{\gamma}_0 F^i Q - (Q + z^k) \equiv 0$ *the sequence stops at* F^i.

Remark 3.3.1. So the sequence F^n is finite if and only if we arrive at a constant of modulus ≥ 1, or at a series of the form $\varepsilon \dfrac{Q + z^k}{Q}$ where Q can be written as $Q = c_0 + c_1 z + \cdots + c_{k-1} z^{k-1} - (\overline{c}_{k-1} z^{k+1} + \cdots + \overline{c}_0 z^{2k})$ with $c_0 \neq 0$ and $|\varepsilon| = 1$.

Remark 3.3.2. In the case

1) we have $\delta_n(F^i) = 0 \quad \forall n \in \mathbf{N}$
2) we have $\delta_n(F^i) = (1 - \gamma_0 \overline{\gamma}_0)^{n+1}$ with $|\gamma_0| > 1$

5b) we have $\delta_n(F^i) = (-1)^{n+1-k} |\gamma_k|^{2(n+1-k)} \quad \forall n \geq 2k$.

3.4 Characterization of certain meromorphic functions on D(0,1)

Definition 3.4. $\mathcal{M}_\infty$ denotes the set of meromorphic functions on $D(0,1)$ with no pole at the origin, that have a finite number of poles in $D(0,1)$ and that verify

$$\limsup_{\substack{z \longrightarrow e^{i\theta} \\ |z|<1}} |f(z)| \leq 1 \qquad \forall \theta \in [0, 2\pi].$$

Remark 3.4.1. A function f belongs to $\mathcal{M}_\infty$ if and only if f can be written as $f = \dfrac{P^*}{P} h$ with $h \in \mathcal{M}$, $P \in \mathbf{C}[z]$, $P(0) \neq 0$ and P has all its zeros in $D(0,1)$.

The generalized Shur algorithm allows us to prove the following theorem.

Theorem 3.4.1. *If F belongs to $\mathbf{C}[[z]]$, then F is the Taylor series at zero of a function belonging to $\mathcal{M}_\infty$ if an only if there exists an integer n_0 such that*

either $\delta_n(F)\delta_{n+1}(F) > 0 \quad \forall n \geq n_0$,

or $\delta_n(F) = 0 \quad \forall n \geq n_0$.

To prove this theorem we need properties relating F to its Schur transform.

Lemma 3.4.1. *Let F be an element of $\mathbf{C}[[z]]$ and $i \in \mathbf{N}$. If F^i is defined and F^{i+l} is the element of the sequence following F^i in the algorithm, then*

a) there exist two polynomials E_l and S_l belonging to $\mathbf{C}[z]$ such that

i) $d^\circ E_l \leq l, \quad d^\circ S_l \leq l$

ii) $F^{i+l} = \dfrac{F^i E_l - S_l}{\widetilde{E}_l - F^i \widetilde{S}_l}$ *with* $\begin{array}{l} \widetilde{E}_l(z) = z^l \overline{E}_l(1/z) \\ \widetilde{S}_l(z) = z^l \overline{S}(1/z) \end{array}$

iii) $E_l \widetilde{E}_l - S_l \widetilde{S}_l = z^l t_l$ *with* $t_l > 0$

iv) $F^i \widetilde{S}_l - \widetilde{E}_l = v_l z^l + \cdots$ *with* $v_l \neq 0$

v) $\mathrm{ord}(F^i E_l - S_l) \geq l$

b) $\delta_n(F^i) = \dfrac{|v_l|^{2n}}{t_l^n} \rho_l \delta_{n-l}(F^{i+l})$ *with* $\rho_l \neq 0 \quad \forall n \geq l$.

Proof. 1) If F^{i+l} is obtained from F^i by the transformation 5a) we take $E_l = z^s(Q - z^k)$, $S_l = z^s \gamma_0 Q$, $l = 2k + s$ and Lemma 3.3.1 finishes the proof.

2) If F^{i+l} is obtained by F^i by the transformation 3) i.e. F^{i+1} is the Schur transform of the preceding paragraph, we take $l = 1$, $S_1 = \gamma_0$, $E_1 = 1$ with $|\gamma_0| < 1$; then $E_1 \widetilde{E}_1 - S_1 \widetilde{S}_1 = (1 - \gamma_0 \overline{\gamma}_0) z$ and by part e) of Lemma 3.1.2

$$\delta_n(F^i) = (1 - \gamma_0 \overline{\gamma}_0)^{n+1} \delta_{n-1}(F^{i+l})$$

3) If F^{i+l} is obtained from F^i by the transformation 4) then we take $l = k$, $S_k = z^k$, $E_k = \overline{\gamma}_0 z^k$, $\widetilde{E}_k = \gamma_0$, $\widetilde{S}_k = 1$; we easily verify that a) holds with $v_l = \gamma_k$, $t_l = \gamma_0 \overline{\gamma}_0 - 1$. Moreover, by Lemma 3.1.1 we have

$$\delta_{n-k}(F^{i+k}) = (-1)^{n+1-k}\left(\frac{1-\gamma_0\overline{\gamma}_0}{|\gamma_k|^2}\right)^{2(n+1-k)}\delta_{n-k}\left(\frac{F^i-\gamma_0}{z^k(1-\overline{\gamma}_0F^i)}\right)$$

and by Remark 3.1.1 and Lemma 3.2.2

$$\delta_{n-k}\left(\frac{F^i-\gamma_0}{z^k(1-\overline{\gamma}_0F^i)}\right) = \delta_{n-1}\left(\frac{F^i-\gamma_0}{z(1-\overline{\gamma}_0F^i)}\right) = \frac{\delta_n(F^i)}{(1-\gamma_0\overline{\gamma}_0)^{n+1}}\,.$$

Hence

$$\delta_{n-k}(F^{i+k}) = (-1)^{n+1-k}\frac{(1-\gamma_0\overline{\gamma}_0)^{n+1-2k}}{|\gamma_k|^{2(n+1-k)}}\delta_n(F^i).$$ ∎

Lemma 3.4.2. *If F belongs to $\mathbf{C}[[z]]$ and if F^i is defined, we have the following properties:*

a) there exist two polynomials A_i and Q_i such that

i) $d^\circ A_i \le i, \quad d^\circ Q_i \le i$

ii) $F = \dfrac{A_i+\widetilde{Q}_iF^i}{Q_i+\widetilde{A}_iF^i}$

iii) $F\widetilde{A}_i - \widetilde{Q}_i = u_iz^i + \cdots$ *with* $u_i \neq 0$

iv) $\operatorname{ord} FQ_i - A_i) \ge i$

v) $Q_i\widetilde{Q}_i - A_i\widetilde{A}_i = \omega_i z^i$ *with* $\omega_i > 0$

b) $\delta_n(F) = \dfrac{|u_i|^{2n}}{\omega_i^n}\lambda_i\delta_{n-i}(F^i)$ *with* $\lambda_i \neq 0 \quad \forall n \ge i.$

Proof. By induction, and making use of the preceding lemma.

Lemma 3.4.3. *If F belongs to $\mathbf{C}[[z]]$ and if the algorithm is finite then:*

either F has finite rank

or $F = A/B$ with A and $B \in \mathbf{C}[z]$, $|A(z)| > |B(z)|)$ if $|z| = 1$ and $\delta_n(F) = \lambda b^n$ with $\lambda \neq 0$ $b < 0$.

Proof. Assume that the algorithm is finite and stops at F^i;

i) if $F^i \equiv \gamma_0$ with $|\gamma_0| = 1$, according to ii) of the preceding lemma, F^i has finite rank.

ii) if $F^i \equiv \gamma_0$ with $|\gamma_0| > 1$ then $F = \dfrac{A_i + \widetilde{Q}_i\gamma_0}{Q_i + \widetilde{A}_i\gamma_0}$ with $|Q_i(z)|^2 - |A_i(z)|^2 = \omega_i$ if $|z| = 1$. Hence $1 - |F(z)|^2 = \dfrac{\omega_i(1 - \gamma_0\overline{\gamma}_0)}{|Q_i(z) + \gamma_0\widetilde{A}_i(z)|^2}$ if $|z| = 1$.

Moreover $\delta_n(F^i) = (1-\gamma_0\overline{\gamma}_0)^{n+1}$ and by the preceding lemma we have $\delta_n(F) = \lambda b^n$ with $b < 0 \quad \forall n \geq i$.

iii) If $F^i = \dfrac{Q + z^k}{\overline{a}_0 Q}$, by Remark 3.3.2 we have $\delta_n(F) = \lambda b^n$ where $\lambda \neq 0$ and $b < 0$ and $1 - |F(z)|^2 = \dfrac{\omega_i(1 - |F^i(z)|^2)}{|Q_i(z) + \widetilde{A}_i(z)F^i|^2}$ if $|z| = 1$.

Moreover we have $|Q(z) + z^k|^2 - |\overline{a}_0 Q(z)|^2 = 1$ if $|z| = 1$. So $|F(z)| > 1$ if $|z| = 1$. ∎

Lemma 3.4.4. *Let g be a function of $\mathcal{M}_\infty$ that has p poles in $D(0,1)$ and G the Taylor series of g at zero, G^l the first Schur transform of G. Then:*

i) G^l is the Taylor series at zero of a function g^l of $\mathcal{M}_\infty$ that has at most p poles in $D(0,1)$

ii) if G^l is obtained from G by the transformations 4) or 5a) g_l has at most $p-1$ poles in $D(0,1)$.

Proof. **i)** By Lemma 3.4.1 we have $G^l = \dfrac{GE_l - S_l}{\widetilde{E}_l - G\widetilde{S}_l}$ where E_l and S_l satisfy i), ii), iii), iv), v).

Set $g_l = \dfrac{gE_l - S_l}{\widetilde{E}_l - g\widetilde{S}_l}$. Then g_l is meromorphic on $D(0,1)$. We easily verify by using the properties iii) that

$$\limsup_{\substack{z \longrightarrow e^{i\theta} \\ |z|<1}} |g_l(z)| \leq 1.$$

Moreover g can be written as $g = \dfrac{P^*}{P} h$, where $h \in \mathcal{M}$, $P \in \mathbf{C}[z]$ has all its zeros in $D(0,1)$, $d^\circ P = p$, $P(0) = 1$.

So g_l can be written as $g_l = \dfrac{P^* h E_l - P S_l}{P\widetilde{E}_l - hP^*\widetilde{S}_l}$;

Properties iv) and v) imply that g_l has no pole at zero; properties iii) and Rouché's theorem imply that $P\widetilde{E}_l - hP^*\widetilde{S}_l$ has $p+s$ zeros in $D(0,1)$ if s denotes the numbers of zeros of $\widetilde{E}_l$ in $D(0,1)$. By iv) $P\widetilde{E}_l - hP^*\widetilde{S}_l$ has a zero of order l at zero; so it has $p+s-l$ zeros in $D(0,1)$ distinct from zero and g_l has at most $p+s-l$ poles in $D(0,1)$. This implies i) because $s \leq d^\circ \widetilde{E}_l \leq l$.

ii) In the case 4), $\widetilde{E}_l = \gamma_0$; so $s = 0$ and $p+s-l = p-l \leq p-1$. In the case 5a), $\widetilde{E}_l$ is a polynomial whose first and last coefficients have the same absolute value, and has no zero on $|z| = 1$; hence $s < l$ and $p + s - l \leq p - 1$. ■

Lemma 3.4.5. *Let f be a function of $\mathcal{M}_\infty$ with no finite rank and F the Taylor series of f at zero. Then there exists $n_0 \in \mathbf{N}$ such that the Schur transform F^{n_0} of F exists and F^{n_0} is the Taylor series at zero of a function belonging to $\mathcal{M}$ and with no finite rank.*

Proof. By Lemma 3.4.3 the algorithm applied to F does not stop. By the preceding lemma, the transformations 4) and 5a) are used only a finite number of times because the number of poles of f in $D(0,1)$ is finite. So there exists n_0 such that the definition of the transforms of F^{n_0} need only use transformation 3), and F^{n_0} is the Taylor series at zero of a function $f_{n_0} \in \mathcal{M}$ with no finite rank. ■

Lemma 3.4.6. *If F belongs to $\mathbf{C}[[z]]$ and if $\delta_n(F) < 0 \quad \forall n \in \mathbf{N}$ then F is the Taylor series at zero of a function f belonging to M_∞ that has exactly one pole in $D(0,1)$.*

Proof. Set $F = \sum_{i\in\mathbf{N}} a_i z^i$. Then $\delta_0(F) = 1 - a_0\bar{a}_0 < 0$, and hence $|a_0| > 1$.

As $\delta_1(F) = (1 - a_0\bar{a}_0)^2 - |a_1|^2 < 0$, we have $a_1 \neq 0$; so $F^1 = z\dfrac{1-\bar{a}_0 F}{F - a_0}$. By Lemma 3.4.2, $\delta_n(F) = h^n\lambda\delta_{n-1}(F^1)$ with $h > 0$, $\lambda \neq 0 \quad \forall n \geq 1$. As $|F^1(0)| = \dfrac{|1 - a_0\bar{a}_0|}{|a_1|}$ we have $\delta_0(F^1) > 0$ and $\lambda < 0$. So $\delta_n(F^1) > 0 \quad \forall n \in \mathbf{N}$ and, by Theorem 3.2.1, F^1 is the Taylor series at zero of a function f_1 of $\mathcal{M}$.

As $F = \dfrac{a_0F^1 + z}{F^1 + \bar{a}_0 z}$, if we set $f = \dfrac{a_0 f_1 + z}{f_1 + \bar{a}_0 z}$, then by Rouché's theorem f has at most one pole in $D(0,1)$.

Moreover as $|a_0| > 1$, we easily see that:

$$\limsup_{\substack{z \to e^{i\theta} \\ |z|<1}} |f(z)| \leq 1 \quad \forall\theta \in [0, 2\pi].$$

So f belongs to $\mathcal{M}_\infty$ and has at most one pole in $D(0,1)$. As $\delta_n(F) < 0 \quad \forall n \in \mathbf{N}$, by Theorem 3.2.1, f has exactly one pole in $D(0,1)$. ∎

Proof of Theorem 3.4.1.

a) Let f be a function of $\mathcal{M}_\infty$, with no finite rank and F the Taylor series of f at zero. By Lemma 3.4.5 there exists n_0 such that F^{n_0} is the Taylor series at zero of a function f_{n_0} of $\mathcal{M}$. Hence $\delta_n(F^{n_0}) > 0 \quad \forall n \in \mathbf{N}$.

By Lemma 3.4.2, $\delta_{n+n_0}(F) = \rho^n \lambda \delta_n(F^{n_0})$ with $\rho > 0$, $\lambda \neq 0$; so

$$\delta_{n+n_0}(F)\delta_{n+1+n_0}(F) = \rho^{2n+1}\lambda^2 \delta_n(F^{n_0})\delta_{n+1}(F^{n_0}) > 0 \qquad \forall n \in \mathbf{N}.$$

b) Assume that $\delta_n(F)\delta_{n+1}(F) > 0 \quad \forall n \geq n_0$. By Lemma 3.4.3 the algorithm does not stop; so consider $i \geq n_0$ such that F^i is defined. By Lemma 3.4.2, $\delta_n(F) = \rho^n \lambda \delta_{n-i}(F^i)$ with $\rho > 0$, $\lambda \neq 0$. So $\delta_n(F^i)\delta_{n+1}(F^i) > 0 \quad \forall n \in \mathbf{N}$ and by Lemma 3.4.6 and Theorem 3.2.1, F^i is the Taylor series of a function f_i of $\mathcal{M}_\infty$ that has at most one pole in $D(0,1)$.

By Lemma 3.4.2 we have

$$F = \frac{A_i + \widetilde{Q}_i F^i}{Q_i + \widetilde{A}_i F^i}; \quad \text{set } f = \frac{A_i + \widetilde{Q}_i f_i}{Q_i + \widetilde{A}_i f_i};$$

then we prove as in Lemma 3.4.4 that f belongs to $\mathcal{M}_\infty$. ∎

Notation. If f is an analytic function in the neighborhood of zero, $\forall n \in \mathbf{N}$, we define $\delta_n(f)$ by $\delta_n(f) = \delta_n(F)$, F being the Taylor series of f at zero.

Corollary 3.4.1. *If f belongs to $\mathcal{M}_\infty$ and has no finite rank, then there exists $n_0 \in \mathbf{N}$ such that the sequence $\left(\dfrac{\delta_{n+1}(f)}{\delta_n(f)}\right)_{n \geq n_0}$ is defined, positive and decreasing.*

Proof. By Lemma 3.4.5, there exists $n_0 \in \mathbf{N}$ such that F^{n_0} is defined and F^{n_0} is the Taylor series at zero of a function f_{n_0} belonging to $\mathcal{M}$ with no finite rank. So $\delta_n(F^{n_0}) > 0 \quad \forall n \in \mathbf{N}$.

Since Lemma 3.4.2 implies that there exists $\rho > 0$, $\mu \neq 0$ such that

$$\delta_{n+n_0}(F) = \rho^n \mu \delta_n(F^{n_0}) \qquad \forall n \geq 0;$$

therefore $\delta_n(f) \neq 0 \quad \forall n \geq n_0$ and $\dfrac{\delta_{n+n_0+1}(f)}{\delta_{n+n_0}(f)} = \rho \dfrac{\delta_{n+1}(f_{n_0})}{\delta_n(f_{n_0})} \quad \forall n \geq 0.$

By Corollary 3.2.1, the sequence $\left(\dfrac{\delta_{n+1}(f)}{\delta_n(f)}\right)_{n\geq n_0}$ is positive and decreasing. ■

Now we determine the limit of the sequence $\left(\dfrac{\delta_{n+1}(f)}{\delta_n(f)}\right)_{n\geq n_0}$ when f belongs to $\mathcal{M}_\infty$.

Theorem 3.4.2. *If f belongs to $\mathcal{M}_\infty$ and has no finite rank and if $\theta_1, \theta_2, \dots, \theta_s$ denote the poles of f in $D(0,1)$, then*

$$\lim_{n\longrightarrow\infty} \frac{\delta_n(f)}{\delta_{n-1}(f)} = \frac{\mathcal{L}(1-|\widehat{f}|^2)}{\prod_{i=1}^{i=s}|\theta_i|^2}$$

with $\widehat{f}(e^{i\theta}) = \lim\limits_{\substack{r\longrightarrow 1\\ r<1}} f(re^{i\theta})$ for almost all points $e^{i\theta} \in T$.

Proof. Let n_0 be an integer such that F^{n_0} is the Taylor series at zero of a function f_{n_0} belonging to M.

We will denote f_n for $n \geq n_0$, the functions of $\mathcal{M}$ whose the Taylor series are F^n. So, with the notation of Lemma 3.4.2 we have

$$f_n = \frac{A_n - Q_n f}{f\widetilde{A}_n - \widetilde{Q}_n}.$$

As in the proof of Theorem 3.2.2 we have

$$1 - |\widehat{f}_n(e^{i\theta})|^2 = \frac{\left(|Q_n(e^{i\theta})|^2 - |A_n(e^{i\theta})|^2\right)\left(1-|\widehat{f}(e^{i\theta})|^2\right)}{|\widehat{f}(e^{i\theta})\widetilde{A}_n(e^{i\theta}) - \widetilde{Q}_n(e^{i\theta})|^2}.$$

Hence we have by Lemma 3.4.2

$\mathcal{L}(1-|\widehat{f}_n|^2) = \dfrac{\omega_n \mathcal{L}(1-|\widehat{f}|^2)}{\mathcal{L}(|\widehat{f}\widetilde{A}_n - \widetilde{Q}_n|^2)}$ with $\omega_n > 0$. We determine $\mathcal{L}(|\widehat{f}\widetilde{A}_n - \widetilde{Q}_n|^2)$.

Set $f = \dfrac{g}{P}$ with $P = \displaystyle\prod_{i=1}^{i=s}(z-\theta_i)$.

Then g is analytic and bounded on $D(0,1)$; moreover

$$f_n = \frac{A_n P - Q_n g}{\widetilde{A}_n g - \widetilde{Q}_n P}.$$

Assume that $A_n P - Q_n g$ and $\widetilde{A}_n g - \widetilde{Q}_n P$ have a common zero α in $D^*(0,1)$; then

$$P(\alpha)\left(A_n(\alpha)\widetilde{A}_n(\alpha) - Q_n(\alpha)\widetilde{Q}_n(\alpha)\right) = \omega_n \alpha^n P(\alpha) = 0$$

$$g(\alpha)\left(A_n(\alpha)\widetilde{A}_n(\alpha) - Q_n(\alpha)\widetilde{Q}_n(\alpha)\right) = \omega_n \alpha^n g(\alpha) = 0.$$

As $\omega_n > 0$, we have $P(\alpha) = g(\alpha) = 0$, which contradicts the definition of g and P. So the function $\widetilde{A}_n g - \widetilde{Q}_n P$ has no zero in $D^*(0,1)$, since f_n is analytic on $D(0,1)$. By Lemma 3.4.2 we have the following equations:

$$A_n P - Q_n g = (\mu_n z^n + \cdots)P$$

$$\widetilde{A}_n g - \widetilde{Q}_n P = (u_n z^n + \cdots)P = \left((-1)^s \prod_{i=1}^{i=s} \theta_i\right)(u_n z^n + \cdots) \quad \text{with } u_n \neq 0.$$

So the function $h = \dfrac{\widetilde{A}_n g - \widetilde{Q}_n P}{z^n}$ is analytic, bounded, and is not zero on $D(0,1)$. So

$$\mathcal{L}(|\widehat{h}|) = |u_n| \prod_{i=1}^{i=s} |\theta_i| = \mathcal{L}(|\widetilde{A}_n \widehat{g} - \widetilde{Q}_n P|) = \mathcal{L}(|\widetilde{A}_n \widehat{f} - \widetilde{Q}_n|)\mathcal{L}(|P|).$$

Hence

$$\mathcal{L}(1 - |\widehat{f}_n|^2) = \frac{\omega_n \mathcal{L}(1 - |\widehat{f}|^2)\mathcal{L}(|P|^2)}{|u_n|^2 \prod_{i=1}^{i=s} |\theta_i|^2}.$$

As $\mathcal{L}(|P|) = \mathcal{L}(|P^*|) = 1$ since $P^*(0) = 1$ and P^* is not zero in $D(0,1)$, we obtain

$$\mathcal{L}(1 - |\widehat{f}_n|^2) = \frac{\omega_n \mathcal{L}(1 - |\widehat{f}|^2)}{|u_n|^2 \prod_{i=1}^{i=s} |\theta_i|^2}. \tag{1}$$

Moreover, by Lemma 3.4.2 we have

$$\frac{\delta_p(f)}{\delta_{p-1}(f)} = \frac{|u_n|^2}{\omega_n} \times \frac{\delta_{p-n}(f_n)}{\delta_{p-n-1}(f_n)} \qquad \forall p \geq n+1.$$

Since f_n belongs to $\mathcal{M}$, we have by Theorem 3.2.2

$$\lim_{p \longrightarrow \infty} \frac{\delta_p(f)}{\delta_{p-1}(f)} = \frac{|u_n|^2}{\omega_n} \mathcal{L}(1 - |\widehat{f}_n|^2). \tag{2}$$

Equations (1) and (2) imply that

$$\lim_{p\longrightarrow\infty}\frac{\delta_p(f)}{\delta_{p-1}(f)}=\frac{\mathcal{L}(1-|\widehat{f}|^2)}{\prod_{i=1}^{i=s}|\theta_i|^2}.$$

■

We end this chapter with Smyth's theorem.

3.5 Smyth's theorem

Theorem 3.5. *Let P be a polynomial contained in $\mathbf{Z}[z]$, different from $\pm P^*$, monic, and with $P(0)\neq 0$. Denote $\theta_1,\theta_2,\ldots,\theta_s$ the roots of P. Then:*

$$\prod_{|\theta_i|>1}|\theta_i|\geq\theta_0$$

where $\theta_0=1.3247\ldots$ is the real root of $\theta^3-\theta-1=0$. (θ_0 is the smallest Pisot number).

Before embarking on the proof of the theorem, we prove two lemmas about the functions of $\mathcal{M}$.

Lemma 3.5.1. *Let f be a function belonging to $\mathcal{M}$, and such that $f(z)=\sum_{n\in\mathbf{N}}a_nz^n\quad\forall z\in D(0,1)$ where a_n are real $\forall n\in\mathbf{N}$. Then:*

i) $|a_0|\leq 1$

ii) $|a_k|\leq 1-a_0^2\quad\forall k\in\mathbf{N}$

iii) $-\left(1-a_0^2-\dfrac{a_k^2}{1+a_0}\right)\leq a_{2k}\leq 1-a_0^2-\dfrac{a_k^2}{1-a_0}\qquad\forall k\in\mathbf{N}^*.$

Proof. By Corollary 3.2.3, the matrices $I_{n+1}\pm J_{n+1}F_n$ are positive hermitian $\forall n\in\mathbf{N}$.

i) If $n=0$ we obtain $1\pm a_0\geq 0$, i.e., proposition i).

ii) If $n=k$, and if we set $\varepsilon=\pm 1$, the matrix

$$\begin{pmatrix}1+\varepsilon a_k & \cdots & \varepsilon a_0\\ \vdots & \ddots & \vdots\\ \varepsilon a_0 & \cdots & 1\end{pmatrix}$$

is positive. Hence the partial matrix

$$\begin{pmatrix} 1+\varepsilon a_k & \varepsilon a_0 \\ \varepsilon a_0 & 1 \end{pmatrix}$$

is positive. Therefore the determinant of this matrix is positive and we obtain ii).

iii) If $n = 2k$ and $\varepsilon = \pm 1$ the matrix

$$\begin{pmatrix} 1+\varepsilon a_{2k} & \cdots & \varepsilon a_k & \cdots & \varepsilon a_0 \\ \vdots & \ddots & \vdots & \ddots & \vdots \\ \varepsilon a_k & \cdots & 1+\varepsilon a_0 & \cdots & 0 \\ \vdots & \ddots & \vdots & \ddots & \vdots \\ a_0 & \cdots & 0 & \cdots & 1 \end{pmatrix}$$

is positive and the partial matrix

$$\begin{pmatrix} 1+\varepsilon a_{2k} & \varepsilon a_k & \varepsilon a_0 \\ \varepsilon a_k & 1+\varepsilon a_0 & 0 \\ \varepsilon a_0 & 0 & 1 \end{pmatrix}$$

is also positive. Hence the determinant of this matrix is positive and we obtain iii). ∎

Lemma 3.5.2. *We keep the hypothesis of Lemma 3.5.1. Let k and l be two integers such that $k < l < 2k$, $\varepsilon = \pm 1$. If we denote by A and B the matrices*

$$A = \begin{pmatrix} a_l & a_k \\ a_k & a_{2k-l} \end{pmatrix}, \quad B = \begin{pmatrix} a_{l-k} & a_0 \\ a_0 & 0 \end{pmatrix},$$

then the matrices $I_2 - B^2 + \varepsilon A$ are positive hermitian .

Proof. By Corollary 3.2.3, the matrix $I_{l+1} + J_{l+1}F_l$ is hermitian and positive. Since

$$I_{l+1} + J_{l+1}F_l = \begin{pmatrix} 1+a_l & \cdots & a_k & \cdots & a_{l-k} & \cdots & a_0 \\ \vdots & \ddots & \vdots & \ddots & \vdots & \ddots & \vdots \\ a_k & \cdots & 1+a_{2k-l} & \cdots & a_0 & \cdots & 0 \\ & \ddots & \vdots & \ddots & \vdots & \ddots & \vdots \\ a_{l-k} & \cdots & a_0 & \cdots & 1 & \cdots & 0 \\ \vdots & \ddots & \vdots & \ddots & \vdots & \ddots & \vdots \\ a_0 & \cdots & 0 & \cdots & 0 & \cdots & 1 \end{pmatrix},$$

the following partial matrix:

$$\begin{pmatrix} I_2 + A & B \\ B & I_2 \end{pmatrix}$$

is also positive.

Since we have the equality

$$\begin{pmatrix} I_2 & -B \\ O_{2,2} & I_2 \end{pmatrix} \begin{pmatrix} I_2 + A & B \\ B & I_2 \end{pmatrix} \begin{pmatrix} I_2 & O_{2,2} \\ -B & I_2 \end{pmatrix} = \begin{pmatrix} I_2 + A - B^2 & O_{2,2} \\ O_{2,2} & I_2 \end{pmatrix},$$

the matrix $\begin{pmatrix} I_2 + A - B^2 & O_{2,2} \\ O_{2,2} & I_2 \end{pmatrix}$ is also positive, and the matrix $I_2 + A - B^2$ is positive. Similarly with $A' = -A$, and we obtain the result. ∎

Proof of Theorem 3.5.

Notation. We set $P(z) = u_0 + \cdots + z^s$, $P^*(z) = 1 + \cdots + u_0 z^s$,

$U(z) = \varepsilon \dfrac{P(z)}{P^*(z)}$ where $\varepsilon = \pm 1$, $\varepsilon u_0 > 0$

$g(z) = \varepsilon' \prod\limits_{|\theta_i|>1} \dfrac{1 - \theta_i z}{\overline{\theta}_i - z}$ where $\varepsilon' = \pm 1$, $\varepsilon' \prod\limits_{|\theta_i|>1} \overline{\theta}_i > 0$

$d(z) = g(z)U(z)$.

The Taylor expansions of U, g, d at zero are denoted by

$U(z) = |u_0| + u_k z^k + u_l z^l + \cdots$ with $u_k \neq 0$, $u_l \neq 0$,

$g(z) = c + c_1 z + \cdots$ with $\dfrac{1}{c} = \prod\limits_{|\theta_i|>1} |\theta_i|$,

$d(z) = g(z)U(z) = d_0 + d_1 z + \cdots = (c + c_1 z + \cdots)(|u_0| + u_k z^k + u_l z^l + \cdots)$.

Remarks. a) Since the coefficients of P are real, the Taylor coefficients of g and d are also real. b) The functions g and d belong to $\mathcal{M}$. c) The function U is not a constant because P is different from $\pm P^*$. d) The Taylor coefficients of U belong to $\mathbf{Z}$.

Case 1: $|u_0| \geq 2$.

By applying Lemma 3.5.1 to the function d we get $c|u_0| \leq 1$; hence

$$\frac{1}{c} \geq |u_0| \geq 2 \geq \theta_0.$$

Case 2: $\begin{cases} |u_0| = 1 \text{ and } |u_k| \geq 2, \text{ or} \\ |u_0| = 1 \text{ and } u_k c_k \geq 0. \end{cases}$

We have $d(z) = c + c_1 z + \cdots + c_{k-1} z^{k-1} + (cu_k + c_k) z^k + \cdots$.

By applying Lemma 3.5.1 to the functions g and d we get

$$|cu_k + c_k| \leq 1 - c^2 \tag{1}$$

$$|c_k| \leq 1 - c^2. \tag{2}$$

a) If $|u_k| \geq 2$, by adding the inequalities (1) and (2) we see that: $2c \leq |cu_k| \leq 2(1 - c^2)$. Hence $1 - c - c^2 \geq 0$, and so $\frac{1}{c}$ is greater than the positive root of the equation $x^2 - x - 1 = 0$. Hence we obtain:

$$\frac{1}{c} \geq \frac{1 + \sqrt{5}}{2} \geq \theta_0.$$

b) If $u_k c_k \geq 0$ it follows from inequality (1) that $|cu_k| \leq 1 - c^2$. Hence $c \leq 1 - c^2$ and we use the same trick as in case a).

Case 3: $|u_0| = 1$, $|u_k| = 1$, $u_k c_k < 0$ and $l > 2k$.

For $z \in D(0, 1)$, we have

$$d(z) = c + c_1 z + \cdots + (cu_k + c_k) z^k + \cdots + (cu_{2k} + c_k u_k + c_{2k}) z^{2k} + \cdots.$$

So we have

$$d_{2k} = cu_{2k} + c_k u_k + c_{2k} = cu_{2k} - |c_k| + c_{2k},$$

$$d_k^2 = (|c_k| - c)^2.$$

By applying Lemma 3.5.1 to the functions g and d we obtain

$$-(1 - c^2 - \frac{c_k^2}{1 + c}) \leq c_{2k} \leq 1 - c^2 - \frac{c_k^2}{1 - c}$$

$$-\left(1 - c^2 - \frac{(|c_k| - c)^2}{1 - c}\right) \leq -c_{2k} + |c_k| - cu_{2k} \leq 1 - c^2 - \frac{(|c_k| - c)^2}{1 + c}.$$

By adding the inequalities we get

$$-2(1-c^2)+\frac{c_k^2}{1+c}+\frac{(|c_k|-c)^2}{1-c} \leq |c_k|-cu_{2k} \leq 2(1-c^2)-\frac{c_k^2}{1-c}-\frac{(|c_k|-c)^2}{1+c}.$$

So

$$-cu_{2k} \leq 2(1-c^2)-\frac{c_k^2}{1-c}-\frac{(|c_k|-c)^2}{1+c}-|c_k| \qquad (3)$$

$$cu_{2k} \leq 2(1-c^2)-\frac{c_k^2}{1+c}-\frac{(|c_k|-c)^2}{1-c}+|c_k|. \qquad (4)$$

From inequalities (1) and (2) we again have $c-|c_k| \leq 1-c^2$ and $|c_k| \leq 1-c^2$. So

$$c-(1-c^2) \leq |c_k| \leq 1-c^2.$$

a) Assume $u_{2k} \leq 0$ and $1-c^2 \leq c$, i.e., $\frac{1}{c} \leq \frac{1+\sqrt{5}}{2}$.

We use inequality (3). The function h defined by

$$h(x) = -\frac{x^2}{1-c}-\frac{(x-c)^2}{1+c}-x$$

has a derivative equal to

$$h'(x) = -\frac{2x}{1-c}-\frac{2x}{1+c}+\frac{c-1}{1+c}.$$

Hence $h'(x) \leq 0$ if $x \geq 0$. Then the second member of inequality (3) attains its maximum when $|c_k| = c-(1-c^2)$. So we obtain

$$0 \leq 2(1-c^2)-\frac{(1-c^2)^2}{1+c}-\frac{((1-c^2)-c)^2}{1-c} = \frac{(1-c^2)(1+c)-c}{1-c^2}$$

So $1-c^2 \geq \frac{c}{1+c}$ and $1/c$ satisfy the inequalities $x^3-x-1 \geq 0$ and $\frac{1}{c} \geq \theta_0$.

b) Assume $u_{2k} \geq 1$ and $1-c^2 \leq c$.

Here we use inequality (4). The function t defined by

$$t(x) = -\frac{x^2}{1+c}-\frac{(x-c)^2}{1-c}+x$$

has a derivative equal to

$$t'(x) = \frac{-4x+(1+c)^2}{(1-c)(1+c)}.$$

So the function t is increasing on $[0, \dfrac{(1+c)^2}{4}]$. Since we have the inequalities

$$1 - c^2 \leq c \leq \frac{(1+c)^2}{4},$$

the second member of inequality (4) attains its maximum for $|c_k| = 1 - c^2$. So we get

$$c \leq 2(1-c^2) - \frac{(1-c^2)^2}{1+c} - \frac{(1-c^2-c)^2}{1-c} + 1 - c^2.$$

Hence $1 - c^2 \geq \dfrac{c}{1+c}$ and $\dfrac{1}{c} \geq \theta_0$.

Case 4: $|u_0| = 1$, $k < l < 2k$, $|u_k| = 1$, and $u_k c_k < 0$.

We have $c_{l-k} = d_{l-k}$ since $l - k < k$, and $c_{2k-l} = d_{2k-l}$ since $0 < 2k - l < k$. Furthermore we have $d_k = c_k + u_k c$, $d_l = cu_l + u_k c_{l-k} + c_l$. Denote D, U, B, A the matrices

$$D = \begin{pmatrix} d_l & d_k \\ d_k & d_{2k-l} \end{pmatrix},\ U = \begin{pmatrix} u_k & 0 \\ u_l & u_k \end{pmatrix},\ B = \begin{pmatrix} c_{l-k} & c \\ c & 0 \end{pmatrix},\ A = \begin{pmatrix} c_l & c_k \\ c_k & c_{2k-l} \end{pmatrix};$$

then we have $D = BU + A$. By applying Lemma 3.5.2 to the functions g and d we obtain that the matrices $I_2 - B^2 + A$, $I_2 - B^2 - A$, $I_2 - B^2 + A + BU$, $I_2 - B^2 - (A + BU)$ are hermitian and positive. So we have $2I_2 - 2B^2 + BU \geq 0$ and $2I_2 - 2B^2 - BU \geq 0$. As

$$B^2 = \begin{pmatrix} c_{l-k} + c^2 & cc_{l-k} \\ cc_{l-k} & c^2 \end{pmatrix}, \quad BU = \begin{pmatrix} u_k c_{l-k} & cu_k \\ cu_k & 0 \end{pmatrix}$$

we obtain the matrices:

$$\begin{pmatrix} 1 - c^2 - c_{l-k} + (\varepsilon/2)(cu_l + u_k c_{l-k}) & (\varepsilon/2)cu_k - cc_{l-k} \\ (\varepsilon/2)cu_k - cc_{l-k} & 1 - c^2 \end{pmatrix}$$

where $\varepsilon = \mp 1$ are hermitian and positive. So the determinants of these matrices are positive, and we obtain

$$(1-c^2)^2 - \frac{c^2}{4} - c_{l-k} \geq |cu_l(\frac{1-c^2}{2}) + u_k c_{l-k}(\frac{1+c^2}{2})|. \tag{5}$$

a) Assume $|c_{l-k}| \geq \dfrac{c(1-c^2)}{1+c^2}$. Since the first member of inequality (5) is positive we obtain

$$(1-c^2)^2 - \frac{c^2}{4} - c^2\left(\frac{1-c^2}{1+c^2}\right)^2 \geq 0; \tag{6}$$

i.e., $4c^8 - 5c^6 - 2c^4 - 5c^2 + 4 \geq 0$. Set $u = c^{-2}$and $v = u + \frac{1}{u}$; so we have

$$4(u^2 + \frac{1}{u^2}) - 5(u + \frac{1}{u}) - 2 = 4(u + \frac{1}{u})^2 - 5(u + \frac{1}{u}) - 10 = 4(v-\alpha)(v-\beta) \geq 0$$

where $\alpha = \frac{5+\sqrt{185}}{8}$, $\beta = \frac{5-\sqrt{185}}{8}$.

Then we obtain $u^2 - \alpha u + 1 \geq 0$ and

$$\frac{1}{c} \geq \left(\frac{\alpha + \sqrt{\alpha^2 - 4}}{2}\right)^{1/2} = 1.3249781\ldots > \theta_0 = 1.3247\ldots.$$

b) Assume $|c_{l-k}| \leq c(\frac{1-c^2}{1+c^2})$ and $1 - c^2 \leq c$. Then

$$c(\frac{1-c^2}{1+c^2}) \leq \frac{c^2}{1+c^2} \leq \frac{1+c^2}{4};$$

by inequality (5) we obtain

$$(1-c^2)^2 - \frac{c}{4} \geq c|u_l|(\frac{1-c^2}{2}) + c_{l-k} - |c_{l-k}|(\frac{1+c^2}{2}). \tag{7}$$

As the function $x \longrightarrow x^2 - x(\frac{1+c^2}{2})$ is decreasing on $[0, \frac{1+c^2}{4}]$, the minimum of the second member of inequality (7) is obtained for $|c_{l-k}| = c(\frac{1-c^2}{1+c^2})$. As $|u_l| \geq 1$, we obtain the inequality

$$(1-c^2)^2 - \frac{c^2}{4} \geq c(\frac{1-c^2}{2}) + c^2(\frac{1-c^2}{1+c^2})^2 - c(\frac{1-c^2}{2}),$$

which is equivalent to inequality (6). As in a), we have $\frac{1}{c} > \theta_0$. ∎

Notes

The first algorithm of Lemma 3.1.2 is due to Schur [6]; it reappears in Wall's article [7]. The criterion of hyper-rationality does not seem to appear in the literature; it can be generalized in a way to $K[[X]]$, where K is an arbitrary field.

C. Chamfy [3] introduced another algorithm that allowed us to prove that there exist necessary and sufficient conditions for a series belonging to $\mathbf{C}[[z]]$ to be the Taylor series at zero of a function of $\mathcal{M}_\infty$; but the conditions were not explicit. In 1958 J. Dufresnoy [4] gave a new algorithm, which is the one of definition 3.3.1. But he too did not succeed in giving an explicit condition.

So Theorem 3.4.1 is original and is due to M. Pathiaux. It depends on Lemma 3.3.1, which gave explicit conditions. By reducing the hermitian matrix $I_{n+1} - F_nF_n^*$ in the case $|a_0| = 1$, she recovers the algorithm of J. Dufresnoy.

Theorems 3.2.2 and 3.4.2 are due to D. Boyd [1], [2] and show the importance of the Schur determinants. The properties of $\mathcal{L}(g)$ can be found in the books [4] and [5].

Many questions remain unanswered: it seems that if F is a rational series then $G = \sum\limits_{n\in\mathbf{N}} \delta_n(F)z^n$ is a rational series, too. How are F and G connected?

Does the Boyd theorem remain true if we replace the condition "f has a finite number of poles θ_i in $D(0,1)$" by "$\prod \theta_i$ converges" and "$\lim\limits_{n\longrightarrow+\infty} \delta_n(f)/\delta_{n-1}(f)$" by "$\limsup\limits_{n\longrightarrow+\infty} |\delta_n(f)|^{1/n}$"? (we can easily prove that if f is meromorphic on $D(0,1)$ and has a bounded characteristic, then $\limsup\limits_{n\longrightarrow+\infty} |\delta_n(f)|^{1/n} < +\infty$).

At the end of this chapter we prove Smyth's theorem [8]. A proof different from that of Smyth can be found in Schinzel's book [7]. The proof given in this book is original, and relies on positive hermitian forms.

References

[1] D.W. Boyd, Pisot numbers and the width of meromorphic functions. Privately circulated manuscript, (1977).

[2] D.W. Boyd, Schur's algorithm for bounded holomorphic functions, *Bull. London Math. Soc.*, 11, (1979), 145-150.

[3] C. Champhy, Fonctions méromorphes dans le cercle unité et leurs séries de Taylor, *Ann. Inst. Fourier (Grenoble)*, 8, (1958), 211-251.

[4] J. DUFRESNOY, Sur le problème des coefficients pour certaines fonctions dans le cercle unité, *Ann. Acad. Sc. Fennicae*, Ser. A (1958).

[5] G.H. HARDY , J.E. LITTLEWOOD , G. POLYA , *Inequalities*, Cambridge University Press, (1934).

[6] K. HOFFMANN, *Banach spaces of analytic functions*, Prentice Hall (Englewood Cliffs, N.J.) (1962).

[7] A. SCHINZEL, *Selected topics on polynomials*, Ann Arbor , The University of Michigan Press.

[8] I. SCHUR, Über Potenzreihen, die im Innern des Einheitskreises beschränkt sind, *J. f. r. u. ang. Math.*, 147, (1917), 205-232.

[9] C.J. SMYTH, On the product of the conjugates outside the unit circle of an algebraic number, *Bull. London Math. Soc.*, 3, (1971), 169-175.

[10] H.S. WALL, Continued fractions and bounded analytic functions, *Bull. Amer. Math. Soc.*, 50, (1944), 110-119.

CHAPTER 4

GENERALITIES CONCERNING DISTRIBUTION MODULO 1 OF REAL SEQUENCES

The purpose of this chapter is to recall the main results on the distribution modulo 1 of real sequences and to prove some theorems that will prove useful in the following chapters.

We will begin by recalling—without proof—some classical theorems (Weyl's, Van der Corput's, Fejer's). These theorems give a good knowledge of slowly increasing sequences (sequences that increase not faster than a polynomial). Few results are known when the growth of the sequence is fast, exponential for instance, except for Koksma's theorem . We shall prove this theorem and develop some consequences. Koksma's theorem yields an exceptional set of real numbers, and we shall see in Chapter 5 that Pisot and Salem numbers belong to this set.

4.0 Notation and examples

Any real number x can be written in one and only one way $x = E(x) + \varepsilon(x)$, with $E(x)$ belonging to the set $\mathbf{Z}$ of integers and $\varepsilon(x)$ to the interval $[-1/2, 1/2[$. The real $\varepsilon(x)$ is the *residue of x modulo 1.* We choose the interval $[-1/2, 1/2[$ instead of $[0, 1[$ because, if $\|x\| = |\varepsilon(x)|$, then $\|x\|$ represents the distance from x to the set $\mathbf{Z}$. This notion will play an important role in the following chapters. *Studying the distribution modulo 1 of a real sequence* (x_n) means studying the sequence $(\varepsilon(x_n))$. The following examples show some of the different behaviors of sequences modulo 1.

Examples.

1. The sequence $(n\alpha)$ with α rational takes on a finite number of values modulo 1.
2. The sequence $((1+\sqrt{5})/2)^n)$ has a single limit point modulo 1, which is 0: the real $((1+\sqrt{5})/2)^n + ((1-\sqrt{5})/2)^n$ is an integer, so we have

$\varepsilon(((1+\sqrt{5})/2)^n) = -((1-\sqrt{5})/2)^n \quad (\forall n \in \mathbf{N})$, and then $\lim_{n\to+\infty} \varepsilon(((1+\sqrt{5})/2)^n) = 0$.

3. Let $p(n)$ denote the smallest prime number in the factorization of n into primes; the sequence $(1/p(n))$ has infinitely many limit points but is not dense in the interval $[-1/2, 1/2[$.

4. The sequence $(\log n)_{n\geq 1}$ is dense modulo 1: let a and b denote arbitrary reals with $-1/2 \leq a < b \leq 1/2$, then there exist integers m and n such that $e^{m+a} \leq n < e^{m+b}$, so $\varepsilon(x_n)$ belongs to the interval $[a, b[$.

4.1 Sequences with finitely many limit points modulo 1

Sequences with finitely many limit points modulo 1 appear occasionally in studying Pisot numbers. The purpose of Theorem 4.1 is to show that if we multiply such a sequence by a suitable integer, then we obtain a sequence with limit points belonging (modulo 1) to an arbitrarily small interval whose center is 0.

We first observe that if a sequence has finitely many rational limit points modulo 1, then the sequence obtained by multiplying by a common denominator of these rationals has only 0 as limit point modulo 1. So in what follows we shall always suppose that all rational limit points are 0.

Theorem 4.1. *Let (x_n) be a sequence of real numbers with finitely many limit points modulo 1; for every positive real η there exists a non-zero integer h such that the sequence (hx_n) satisfies, for n large enough, $\|hx_n\| \leq \eta$. Moreover let k denote the irrational limit points of the sequence (x_n) modulo 1 and q an integer with $q \geq \max(4, 2/\eta)$. Then h can be chosen such that $0 < h \leq q^k$.*

Proof. Suppose that the sequence $(\varepsilon(x_n))$ has k irrational limit points $(\gamma_i)_{1\leq i\leq k}$ and possibly also the limit point $\gamma_0 = 0$.

For every integer n, x_n can be written in the form $x_n = E(x_n) + \gamma_i + \eta_n$, with i taking one of the values $0, 1, \ldots, k$ and the sequence (η_n) converging to 0. By applying Dirichlet's theorem on simultaneous rational approximations, one shows that, for every integer $q > 1$, there exist integers h and $\ell_i \quad (i = 1, \ldots, k)$ such that $h\gamma_i = \ell_i + \xi_i$, with $0 < h \leq q^k$ and $|\xi_i| \leq 1/q \quad (i = 1, \ldots, k)$. Therefore we have, for every integer n,

$$hx_n = hE(x_n) + \ell_i + \xi_i + h\eta_n$$

with $i \in \{0, 1, \ldots, k\}$ and $\ell_0 = \xi_0 = 0$.

The sequence (η_n) converges to 0 and the integer q satisfies $q \geq \max(4, 2/\eta)$, so there exists some n_0 such that, for every $n \geq n_0$, $|q^k \eta_n| < 1/q$ and therefore $|\xi_i + h\eta_n| < 2/q < 1/2$. It follows that $|\varepsilon(hx_n)| = |\xi_i + h\eta_n| \leq \eta, \quad \forall n \geq n_0$. ∎

4.2 Uniform distribution of sequences

The notion of uniform distribution modulo 1 can be considered as a notion of "good" or "regular" distribution.

Let (x_n) be a sequence of real numbers, a and b two reals with $-1/2 \leq a < b \leq 1/2$, and N a positive integer. We note $\nu(a, b, N)$ the number of terms x_n for which $\varepsilon(x_n)$ belongs to the interval $[a, b[$.

Definition 4.2. *The sequence (x_n) is said to be uniformly distributed modulo 1 (abbreviated u.d. mod 1) if for every pair a, b of real numbers with $-1/2 \leq a < b \leq 1/2$ we have*

$$\lim_{N \to +\infty} \frac{\nu(a, b, N)}{N} = b - a.$$

The notion of uniform distribution modulo 1 can be extended to compact and locally compact groups ($\mathbf{R}^p$, adeles, fields of formal series). This notion is close to that of the probability of a term of a sequence belonging to a given interval, but it is different from the notion of density as is shown in the following example.

Example.

The sequence $(\log n)_{n \geq 1}$ is not u.d. mod 1; for we take $a = 0$ and b with $0 < b \leq 1/2$. The real $\varepsilon(\log n)$ belongs to the interval $[0, b[$ if and only ifthere exists an integer m such that n belongs to the interval $[e^m, e^{m+b}[$. We have then

$$\nu(0, b, E(e^{m+b})) = \sum_{k=0}^{m} (e^{k+b} - e^k) + O(m) = \frac{(e^b - 1)(e^{m+1} - 1)}{e - 1} + O(m).$$

From the equality $\lim_{m \to +\infty} E(e^{m+b})/e^{m+b} = 1$, it follows that

$$\lim_{m \to +\infty} \frac{\nu(0, b, E(e^{m+b}))}{E(e^{m+b})} = \frac{(e^b - 1)e^{1-b}}{e - 1} \neq b.$$

4.3 Weyl's theorems

Weyl's Theorems are the earliest criteria of uniform distribution modulo 1; the first one can be considered as the starting point for the whole study of uniform distribution modulo 1.

Theorem 4.3.1. *A sequence (x_n) of real numbers is uniformly distributed modulo 1 if and only if, for every Riemann-integrable function f on the interval $[-1/2, 1/2]$, we have*

$$\text{(1)} \qquad \lim_{N\to+\infty} \frac{1}{N} \sum_{n=1}^{N} f(\varepsilon(x_n)) = \int_{-1/2}^{1/2} f(t)dt.$$

If f is only Lebesgue-integrable and not Riemann-integrable, the relation (1) might not be satisfied for a u.d. mod 1 sequence, as is shown in the following example. Let (x_n) be an arbitrary real sequence, $J_{(x_n)}$ the set of the points $\varepsilon(x_n)$ and let χ be the characteristic function of $J_{(x_n)}$ on the interval $[-1/2, 1/2]$; the function χ is Lebesgue-integrable on $[-1/2, 1/2]$ and its integral is 0. We have then $\frac{1}{N} \sum_{n=1}^{N} \chi(\varepsilon(x_n)) = 1$ for every integer N.

The functions $f_h: t \mapsto e^{2i\pi ht}$ $(h \in \mathbf{Z}^\star)$ satisfy the conditions of Theorem 4.3.1. It is an important fact, shown in Theorem 4.3.2, that these functions suffice to determine the uniform distribution modulo 1 of a sequence.

Theorem 4.3.2. *A sequence (x_n) of real numbers is u.d. mod 1 if and only if*

$$\text{(2)} \qquad \lim_{N\to+\infty} \frac{1}{N} \sum_{n=1}^{N} \exp(2i\pi h x_n) = 0$$

for every non-zero integer h.

For $h \in \mathbf{Z}^\star$ and $N \in \mathbf{N}^\star$ the expressions $\sigma_h(N) = \frac{1}{N} \sum_{n=1}^{N} \exp(2i\pi h x_n)$ are called Weyl sums and the condition (2) is written $\lim_{N\to+\infty} \sigma_h(N) = 0 \quad \forall h \in \mathbf{Z}^\star$.

Corollary. *The sequence $(n\alpha)$ is u.d. mod 1 if and only if α is irrational.*

Proof. If α is rational, then we have seen that the sequence $(n\alpha)$ takes a finite number of values modulo 1; so we assume that α is irrational. We have then, for $N \in \mathbf{N}^\star$ and $h \in \mathbf{Z}^\star$,

$$\sigma_h(N) = \frac{1}{N}\sum_{n=1}^{N}\exp(2i\pi hn\alpha) = \frac{\exp(2i\pi h(N+1)\alpha) - \exp(2i\pi h\alpha)}{N(\exp(2i\pi h\alpha) - 1)}$$

and therefore $\sigma_h(N) \leq 1/N|\sin\pi h\alpha|$. For every non-zero integer h, $\sin\pi h\alpha$ is not 0 and therefore, $\lim_{N\to+\infty}\sigma_h(N) = 0 \quad (\forall h \in \mathbf{Z}^\star)$. ■

As a consequence of Weyl's criterions we obtain the following theorem, which we will later extend to $\mathbf{R}^p \quad (p \geq 2)$.

Theorem 4.3.3. *Let (x_n) be a uniformly distributed sequence of reals modulo 1 and φ a continuous function on the interval $[-1/2, 1/2]$; the sequence $(\varphi(\varepsilon(x_n)))$ is uniformly distributed modulo 1 if and only if*

$$\int_{-\frac{1}{2}}^{\frac{1}{2}}\exp(2i\pi h\varphi(t))dt = 0 \qquad \forall h \in \mathbf{Z}^\star.$$

Proof. By applying Theorem 4.3.1 to the functions $t \mapsto \exp(2i\pi h\varphi(t))$ for $h \in \mathbf{Z}^\star$ we obtain

$$\lim_{N\to+\infty}\frac{1}{N}\sum_{n=1}^{N}\exp(2i\pi h\varphi(\varepsilon(x_n))) = \int_{-1/2}^{1/2}\exp(2i\pi h\varphi(t))dt.$$

The conclusion follows then from Theorem 4.3.2. ■

4.4 Van des Corput's and Fejer's theorems. Applications

The theorems mentioned in this section give, contrary to Weyl's theorems, only sufficient conditions for uniform or non-uniform distribution modulo 1.

Theorem 4.4.1. (Van der Corput's theorem) *Let (x_n) be a sequence of real numbers. If, for every positive integer h, the sequence $(x_{n+h} - x_n)_{n\in\mathbf{N}}$ is uniformly distributed modulo 1, then (x_n) is uniformly distributed modulo 1.*

Theorem 4.4.2. (Fejer's theorem) *Let g be a function with a continuous derivative on the interval $[1, +\infty[$, satisfying the following conditions:*

(i) g is an increasing function with $\lim_{t \to +\infty} g(t) = +\infty$,

(ii) g' is a decreasing function with $\lim_{t \to +\infty} g'(t) = 0$ and $\lim_{t \to +\infty} tg'(t) = +\infty$,

then the sequence $(g(n))$ is uniformly distributed modulo 1.

Theorem 4.4.3. *Let g be a function with a continuous derivative on the interval $[1, +\infty[$, satisfying the following conditions:*

(i) g is an increasing function with $\lim_{t \to +\infty} g(t) = +\infty$,

(ii) g' is a decreasing function with $\lim_{t \to +\infty} tg'(t) = 0$;

then the sequence $(g(n))$ is dense but not uniformly distributed modulo 1.

It follows from Theorem 4.4.1 that if P is a polynomial with real coefficients, the sequence $(P(n))$ is u.d. mod 1 if and only ifat least one of the coefficients of $P - P(0)$ is irrational.

By applying Theorems 4.4.1, 4.4.2 and 4.4.3 one deduces the following results.

Theorem 4.4.4. *Let a, α and β be real positive numbers.*

(i) The sequence (an^α) is u.d. mod 1 if α is not an integer.

(ii) The sequence $(a \log^\beta n))$ with $\beta > 1$ is u.d. mod 1, it is dense but not u.d. mod 1 for β satisfying $0 < \beta \leq 1$.

(iii) The sequence $(an^\alpha \log^\beta n)_{n \geq 2})$ with α not an integer is u.d. mod 1.

(iv) The sequence $(\log(\log n))_{n \geq 2}$ is not u.d. mod 1.

Here again we see that the sequence $(\log n)_{n \geq 1}$ is not u.d. mod 1. This is also true for sequences whose growth is slower than $(\log n)_{n \geq 1}$

On the other hand sequences increasing faster than $(\log n)_{n \geq 1}$, but not faster than a polynomial are u.d. mod 1. The rapidity of growth of a sequence seems to play a major role in its distribution.

4.5 Koksma's theorem

With Koksma's theorem we arrive at exponentially increasing sequences.

Theorem 4.5.1. *Let (f_n) be a sequence of functions with continous derivative on the interval $[a,b]$.*

For m and n integers, $m \neq n$, we define

$$F_{m,n}(t) = f_m(t) - f_n(t).$$

Suppose the following conditions are satisfied for every pair (m,n), $m \neq n$.

(i) The derivative function $F'_{m,n}$ is monotone and never equal to 0 on the interval $[a,b]$.

(ii) There exists an increasing sequence (N_ν) of integers, satisfying $\lim_{\nu\to+\infty} N_{\nu+1}/N_\nu = 1$, such that if we set for $N \geq 2$

$$A_N = \frac{1}{N^2}\sum_{n=2}^{N}\sum_{m=1}^{n-1} \max\left[\frac{1}{|F'_{(m,n)}(a)|}\ ,\ \frac{1}{|F'_{(m,n)}(b)|}\right]$$

the series $\sum_{\nu\in\mathbf{N}} A_{N_\nu}$ converges.

Then the sequence $(f_n(t))$ is uniformly distributed modulo 1 for almost all $t \in [a,b]$ (i.e., for every $t \in [a,b]$ apart from a set that has Lebesgue measure 0).

Koksma's theorem is more often used with condition (iii) rather than (ii):

(iii) There exists a real K such that

$$|F'_{m,n}(t)| \geq K > 0 \qquad \forall t \in [a,b].$$

We will first prove the theorem with the conditions (i) and (ii) and then we will show that we can deduce condition (ii) from condition (iii).

Two lemmas are needed in the proof.

Lemma 1. *Let (N_ν) be a strictly increasing sequence of integers with $\lim_{\nu\to+\infty} \frac{N_{\nu+1}}{N_\nu} = 1$; then one can extract a subsequence $(N_{\nu_k})_k$ with $\lim_{k\to+\infty} \frac{N_{\nu_{k+1}}}{N_{\nu_k}} = 1$ such that the series $\sum_{k\in\mathbf{N}} \frac{1}{N_{\nu_k}}$ is convergent.*

Proof. For every $m \in \mathbf{N}$, let I_m denote the interval $[m^2, (m+1)^2]$; we define the sequence (N_{ν_k}) as follows: if the interval I_m contains at least two terms of the sequence (N_ν), we let N_{ν_k} be the smallest and $N_{\nu_{k+1}}$ the greatest. If it contains only one term, we let N_{ν_k} be that term. Then when we consider two terms N_{ν_k} and $N_{\nu_{k+1}}$, either they are consecutive in (N_ν), or they satisfy $1 \leq N_{\nu_{k+1}}/N_{\nu_k} \leq (m+1)^2/m^2$, so that we have $\lim_{k\to+\infty} N_{\nu_{k+1}}/N_{\nu_k} = 1$. From the inequality $N_{\nu_k} \geq (k/2)^2$ it follows that the series $\sum_{k\in\mathbf{N}} 1/N_{\nu_k}$ is convergent.∎

Lemma 2. *Let (u_n) be a sequence of real positive numbers such that the series $\sum_{n\in\mathbf{N}} u_n$ is convergent. Then there exists an increasing sequence (γ_n) with $\lim_{n\to+\infty} \gamma_n = +\infty$ such that the series $\sum_{n\in\mathbf{N}} u_n\gamma_n$ is convergent.*

Proof. We set $S = \sum_{n=0}^{+\infty} u_n$ and, for all integers n, $R_n = \sum_{k=n+1}^{+\infty} u_k$. The sequence (γ_n) defined by $\gamma_n = \sqrt{S}/\sqrt{R_{n-1}} + \sqrt{R_n} \quad (n \geq 1)$ is an increasing sequence with $\lim_{n\to+\infty} \gamma_n = +\infty$ and $\sum_{k=1}^{n} u_k\gamma_k = \sqrt{S}(\sqrt{S} - \sqrt{R_n})$. Therefore the series $\sum_{n\in\mathbf{N}} u_n\gamma_n$ is convergent. ∎

Proof of Theorem 4.5.1. Suppose conditions (i) and (ii) are satisfied. We consider the Weyl sums $\sigma_h(N,t) = \dfrac{1}{N}\sum_{n=1}^{N} \exp(2i\pi h\, f_n(t))$ associated to the sequence $(f_n(t))$ $t \in [a,b]$. We now construct a 0-Lebesgue measure set E such that, for every t belonging to the complement of E in $[a,b]$, we have $\lim_{N\to+\infty} \sigma_h(N,t) = 0 \quad (\forall h \in \mathbf{Z}^\star)$.

The equalities

$$\begin{aligned}|\sigma_h(N,t)|^2 &= \frac{1}{N} + \frac{1}{N^2}\sum_{n=2}^{N}\sum_{m=1}^{n-1}\big[\exp 2i\pi h(f_m(t) - f_n(t)) \\ &\qquad\qquad + \exp 2i\pi h(f_n(t) - f_m(t))\big] \\ &= \frac{1}{N} + \frac{2}{N^2}\sum_{n=2}^{N}\sum_{m=1}^{n-1}\cos(2\pi h F_{m,n}(t))\end{aligned}$$

imply

$$\int_a^b |\sigma_h(N,t)|^2 dt = \frac{b-a}{N} + \frac{2}{N^2}\sum_{n=2}^{N}\sum_{m=1}^{n-1}\int_a^b \cos(2\pi h F_{m,n}(t))dt.$$

We set $I_{m,n}(h) = \int_a^b \cos(2\pi h F_{m,n}(t))dt$; the sign of the function $F'_{m,n}$ stays constant on the interval $[a,b]$, hence the function $F_{m,n}$ has an inverse function $\phi_{m,n}$, and we have $I_{m,n}(h) = \int_\alpha^\beta \phi'_{m,n}(u)\cos 2\pi hu\,du$ with $\alpha = F_{m,n}(a)$ and $\beta = F_{m,n}(b)$.

Since the function $\phi'_{m,n}$ is monotone, the Second Mean Value theorem yields

$$|I_{m,n}(h)| \leq \frac{1}{\pi|h|}\max(|\phi'_{m,n}(\alpha)|, |\phi'_{m,n}(\beta)|) = \frac{1}{\pi|h|}\max(\frac{1}{|F'_{m,n}(a)|}, \frac{1}{|F'_{m,n}(b)|})$$

and then

$$\int_a^b |\sigma_h(N,t)|^2 dt \leq \frac{b-a}{N} + A_N.$$

From condition (ii) it follows, by Lemma 1, that one can extract from the sequence (N_ν) a subsequence, still denoted by (N_ν), such that the series $\sum_{\nu\in\mathbf{N}} ((b-a)/N_\nu + A_{N_\nu})$ converges. According to Lemma 2 there exists an increasing sequence such that the series

$$\sum_{\nu\in\mathbf{N}} \left(\frac{b-a}{N_\nu} + A_{N_\nu}\right)\gamma_\nu$$

still converges. We now define the set E. Let us assume that

$$E_\nu(h) = \{t \in [a,b] / |\sigma_h(N_\nu,t)| > \frac{1}{\sqrt{\gamma_\nu}}\} \qquad (\forall\nu\in\mathbf{N}^\star,\ \forall h\in\mathbf{Z}^\star).$$

The Lebesgue measure of $E_\nu(h)$ satisfies

$$m_\nu(h) \leq \gamma_\nu\int_a^b |\sigma_h(N_\nu,t)|^2 dt \leq \gamma_\nu(\frac{b-a}{N_\nu} + A_{N_\nu}).$$

The set $F_\mu(h)$ defined by $F_\mu(h) = \bigcup_{\nu=\mu+1}^{+\infty} E_\nu(h)$ $(\forall\mu\in\mathbf{N}^\star,\ \forall h\in\mathbf{Z}^\star)$ is measurable since it is a countable union of measurable sets, and the sequence

$(F_\mu(h))_\mu$ is a decreasing sequence of sets. Let $M_\mu(h)$ be the measure of $F_\mu(h)$. We have

$$M_\mu(h) \le \sum_{\nu=\mu+1}^{+\infty} m_\nu(h) \le \sum_{n=\mu+1}^{+\infty} (\frac{b-a}{N_\nu} + A_{N_\nu})\gamma_\nu$$

and therefore $\lim_{\mu\to+\infty} M_\mu(h) = 0$. The set $\bigcap_{\mu=1}^{+\infty} F_\mu(h)$ has Lebesgue measure zero and the same is true for the set E defined by $E = \bigcup_{h\in\mathbf{Z}^\star} [\bigcap_{\mu=1}^{+\infty} F_\mu(h)]$, since E is a countable union of sets of measure zero.

If t does not belong to E, then for every $h \in \mathbf{Z}^\star$, t does not belong to $\bigcap_{\mu=1}^{+\infty} F_\mu(h)$ and there exists some $\mu_1(h)$ such that for $\mu \ge \mu_1(h)$ we have $|\sigma_h(N_\mu, t)| \le 1/\sqrt{\gamma_\mu}$; it follows that $\lim_{\mu\to+\infty} \sigma_h(N_\mu, t) = 0$. For every integer N, there exists an integer μ such that $N_\mu \le N < N_{\mu+1}$; we then have

$$|\sigma_h(N,t)| \le \frac{1}{N_\mu} |\sum_{n=1}^{N_\mu} \exp 2i\pi h f_n(t)| + \frac{N_{\mu+1} - N_\mu}{N_\mu}.$$

Hence $\lim_{N\to+\infty} \sigma_h(N,t) = 0 \quad (\forall h \in \mathbf{Z}^\star)$; the sequence $(f_n(t)$ is therefore u.d. mod 1.

We will now show that it is possible to use condition (iii) instead of condition (ii). Let t be a fixed number belonging to the interval $[a,b]$; for every integer $N > 1$ we can order the numbers $f'_n(t), \quad (n = 1, \dots, N)$, according to their magnitude. We have in the new ordering $f'_n(t) - f'_{n-1}(t) \ge K \quad (1 \le n \le N)$ and therefore $f'_m(t) - f'_n(t) \ge (m-n)K$, $(1 \le n < m \le N)$. Hence we obtain

$$\sum_{n=1}^{m-1} \frac{1}{|F'_{m,n}(t)|} \le \frac{1}{K} \sum_{n=1}^{m-1} \frac{1}{m-n} \le \frac{1}{K}(1 + \log N)$$

and then $A_N \le N(1+\log N)/KN^2 = (1+\log N)/NK$.

We now define the sequence (N_ν) by $N_\nu = \nu^2$. We then have $\lim_{\nu\to+\infty} \frac{N_{\nu+1}}{N_\nu} = 1$. It follows that $A_{N_\nu} \le (1 + 2\log\nu)/K\nu^2$. The series $\sum_{\nu\in\mathbf{N}} A_{N_\nu}$ is convergent, so condition (ii) is satisfied. ■

In succeeding chapters we will apply Koksma's theorem in the following way.

Theorem 4.5.2.

a) Let α be a real number with $\alpha > 1$; the sequence $(\lambda\alpha^n)$ is uniformly distributed modulo 1 for almost all real λ.

b) Let λ be a non-zero real number; the sequence $(\lambda\alpha^n)$ is uniformly distributed modulo 1 for almost all $\alpha > 1$.

Proof.

a) We set $f_n(t) = t\alpha^n \quad (n \in \mathbf{N})$; we have then, for $n \neq m \quad f'_m(t) - f'_n(t) = \alpha^m - \alpha^n$. Hence $|f'_m(t) - f'_n(t)| > \alpha(\alpha - 1) > 0$. Applying Theorem 4.5.1. we deduce that the sequence $(t\alpha^n)$ is u.d. mod 1 for almost all $t \in [k, k+1] \quad (k \in \mathbf{Z})$. The countable union of sets having Lebesgue measure zero has Lebesgue measure zero, so the sequence $(t\alpha^n)$ is u.d. mod 1 for almost all $t \in \mathbf{R}$.

b) We set $g_n(t) = \lambda t^n \quad (n \in \mathbf{N})$; we then have, for $n \neq m$ and $t > 1$, $g'_m(t) - g'_n(t) = \lambda(mt^{m-1} - nt^{n-1})$, and therefore $|g'_m(t) - g'_n(t)| \geq |\lambda| > 0$. The proof is then completed as in a). ∎

Koksma's theorem shows in particular that the sequence (α^n) is u.d. mod 1 for almost all $\alpha > 1$. The set of reals $\alpha > 1$ such that the sequence (α^n) is not u.d. mod 1 has Lebesgue measure 0; it is called the *exceptional set of Koksma's theorem.* It is interesting to note that although one knows many reals belonging to the exceptional set, no sequence (α^n) u.d. mod 1 is known. The sequence $(((1+\sqrt{5})/2)^n)$ (cf. § 4.0) is not u.d. mod 1 and the number $(1+\sqrt{5})/2$ belongs to the exceptional set, but one does not know if sequences such as $((3/2)^n)$, (e^n) are u.d. mod 1 or not. Computer calculations have been made that show a very regular distribution of the sequence $((3/2)^n)$.

4.6 Some notions about uniform distribution modulo 1 in $\mathbf{R}^p$.

The notion of uniform distribution modulo 1 can be extended to $\mathbf{R}^p$, $p \geq 2$. Here we only give definitions and state theorems that will be used later.

Let $\mathbf{a} = (a_k)_{1\leq k\leq p}$ and $\mathbf{b} = (b_k)_{1\leq k\leq p}$ belong to $\mathbf{R}^p$. We say that $\mathbf{a} < \mathbf{b}$ (resp. $(\mathbf{a} \leq \mathbf{b})$ if $a_k < b_k$ (resp. $a_k \leq b_k$), $(\forall k = 1, \ldots, p)$. The set of points $\mathbf{x} \in \mathbf{R}^p$ such that $\mathbf{a} \leq \mathbf{x} < \mathbf{b}$ will be denoted $[\mathbf{a}, \mathbf{b}[= \prod_{k=1}^{p} [a_k, b_k[$. The other p-dimensional intervals $[\mathbf{a}, \mathbf{b}],]\mathbf{a}, \mathbf{b}],]\mathbf{a}, \mathbf{b}[$ have similar interpretations. For $r \in \mathbf{R}$ we set $\mathbf{r} = (r, \ldots, r)$, and we denote Π^p (resp. $\overline{\Pi}^p$) the s-dimensional unit cube $[-\mathbf{1/2}, \mathbf{1/2}[$ (resp. $[-\mathbf{1/2}, \mathbf{1/2}]$).

Every $\mathbf{x} \in \mathbf{R}^p$ can be written in one and only one way $\mathbf{x} = E(\mathbf{x}) + \varepsilon(\mathbf{x})$ with $E(\mathbf{x})$ belonging to $\mathbf{Z}^p$ and $\varepsilon(\mathbf{x})$ to Π^p. Studying the distribution modulo 1 of a sequence $(\mathbf{x}_n)$ in $\mathbf{R}^p$ means studying the sequence $(\varepsilon(\mathbf{x}_n))$ of the residues modulo 1. For $\mathbf{a}$, $\mathbf{b} \in \Pi^p$ $(\mathbf{a} < \mathbf{b})$ and $N \in \mathbf{N}^\star$ we denote $\nu(\mathbf{a}, \mathbf{b}, N)$ the number of terms $\mathbf{x}_n$ for which $\varepsilon(\mathbf{x}_n)$ belongs to $[\mathbf{a}, \mathbf{b}[$.

Definition 4.6. *The sequence $(\mathbf{x}_n)$ is said to be uniformly distributed modulo 1 (u.d. mod 1) in $\mathbf{R}^p$ if and only iffor every pair $(\mathbf{a}, \mathbf{b})$ with $\mathbf{a}$ and $\mathbf{b}$ belonging to $\overline{\Pi}^p$, $\mathbf{a} < \mathbf{b}$ we have*

$$\lim_{N \to +\infty} \frac{\nu(\mathbf{a}, \mathbf{b}, N)}{N} = \prod_{k=1,\ldots,p} (b_k - a_k).$$

Let $<\ >$ be the standard inner product in $\mathbf{R}^p$; we then have the following analogues of the one-dimensional results, in particular Weyl's results.

Theorem 4.6.1. *A sequence $(\mathbf{x}_n)$ is uniformly distributed modulo 1 in $\mathbf{R}^p$ if and only iffor every Riemann-integrable function f on $\overline{\Pi}^p$ we have*

$$\lim_{N \to +\infty} \frac{1}{N} \sum_{k=1}^{N} f(\varepsilon(\mathbf{x}_n)) = \int_{\overline{\Pi}^p} f(\mathbf{t}) d\mathbf{t}.$$

Theorem 4.6.2. *A sequence $(\mathbf{x}_n)$ is uniformly distributed modulo 1 in $\mathbf{R}^p$ if and only if*

$$\lim_{N \to +\infty} \frac{1}{N} \sum_{k=1}^{N} \exp(2i\pi < \mathbf{h}, \mathbf{x}_n >) = 0$$

for every lattice point $\mathbf{h} \in \mathbf{Z}^p \setminus \{0\}$.

The following results are consequences of Weyl's theorems; they will be used in § 5.3 for studying the distribution modulo 1 of sequences (τ^n), where τ is a Salem number.

Theorem 4.6.3. *Suppose $\boldsymbol{\alpha} = (\alpha_k)_{1 \le k \le p} \in \mathbf{R}^p$ has the property that the reals $1, \alpha_1, \ldots, \alpha_p$ are $\mathbf{Q}$-linearly independent. Then the sequence $(n\boldsymbol{\alpha})$ is uniformly distributed modulo 1 in $\mathbf{R}^p$.*

Kronecker's theorem is a consequence of this result.

Theorem 4.6.4. *Suppose $\boldsymbol{\alpha} = (\alpha_k)_{1\le k\le p} \in \mathbf{R}^p$ has the property that the reals $1, \alpha_1, \ldots, \alpha_p$ are $\mathbf{Q}$-linearly independent, and let $\boldsymbol{\mu}$ denote an arbitrary vector in $\mathbf{R}^p$, N an integer and ε a positive real number. Then there exists an integer $n > N$ such that $|\varepsilon_k(n\boldsymbol{\alpha} - \boldsymbol{\mu})| < \varepsilon \quad (k = 1, \ldots, p)$.*

Theorem 4.6.5. *Let $(\mathbf{x}_n)$ be a sequence uniformly distributed modulo 1 in $\mathbf{R}^p$, and φ a function continuous in the interval $[-1/2, 1/2]$; the real sequence defined by $y_n = \sum_{k=1}^{p} \varphi(\varepsilon_k(\mathbf{x}_n))$ is uniformly distributed modulo 1 if and only ifwe have*

$$\int_{-1/2}^{1/2} \exp(2i\pi h\varphi(t))dt = 0 \qquad (\forall h \in Z^\star).$$

This theorem is the analogue of Theorem 4.3.3 in one dimension; the proof uses Theorem 4.6.3.

Notes

The notion of uniform distribution modulo 1 for real sequences was introduced by Weyl in 1914. Theorems recalled in § 4.2 as well as their generalizations to $\mathbf{R}^p$ are also due to him [14, 15]. The 1930s produced the classical theorems of Van der Corput [12] and Koksma [6]. Pisot's first results followed soon after. These began as a search for elements of the exceptional set of Koksma's theorem. In their work Pisot and Salem were led to consider distribution modulo 1 theorems, such as Theorem 4.1 [9] and Theorems 4.3.3 and 4.6.5 [11].

A few years before Weyl's papers, Borel introduced in 1909 the notion of a normal number in base g by using g-adic expansions [1,2]. It soon appeared that there was a relation between normal numbers and uniformly distributed sequences modulo 1. This relation was made explicit in 1948 when Wall proved that a normal number in base g may be characterized by the uniform distribution of the sequence $(g\alpha^n)$ [13]. Following Niven and Zuckerman, and then Cassels [3], Mendès-France in the late 1960s used Weyl's characterization as a means of generalizing the notion of normal number [8]. In 1980 Choquet applied new methods taken from measure theory and dynamical systems to the study of the distribution of sequences $(k(3/2)^n)$ [5]. It will be noticed that the notion of uniform distribution modulo 1 can be expressed in probabilistic terms by noting that the sequence of probability measures $(1/N)\sum_{i=1}^{N} \delta_{\varepsilon(x_i)}$ is weakly convergent to the Lebesgue measure of $[-1/2, 1/2]$ when $N \to +\infty$. The Weyl criterion is then the transposition of this convergence in terms of Fourier transform measures. The condition of Theorems 4.3.3 and 4.6.5 on the function φ expresses the fact that the Lebesgue measure of $[-1/2, 1/2]$ is invariant by the transformation $t \mapsto \varepsilon_0 \varphi(t)$.

Kuipers and Niederreiter 's book [7] can be considered as the best reference concerning uniform distribution modulo 1. Cassels's [4] and Rauzy's [10] works should also be consulted.

References

[1] E. Borel, Les probabilités dénombrables et leurs applications arithmétiques, *Rend. Circ. Math. Palermo* 27, (1909), 247-271.

[2] E. Borel, *Leçons sur la théorie des fonctions*, Gauthiers-Villars, 2nde édition, Paris (1914).

[3] J.W.S. Cassels, On a paper of Niven and Zuckerman, *Pacific J. Math.* 2, (1952), 555-557.

[4] J.W.S. Cassels, An introduction to diophantine approximation, *Cambridge Tracts* n° 45, (1957).

[5] G. Choquet, Répartition des nombres $k(3/2)^n$; mesures et ensembles associés *C.R.A.S.* Vol. 290, (1980), 575-580.

[6] J.F. Koksma, Ein mengentheoretischer Satz über die Gleichverteilung modulo Eins, *Compositio Math.* 2, (1935), 250-258.

[7] L. Kuipers and H. Niederreiter, *Uniform Distribution of Sequences*, J. Wiley and Sons, (1974).

[8] M. Mendes-France, Nombres normaux. Applications aux fonctions pseudoaléatoires, *J. Analyse Math.* 20, (1967), 1-56.

[9] C. Pisot, Répartition modulo 1 des puissances successives des nombres réels, *Comment. Math. Helv.* (19, (1946), 153-160.

[10] G. Rauzy, Propriétés des suites arithmétiques, *Collection Sup. P.U.F.*, Paris, (1976).

[11] R. Salem, *Algebraic Numbers and Fourier Analysis*, Heath Math. Monographs, Boston, Mass., (1963).

[12] J.G. Van Der Corput, Diophantische Ungleichungen I. Zur Gleichverteilung modulo Eins, *Acta Math.* Vol. 56, (1931), 373-456.

[13] D.D. Wall, Normal numbers, *Ph.D. Thesis* (1949), University of California, Berkeley.

[14] H. Weyl, Über ein Problem aus dem Gebiete der diophantischen Approximationen, *Nachr. Wiss. Ges. Göttingen, Math. Phys. Kl.* (1914), 234-244.

[15] H. Weyl, Über die Gleichverteilung von Zahlen mod Eins. *Math. Ann.* 77, (1916), 313-352.

CHAPTER 5

PISOT NUMBERS, SALEM NUMBERS AND DISTRIBUTION MODULO 1

This first chapter on Pisot and Salem numbers deals mainly with properties of distribution modulo 1 of certain sequences $(\lambda\alpha^n)$. In particular we will show that Pisot and Salem numbers belong to the exceptional set of Koksma's theorem. In order to display similarities and differences we will study the two sets together as often as posssible.

5.0 Notation

We first define the notation used in this chapter and in the following ones.

We denote by α (eventually θ or τ) a real number greater than 1. Then if λ is a non-zero real, we set $u_n = E(\lambda\alpha^n)$ and $\varepsilon_n = \varepsilon(\lambda\alpha^n)$ $(\forall n \in \mathbf{N})$, hence $\|\lambda\alpha^n\| = |\varepsilon_n|$. Thus a sequence (u_n) of integers is associated to a pair (λ, α). Two different pairs (λ, α) and (λ', α') cannot have the same sequence (u_n); otherwise we would have for every integer n, $|\lambda\alpha^n - \lambda'\alpha'^n| > 1$, and suppose for instance $\alpha > \alpha' > 1$, $\alpha^n \left|\lambda - \lambda'\left(\frac{\alpha'}{\alpha}\right)^n\right| < 1$. This inequality cannot hold if n is large enough and we have $\alpha' = \alpha$ and $\lambda' = \lambda$.

In this chapter many proofs are based on the properties of the formal series $\sum\limits_{n\in\mathbf{N}} u_n X^n$ so defined. Thus, in order to prove that a real α is algebraic, one shows that there exists a non-zero real λ such that the associated series $\sum\limits_{n\in\mathbf{N}} u_n X^n$ is rational. Here we will make use of criteria proved in Chapter 1.

The power series $\sum\limits_{n=0}^{+\infty} u_n z^n$ and $\sum\limits_{n=0}^{+\infty} \varepsilon_n z^n$ are Taylor expansions of functions respectively denoted f and ε. These functions are analytic respectively in the disks $D(0, 1/\alpha)$ and $D(0, 1)$. When the series $\sum\limits_{n\in\mathbf{N}} u_n X^n$ is rational the functions f and ε are rational functions.

Finally if α is an algebraic number of degree s we denote $\alpha^{(j)}$ $(j = 2, \ldots, s)$ the remaining conjugates of α.

5.1 Some sequences $(\lambda\alpha^n)$ non-uniformly distributed modulo 1 and a set of algebraic integers

The following theorem shows that if a certain condition of non-uniform distribution is satisfied by a sequence $(\lambda\alpha^n)$, then α belongs to a remarkable set of algebraic integers.

Theorem 5.1.1. *Let α be a real greater than 1. Suppose there exists a real λ with $\lambda \geq 1$ such that*

(1) $$\|\lambda\alpha^n\| \leq \frac{1}{2e\alpha(\alpha+1)(1+\log\lambda)} \qquad (\forall n \in \mathbf{N}).$$

Then α is an algebraic integer, its remaining conjugates have modulus at most equal to 1 and λ belongs to the field $\mathbf{Q}(\alpha)$.

Proof. Using notations defined in §5.0 we will prove that the series $\sum\limits_{n\in\mathbf{N}} u_n X^n$ is rational by applying Proposition 1.1. Our purpose is to show that there exists an integer $s \geq 1$ and an element $\mathbf{a} = (a_i)_{0\leq i\leq s} \in \mathbf{Z}^{s+1} \setminus \{\mathbf{0}\}$ such that, if V_n $(n \in \mathbf{N})$ denotes the linear form defined on $\mathbf{R}^{s+1}$ by $V_n(\mathbf{x}) = \sum\limits_{i=0}^{s} u_{n+i}x_i$, then $V_n(\mathbf{a}) = 0$ for every integer n.

There are three steps to the proof:

First, assuming there exist numbers $A \in \mathbf{N}^\star$ and $\varepsilon \in \mathbf{R}_+^\star$ such that $|a_i| \leq A$ $(i = 0, \ldots, s)$ and $|\varepsilon_n| \leq \varepsilon$ $(\forall n \in \mathbf{N})$, we state a condition (2) for A, ε and s under which the equality $V_n(\mathbf{a}) = 0$ implies $V_{n+1}(\mathbf{a}) = 0$. Then, using another condition (3) we show that it is possible to find $\mathbf{a}$ such that $V_0(\mathbf{a}) = 0$. It remains to prove that by using condition (1) we can determine A and s such that conditions (2) and (3) are simultanously satisfied. The algebraicity of α will then be proved, and certain properties of its conjugates follow from analytic considerations.

a) *The equality $V_n(\mathbf{a}) = 0$ $(\mathbf{a} \in \mathbf{Z}^{s+1})$ and the condition*

(2) $$|\varepsilon_n| < \frac{1}{(s+1)(\alpha+1)A}$$

imply $V_{n+1}(\mathbf{a}) = 0$.

Thus we have $|V_{n+1}(\mathbf{a}) - \alpha V_n(\mathbf{a})| \leq \sum_{i=0}^{s} |a_i(\varepsilon_{n+i+1} - \alpha\varepsilon_{n+i})|$. Hence

$$|V_{n+1}(\mathbf{a}) - \alpha V_n(\mathbf{a})| \leq (s+1)(\alpha+1)A\varepsilon.$$

It follows from the inequality (2) that $V_n(\mathbf{a}) = 0$ implies $V_{n+1}(\mathbf{a}) = 0$.

b) *Conditions* (2) *and*

(3) $$A \geq 2\lambda^{1/s}\alpha - 1$$

imply that for every integer s *there exists* $\mathbf{a} \in \mathbf{Z}^{s+1} \setminus \{0\}$ *such that* $V_0(\mathbf{a}) = 0$.

Let W_0 be the linear form defined by $W_0(\mathbf{x}) = \sum_{i=0}^{s} |u_i| x_i$. Consider the set of points belonging to $\mathbf{Z}^{s+1}$ such that $0 \leq x_i \leq A$ $(i = 0, \ldots, s)$. There are $(A+1)^{s+1}$ such points and the values taken by W_0 on this set satisfy $0 \leq W_0(\mathbf{x}) \leq (s+1)A(\lambda\alpha^s + \varepsilon)$. Hence using condition (2) we have $0 \leq W_0(\mathbf{x}) \leq (s+1)(A+1)\lambda\alpha^s - 1$. Then it follows from condition (3) that

$$(A+1)^{s+1} \geq 2^s(A+1)\lambda\alpha^s \geq (s+1)(A+1)\lambda\alpha^s.$$

According to the box or pigeonhole principle there exist two different points $\mathbf{b}$ and $\mathbf{b}'$ in $\mathbf{Z}^{s+1}$ with $0 \leq b_i \leq A$, $0 \leq b'_i \leq A$ $(i = 0, \ldots, s)$ such that $W_0(\mathbf{b}) = W_0(\mathbf{b}')$. Then we set $\mathbf{a} = \mathbf{b} - \mathbf{b}'$, $\mathbf{a}$ belongs to $\mathbf{Z}^{s+1} \setminus \{0\}$ with $|a_i| \leq A$ $(i = 0, \ldots, s)$, and $V_0(\mathbf{a}) = 0$.

c) *By using condition* (1) *we can determine non-zero integers* s *and* A *such that there exists* $\mathbf{a} \in \mathbf{Z}^{s+1} \setminus \{0\}$ *satisfying* $V_n(\mathbf{a}) = 0$ *for every integer* $n \in \mathbf{N}$.

Let s and A be integers defined by $s - 1 \leq \log\lambda < s$ and $A < 2\lambda^{1/s}\alpha \leq A + 1$. It follows from the properties of the function φ defined by $\varphi(x) = 1 - \frac{x}{s} + \log\frac{1+x}{1+s}$ that $\varphi(\log\lambda) > 0$, hence $\frac{\log\lambda}{s} + \log(1+s) < 1 + \log(1+\log\lambda)$ and finally $\lambda^{1/s}(s+1) < e(1+\log\lambda)$.

Condition (1) and the definition of A imply that conditions (2) and (3) are satisfied. Thus there exists $\mathbf{a} \in \mathbf{Z}^{s+1} \setminus \{0\}$ such that $V_n(\mathbf{a}) = 0$ for every integer n.

d) *The conjugates of* α *(except* α *itself) have modulus at most equal to 1.*

The series $\sum_{n \in \mathbf{N}} u_n X^n$ is rational and so equals A/Q, where A and Q are polynomials with integer coefficients and relatively prime, so that by Fatou's lemma we have $Q(0) = 1$ (cf. §1.3).

The function ε being analytic in the disk $D(0,1)$ the equality $f(z) = \dfrac{A(z)}{Q(z)} = \dfrac{\lambda}{1-\alpha z} + \sum_{n=0}^{+\infty} \varepsilon_n z^n$ $(\forall z \in D(0, 1/\alpha))$ implies that the polynomial Q has a single zero $1/\alpha$ in the disk $D(0,1)$.

The number α is an algebraic integer and its remaining conjugates have modulus at most equal to 1. We note that the degree s of α is bounded by $1 + \log \lambda$. The residue of the function f at $1/\alpha$ is $-\lambda/\alpha$. Then $\lambda = -\alpha A(1/\alpha)/Q'(1/\alpha)$ and thus the number λ belongs to the field $\mathbf{Q}(\alpha)$. ∎

Theorem 5.1.1 defines a set of algebraic integers.

Definition 5.1. *The set U is the set of real algebraic integers α greater than 1 whose remaining conjugates have modulus at most equal to 1.*

By using instead of Proposition 1.1 a more subtle criterion of rationality we can obtain instead of Condition (1) a weaker condition. This is the purpose of Theorems 5.1.2 and 5.1.3.

The stated conditions imply, as for Theorem 5.1.1 , that λ belongs to the field $\mathbf{Q}(\alpha)$. This will be assumed henceforth.

Theorem 5.1.2. *Let α be a real greater than 1; assume there exists a real λ with $\lambda \geq \frac{1}{2}$ such that*

$$(4) \qquad (\alpha+1)^2 \sum_{i=m}^{m+n} (\varepsilon_{i+1} - \alpha\varepsilon_i)^2 < \rho(n+1) \qquad (\forall (m,n) \in \mathbf{N}^2)$$

with

$$(5) \qquad \rho \leq \frac{1}{e[\log(u_0^2 + \frac{1}{(\alpha+1)^3}) + 2]};$$

then α belongs to U.

The following lemma will be used in the proof.

Lemma *Let ρ be real with $0 < \rho < 1/e$ and let r be the integer defined by* $\frac{1}{\rho e} - 1 \leq r < \frac{1}{\rho e}$.

We have then the inequalities

$$(r+1)\log\rho(r+1) < \rho e - \frac{1}{\rho e} \tag{6}$$

$$r\log\rho(r+1) < \rho e - \frac{1}{\rho e} + 1. \tag{7}$$

Proof. The function $x \mapsto x\log\rho x$ has a minimum at $x = \frac{1}{\rho e}$; thus the inequalities $\frac{1}{\rho e} \leq r+1 < \frac{1}{\rho e} + 1$ imply

$$(r+1)\log\rho(r+1) \leq (\frac{1}{\rho e}+1)\log\left(\rho(\frac{1}{\rho e}+1)\right) = (\frac{1}{\rho e}+1)\left(\log(\rho e+1)-1\right).$$

Hence, by the inequality $\log x < x - 1$ $(\forall x > 0)$

$$(r+1)\log\rho(r+1) < (\frac{1}{\rho e}+1)(\rho e - 1) = \rho e - \frac{1}{\rho e},$$

we deduce that

$$r\log\rho(r+1) < \rho e - \frac{1}{\rho e} - \log\rho(r+1) < \rho e - \frac{1}{\rho e} + 1.$$

∎

Proof of the theorem. We use the same notation as in Theorem 1.1.2. The sequence (t_n) is defined by $t_0 = 1,\ t_1 = -\alpha,\ t_n = 0\ \ (n \geq 2)$.

For every integer $r \geq 1$ and every sequence $L_r \in \mathcal{L}_r$ we write the matrix $A(L_r)$ as

$A(L_r) = (x_{\ell_i,j})$ $(i = 0,\dots,r;\ \ j = 0,\dots,r)$ with $x_{m,n} = \sum_{h=0}^{m}\sum_{k=0}^{n} t_h t_k u_{m+n-(h+k)}$.

We set $s_n = \varepsilon_n - \alpha\varepsilon_{n-1}$ for every integer $n \geq 1$. We have then

$$x_{0,0} = u_0,\ x_{0,n} = x_{n,0} = -s_n \quad (n \geq 1),$$
$$x_{m,n} = x_{n,m} = \alpha s_{m+n-1} - s_{m+n} \quad (m \geq 1,\ n \geq 1).$$

We obtain the following inequalities:

$$x_{m,n}^2 \le (\alpha+1)(\alpha s_{m+n-1}^2 + s_{m+n}^2) \qquad (m \ge 1\ n \ge 1).$$

We transform the matrix $A(L_r)$ into a matrix with the same determinant. Let $B(L_r) = (y_{i,j})$ $(i = 0, \ldots, r;\ \ j = 0, \ldots, r)$ with

$$y_{0,0} = x_{\ell_0,0}, \qquad y_{0,j} = x_{\ell_0,j}/\sqrt{\alpha+1} \quad (j = 1, \ldots, r),$$
$$y_{i,0} = \sqrt{\alpha+1}x_{\ell_i,0} \quad (i = 1, \ldots r), \qquad y_{i,j} = x_{\ell_i,j} \quad (i = 1, \ldots, r;\ j = 1, \ldots, r).$$

We apply the Hadamard inequality to $\det B(L_r)$; it suffices to consider sequences L_r such that $\ell_0 = 0$.

The equations $\sum_{j=0}^{r} y_{0,j}^2 = x_{0,0}^2 + \frac{1}{\alpha+1}\sum_{j=1}^{r} x_{0,j}^2 = u_o^2 + \frac{1}{\alpha+1}\sum_{j=1}^{r} s_j^2$ imply

$$\sum_{j=0}^{r} y_{0,j}^2 \le u_0^2 + \frac{\rho(r+1)}{(\alpha+1)^3}. \tag{8}$$

In the same way we get for $i = 1, \ldots, r$

$$\sum_{j=0}^{r} y_{i,j}^2 \le (\alpha+1)x_{\ell_i,0}^2 + \sum_{j=1}^{r} x_{\ell_i,j}^2 \le (\alpha+1)\left[s_{\ell_i}^2 + \sum_{j=1}^{r}(s_{\ell_i+j}^2 + \alpha s_{\ell_i+j-1}^2)\right].$$

Hence by inequality (4)

$$\sum_{j=0}^{r} y_{i,j}^2 \le (\alpha+1)^2 \sum_{j=0}^{r} s_{\ell_i+j}^2 \le \rho(r+1). \tag{9}$$

It follows from inequalities (8) and (9) that

$$\det{}^2 A(L_r) \le \left(u_0^2 + \frac{\rho(r+1)}{(\alpha+1)^3}\right)(\rho(r+1))^r. \tag{10}$$

The inequality $\lambda \ge \frac{1}{2}$ implies $u_0 \ge 1$ and thus $\log\left(u_0^2 + \frac{1}{(\alpha+1)^3}\right) > 0$. From condition (5) we deduce $\rho e < 1$. Then let r be the integer defined by $\frac{1}{\rho e} - 1 \le$

$r < \dfrac{1}{\rho e}$, we have $r \geq 1$ and $\rho(r+1) < 1$. It follows from the lemma that the inequalities (6) and (7) are satisfied.

Thus we have

$$\log\left(u_0^2 + \frac{\rho(r+1)}{(\alpha+1)^3}\right) + 1 < \log\left(u_0^2 + \frac{1}{(\alpha+1)^3}\right) + 1 < \frac{1}{\rho e} - 1 < \frac{1}{\rho e} - \rho e.$$

By inequality (7) we have

$$\log\left(u_0^2 + \frac{\rho(r+1)}{(\alpha+1)^3}\right) + r\log\rho(r+1) < 0,$$

hence $|\det A(L_r)| < 1 \quad (\forall L_r \in \mathcal{L}_r)$. It follows from Theorem 1.1.2 that the series $\sum_{n\in\mathbf{N}} u_n X^n$ is rational and the end of the proof is the same as for Theorem 5.1.1. ∎

Theorem 5.1.3. *Let α be a real greater than 1. Assume there exists a real λ with $\lambda \geq 1$ such that*

$$\|\lambda\alpha^n\| \leq \frac{1}{e(\alpha+1)^2(2+\sqrt{\log\lambda})} \qquad (\forall n \in \mathbf{N}); \tag{11}$$

then α belongs to U.

Proof. According to condition (11) we have

$$(\alpha+1)^2 \sum_{i=m}^{m+n} (\varepsilon_{i+1} - \alpha\varepsilon_i)^2 \leq \frac{n+1}{e^2(2+\sqrt{\log\lambda})^2} \qquad (\forall (m,n) \in \mathbf{N}^2) \qquad \text{and}$$

$\varepsilon_0 < \dfrac{1}{8}$.

We deduce from these inequalities $0 < u_0 < \lambda + \dfrac{1}{8}$ and $u_0^2 + \dfrac{1}{(1+\alpha)^3} < (\lambda + \dfrac{1}{2})^2$.
We have then

$$\log\left(u_0^2 + \frac{1}{(\alpha+1)^3}\right) + 2 \leq 2\log(\lambda + \frac{1}{2}) + 2 \leq 2(\log\lambda + 2).$$

Since the inequalities $2e < e^2$, $\sqrt{\log\lambda + 2} \leq 2 + \sqrt{\log\lambda}$ hold we can apply Theorem 5.1.2 with $\rho = \dfrac{1}{2e(2+\log\lambda)}$. ∎

The interest of Theorem 5.1.3 with regard to Theorem 5.1.1 concerns large values of λ, because of the term $\sqrt{\log\lambda}$ instead of $\log\lambda$. This can be easily seen when we consider $\lambda\alpha^{n_0}$ instead of λ.

We should mention that Boyd has shown that there exist pairs (λ,α) with α trancendental such that $\|\lambda\alpha^n\| \leq \dfrac{5}{e\alpha(\alpha+1)(1+\log\lambda)}$, $\forall n \in \mathbf{N}$. According to Theorem 5.1.3 this is possible only for small values of λ.

All theorems in this paragraph have reciprocals. This will be shown in § 5.2.

5.2 Pisot numbers and Salem numbers. Definitions and algebraic properties

The conjugates of a U-number α have (except α itself) modulus at most equal to 1. Now we distinguish between U-numbers that have at least one conjugate of modulus 1 and those that have no conjugate of modulus 1.

Definition 5.2.1. *The set S of Pisot numbers is the set of real algebraic integers θ greater than 1 whose other conjugates have modulus strictly smaller than 1.*

Definition 5.2.2. *The set T of Salem numbers is the set of real algebraic integers τ greater than 1 whose other conjugates have modulus at most equal to 1, one at least having a modulus equal to 1.*

The sets S and T define a partition of U. This partition remains valid in some generalizations but in a somewhat more intricate way.

All rational integers greater than 1 belong to S. The quadratic numbers in S are zeros of polynomials with integer coefficients $X^2+q_1X+q_0$ with $q_1+|1+q_0|<0$. More generally polynomials $X^s+q_{s-1}X^{s-1}+\cdots+q_0$ whose integer coefficients satisfy $1+\sum_{i=0}^{s-1} q_i < 0$ and $|q_{s-1}| > 1+\sum_{i=0}^{s-2}|q_i|$ have by Rouché's theorem a zero greater than 1. This zero belongs to S. We also note that if θ is an S-number then θ^n is an S-number.

There are no examples of Salem numbers as simple as the ones given for Pisot numbers because there exist no Salem numbers of degree less than 4. Thus if τ belongs to T at least one of its conjugates has modulus equal to 1 and this number's inverse is also a conjugate of τ. Then the minimal polynomial of τ is reciprocal and has a single zero in the disk $D(0,1)$; its other zeros belong to $C(0,1)$, and are pairs of imaginary conjugates. Thus a Salem number is necessarily a unit of even degree at least equal to 4. These properties lead us to introduce the following notation, which will be used in this and the following chapters. If τ is a T-number we denote $1/\tau, \tau^{(j)}, 1/\tau^{(j)} = \overline{\tau}^{(j)}$ $(j = 2, \dots, s)$ the remaining conjugates of τ; we set $\tau^{(j)} = \exp(2i\pi\omega^{(j)})$ with $\omega^{(j)} \in [-1, +1]$ $(j = 2, \dots, s)$. Furthermore, define the reals $\gamma = \tau + 1/\tau, \gamma^{(j)} = \tau^{(j)} + 1/\tau^{(j)}$ $(j = 2, \dots, s)$. These are zeros of the polynomial obtained from the minimal polynomial of τ by transformation $X \mapsto Y = X + 1/X$; these reals satisfy $\gamma > 2$ and $-2 < \gamma^{(j)} < 2$ $(j = 2, \dots, s)$.

One can easily determine the Salem numbers of degree 4. They are the zeros greater than 1 of the polynomials with integer coefficients of the following form $X^4 + q_1X^3 + q_2X^2 + q_1X + 1$ with $2(q_1 - 1) < q_2 < -2(q_1 + 1)$. The smallest is a zero of the polynomial $X^4 - X^3 - X^2 - X + 1$.

The basic properties so far mentioned show that differences exist between Pisot and Salem numbers as regards their algebraic properties. More precisely we will show that every real extension of finite degree of $\mathbf{Q}$ can be generated by S-numbers, whereas if τ is a T-number the extension $\mathbf{Q}(\tau)$ is necessarily a real quadratic extension of a totally real field.

The existence of S-numbers in every real extension of $\mathbf{Q}$ of finite degree follows from the theorem of Minkowski, which we now recall.

Theorem 5.2.1. *Let $(L_i)_{1\leq i\leq n}$ be n linear forms with real coefficients whose determinant Δ is not zero. Let $(\gamma_i)_{1\leq i\leq n}$ be n positive reals with $\prod_{i=1,\dots,n} \gamma_i \geq |\Delta|$. Then there exists $\mathbf{u} \in \mathbf{Z}^n \setminus \{\mathbf{0}\}$ such that $|L_i(\mathbf{u})| \leq \gamma_i$ $(i = 1, \dots, n)$.*

Minkowski's theorem is still valid if the forms L_i have complex coefficients, provided we consider simultaneously the forms L_i and $\overline{L}_i$. We have then $|L_i(\mathbf{u})| \leq \gamma_i$ and $|\overline{L}_i(\mathbf{u})| \leq \gamma_i$ $(i = 1, 2, \dots, n)$.

Theorem 5.2.2. *Every real algebraic extension of finite degree of the field $\mathbf{Q}$ contains infinitely many S-numbers whose degree is that of the extension; some of these numbers are units.*

Proof. Let K be a real algebraic extension of $\mathbf{Q}$; we designate by s its degree and by D its discriminant.

a) The field K contains S-numbers of degree s.

Let $(\omega_i)_{1\le i\le s}$ be a basis of integers of K and denote by $(\omega_i^{(j)})_{1\le i\le s}$ $(j=2,\dots,s)$ the conjugate basis.

By Minkowski's theorem there exists $\mathbf{u} = (u_i)_{1\le i\le s} \in \mathbf{Z}^s \setminus \{0\}$ such that $|\sum_{i=1}^{s} u_i\omega_i| \le M$ and $|\sum_{i=1}^{s} u_i\omega_i^{(j)}| \le \delta < 1$, $(j=2,\dots,s)$ provided the inequality $M\delta^{s-1} \ge \sqrt{|D|}$ holds. We set $\theta = \sum_{i=1}^{s} u_i\omega_i$: the number θ is an algebraic integer whose conjugates $\theta^{(j)} = \sum_{i=1}^{s} u_i\omega_i^{(j)}$ $(j=2,\dots,s)$ satisfy $|\theta^{(j)}| \le \delta < 1$. Thus we have $1 \le |N(\theta)| = |\theta \prod_{j=2}^{s} \theta^{(j)}| \le |\theta|\delta^{s-1}$, hence $|\theta| > \delta^{1-s}$ and $|\theta^{(j)}| < 1$ $(j=2,\dots,s)$.

It follows from these inequalities that θ (or $-\theta$) belongs to S and that the monic polynomial whose θ and $\theta^{(j)}$ $(j=2,\dots,s)$ are zeros is irreducible. Furthermore, θ is of degree s.

b) The field K contains units of degree s that belong to S.

From part a), there exists a sequence (δ_n) of reals and a sequence (θ_n) of numbers of $K\cap S$ all distinct, of degree s, such that $1 < \theta_1 \le \sqrt{|D|}\delta_1^{1-s}$, $0 < \delta_n < \inf_{j=2,\dots,s} |\theta_{n-1}^{(j)}|$ and $|\theta_n^{(j)}| \le \delta_n$ $(j=2,\dots,s)$ $(n\ge 2)$ where δ_1 is an arbitrary real with $0<\delta_1<1$.

The sequence $(|N(\theta_n)|)$ is bounded by $\sqrt{|D|}$, so one can find $m\in\mathbf{Z}$ such that for infinitely many n, $N(\theta_n)=m$. Then there exist in the sequence (θ_n) two terms respectively denoted θ' and θ'' and defined by $\theta' = \sum_{i=1}^{s} u_i'\omega_i$ and $\theta'' = \sum_{i=1}^{s} u_i''\omega_i$, with $\delta' > \delta''$, such that $N(\theta') = N(\theta'') = m$ and $u_i' = u_i''$ (mod m), $(i=1,\dots,s)$. Thus we have $\theta''-\theta' = m\alpha$ with α an algebraic integer in K. We set $\xi = \theta''/\theta'$. ξ is real and $\xi = 1+\alpha\prod_{j=2}^{s}\theta^{(j)}$; it is an algebraic integer that satisfies $N(\xi)=1$, and therefore a unit of K. The inequality $\delta'>\delta''$ implies that $|\xi^{(j)}| = |\theta''^{(j)}/\theta'^{(j)}| < 1$, $(j=2,\dots,s)$, hence $|\xi|>1$. Thus ξ (or $-\xi$) belongs to S. ∎

Theorem 5.2.3. *Let τ be a T-number; the extension $K=\mathbf{Q}(\tau)$ is a real quadratic extension of a totally real field. The numbers of $K\cap T$ can be written τ_0^n, where τ_0 belongs to $K\cap T$ and generates K. Every number in $K\cap T$ is quotient of two numbers of $K\cap S$.*

Proof. It follows from the definiton of γ that $\mathbf{Q}(\gamma)$ is totally real. K is therefore a real quadratic extension of $\mathbf{Q}(\gamma)$.

On the other hand, let τ and τ' be two numbers of $K \cap T$; then $\tau\tau'$ belongs to $K \cap T$. There are finitely many units of K belonging to a bounded part of $\mathbf{C}$, so the set $\{\log \tau \mid \tau \in K \cap T\} \cup \{0\}$ is an additive and discrete subgroup of $\mathbf{R}$, and there exists $\tau_0 \in K \cap T$ such that the degree of τ_0 is the same and that K and every number of $K \cap T$ can be written τ_0^n, $n \in \mathbf{N}^\star$. Finally if θ belongs to $K \cap S$ then the number $\theta' = \theta\tau$ is an algebraic integer; its remaining conjugates have modulus smaller than 1; therefore θ belongs to $K \cap S$. ∎

We end this section by showing that Theorem 5.2.2 allows us to state the necessary condition in some cases. By using Theorem 5.2.2 we find a suitable real $\lambda \in \mathbf{Q}(\alpha)$ and prove the converse of Theorems 5.1.1, 5.1.2 and 5.1.3. We obtain characterizations of U. We will prove the converse of Theorem 5.1.1; the other proofs are analogous.

Theorem 5.2.4. *A real α greater than 1 belongs to the set U if and only if there exists a real λ with $\lambda \geq 1$ such that*

(1) $$\|\lambda\alpha^n\| \leq \frac{1}{2e\alpha(\alpha+1)(1+\log\lambda)} \qquad (\forall n \in \mathbf{N}).$$

Proof of the necessary condition. Let α be a U-element and μ an element of $\mathbf{Q}(\alpha) \cap S$. We wish to find λ of the form $\lambda = \mu^\nu$ $(\nu \in \mathbf{N}^\star)$. The remaining conjugates of α, μ and λ satisfy $|\alpha^{(j)}| \leq 1$, $|\mu^{(j)}| \leq \delta < 1$, $|\lambda^{(j)}| \leq \delta^\nu$ $(j = 2, \dots s)$. The number $\lambda\alpha^n + \sum_{j=2}^{s} \lambda^{(j)}{\alpha^{(j)}}^n$ is a rational integer and the inequality $|\sum_{j=2}^{s} \lambda^{(j)}{\alpha^{(j)}}^n| \leq (s-1)\delta^\nu$ implies that one can find ν such that for every $n \in \mathbf{N}$ we have $\|\lambda\alpha^n\| = |\sum_{j=2}^{s} \lambda^{(j)}{\alpha^{(j)}}^n|$ and $(s-1)\delta^\nu \leq \dfrac{1}{2e\alpha(\alpha+1)(1+\nu\log\lambda)}$; thus condition (1) is satisfied. ∎

5.3 Distribution modulo 1 of the sequences (α^n) with α a U-number

We will show in this paragraph that the set U is included in the exceptional set of Koksma's theorem. If α is a Pisot number then the sequence (α^n) converges to 0 modulo 1, whereas if α is a Salem number the sequence (α^n) is dense but not uniformly distributed modulo 1.

Theorem 5.3.1. *Let θ be an S-number; the sequence (θ^n) converges to 0 modulo 1.*

Proof. Let θ be an S-number; we set $\delta = \sup_{j=2,\dots,s} |\theta^{(j)}|$. The number $\theta^n + \sum_{j=2}^{s} {\theta^{(j)}}^n$ being a rational integer, the inequality $|\sum_{j=2}^{s} {\theta^{(j)}}^n| \le (s-1)\delta^n$ implies that we have for n large enough $\|\theta^n\| = |\sum_{j=2}^{s} {\theta^{(j)}}^n|$. Therefore the sequence $(\|\theta^n\|)$ converges to 0 geometrically. ∎

Theorem 5.3.2. *Let τ be a T-number; the sequence (τ^n) is then dense but not uniformly distributed modulo 1.*

Proof. Let τ be a T-number. The number $\tau^n + \tau^{-n} + \sum_{j=2}^{s} [{\tau^{(j)}}^n + {\tau^{(j)}}^{-n}]$ being a rational integer, the distribution of the sequence (τ^n) is the same as that of the sequence $\left(\tau^{-n} + \sum_{j=2}^{s} \left({\tau^{(j)}}^n + {\tau^{(j)}}^{-n}\right)\right)$. The sequence (τ^{-n}) converges to 0, so we will only consider the distribution of the sequence $\left(\sum_{j=2}^{s} \left[{\tau^{(j)}}^n + {\tau^{(j)}}^{-n}\right]\right)$, in other words of $\left(2 \sum_{j=2}^{s} \cos 2n\pi\omega^{(j)}\right)$.

a) The sequence (τ^n) is dense modulo 1.

Our first step is to show that the numbers $1, \omega^{(2)}, \dots, \omega^{(s)}$ are $\mathbf{Q}$-linearly independent, the second to apply Theorem 4.6.4 to prove that the sequence $\left(2 \sum_{j=2}^{s} \cos 2n\pi\omega^{(j)}\right)$ is dense modulo 1.

Suppose there exist integers $\ell_1, \dots, \ell_s$ of $\mathbf{Z}$ such that $\ell_1 + \sum_{j=2}^{s} \ell_j\omega^{(j)} = 0$. It follows that $\exp(2\pi i \sum_{j=2}^{s} \ell_j\omega^{(j)}) = 1$, hence $\prod_{j=2}^{s} {\tau^{(j)}}^{\ell_j} = 1$. In the Galois extension $\mathbf{Q}(\tau, \tau^{-1}, \tau^{(2)}, \overline{\tau}^{(2)}, \dots, \tau^{(s)}, \overline{\tau}^{(s)})$ there exists an automorphism σ such that $\sigma(\tau^{(2)}) = \tau$. Hence we have $\tau^{\ell_2} \prod_{j=3}^{s} (\sigma(\tau^{(j)})^{\ell_j} = 1$ $(j = 3, \dots, s)$. This equation cannot hold if $\ell_2 \neq 0$. In the same way the integers $\ell_3, \dots, \ell_s$ are zero and the reals $1, \omega^{(2)}, \dots, \omega^{(s)}$ are $\mathbf{Q}$-linearly independent. Let then ρ be a real $-1/2 \leq \rho \leq 1/2$. We set $\rho = 2\cos 2\pi\beta$ with $\beta \in [-1, 1]$. According to Theorem 4.6.4 consider $(\omega^{(j)})_{2\leq j\leq s} \in \mathbf{R}^{s-1}$ with $\boldsymbol{\mu} = (\beta, 1/4, \ldots, 1/4)$. It is clear that for every $\varepsilon > 0$ there exists an arbitrarily large integer m such that $|m\omega^{(2)} - \beta| < \varepsilon \pmod 1$, $|m\omega^{(j)} - 1/4| < \varepsilon \pmod 1$ $(j = 3, \ldots, s)$. Thus we can find a real $2\sum_{j=2}^{s} \cos 2\pi m\omega^{(j)}$ arbitrarily close to ρ.

b) The sequence (τ^n) is not uniformly distributed modulo 1.

The reals $1, \omega^{(2)}, \ldots, \omega^{(s)}$ being linearly independent, it follows from Theorem 4.6.3 that the sequence $\left((n\omega^{(j)})_{j=2,\ldots,s}\right)$ is uniformly distributed modulo 1 in $\mathbf{R}^{s-1}$. Consider the function $x \mapsto 2\cos 2\pi x$. The integral $\int_0^1 \exp(4i\pi h\cos 2\pi t)dt = J_0(4h\pi)$ (with J_0 designating the Bessel function of order 0) is not zero for all $h \in \mathbf{Z}^\star$. We complete the proof by using Theorem 4.6.5. ∎

5.4 Pisot numbers and distribution modulo 1 of certain sequences $(\lambda\theta^n)$

Let θ be an S-number and λ an algebraic integer of $\mathbf{Q}(\theta)$; the real $\lambda\theta^n + \sum_{j=2}^{s} \lambda^{(j)}\theta^{(j)^n}$ is then a rational integer. One then proves as in Theorem 5.3.1 that the sequence $(\|\lambda\theta^n\|)$ converges to zero geometrically.

In this section our object is to study the converse problem, which can be stated as follows: *Let θ be a real greater than 1. Suppose there exists a non-zero real λ such that $\lim_{n\to+\infty} \|\lambda\theta^n\| = 0$. Does θ belong to the set S?*

This problem has so far not been solved. We do however have an answer when a supplementary condition is imposed. One supposes either that θ is algebraic or that the sequence $(\|\lambda\theta^n\|)$ converges to zero sufficiently rapidly.

The following theorems give conditions for a number to be an S-number. In each case the condition given is clearly necessary while the proof of its sufficiency depends on an analytic or algebraic criterion. For all these criteria we refer the reader to Chapter 1. The weaker the convergence condition, the subtler the rationality criterion employed. In each theorem the conditions imply that λ belongs to $\mathbf{Q}(\theta)$. The proof of Theorem 5.4.1 uses an analytic lemma whose purpose is to show that the minimal polynomial of θ has no zeros on the circle $C(0, 1)$.

Lemma 5.4. *Let φ be a meromorphic function in an open set that contains the disk $\overline{\mathrm{D}}(0,1)$. Assume that φ has no poles in 0 and that the coefficients η_n of its Taylor expansion satisfy $\lim\limits_{n\to+\infty} \eta_n = 0$. Then φ, which is analytic in the disk $D(0,1)$, has no poles on the circle $C(0,1)$.*

Proof. The radius of convergence of the series $\sum\limits_{n=0}^{+\infty} \eta_n z^n$ is at least equal to 1. Hence the function φ is analytic in the disk $D(0,1)$. Assume that $R = 1$; then φ has at least one singular point on the circle $C(0,1)$. Suppose without loss of generality that this pole is at the point $z = 1$. Then let ε be a real positive number, there exists an integer n_0 such that for $n \geq n_0$, we get $|\eta_n| < \varepsilon$. Hence for $0 < r < 1$, $|\varphi(z)| \leq \left|\sum\limits_{n=0}^{n_0-1} \eta_n r^n\right| + \left|\sum\limits_{n=n_0}^{+\infty} \eta_n r^n\right| \leq M + \varepsilon\dfrac{r^{n_0}}{1-r}$ with M a constant. Therefore we have $(1-r)|\varphi(r)| \leq M(1-r) + \varepsilon$, and hence $\lim\limits_{r\to 1, r<1} (1-r)\varphi(r) = 0$. This last equation contradicts the fact that 1 is a pole of φ. Therefore $R > 1$. ■

Theorem 5.4.1. *An algebraic real θ greater than 1 belongs to the set S if and only if there exists a non-zero real λ such that*

$$\lim_{n\to+\infty} \|\lambda\theta^n\| = 0.$$

Proof. The number θ being algebraic, it is a zero of a polynomial $\sum\limits_{i=0}^{s} q_i X^i$ with integer coefficients. Thus we have $\lambda \sum\limits_{i=0}^{s} q_i \theta^{i+n} = 0$ for every integer n, hence $\sum\limits_{i=0}^{s} q_i u_{n+i} = -\sum\limits_{i=0}^{s} q_i \varepsilon_{n+i}$. The hypothesis $\lim\limits_{n\to+\infty} \varepsilon_n = 0$ implies $|\sum\limits_{i=0}^{s} q_i \varepsilon_{n+i}| < 1$ for n large enough. Therefore $\sum\limits_{i=0}^{s} q_i u_{n+i} = 0$. Consequently the series $\sum\limits_{n\in\mathbf{N}} u_n X^n$ is rational. As in Theorem 5.1.1 the series equals A/Q where A and Q are polynomials with integral coefficients, relatively prime and such that $Q(0) = 1$. The functions f and ε satisfy $f(z) = \dfrac{A(z)}{Q(z)} = \dfrac{\lambda}{1-\theta z} + \varepsilon(z)$, $\forall z \in D(0, 1/\theta)$. They are rational functions. By applying Lemma 5.4 it follows from the condition $\lim\limits_{n\to+\infty} \varepsilon_n = 0$ that the function ε has no pole on the disk $\overline{\mathrm{D}}(0,1)$. The polynomial Q has a single zero in the disk $\overline{\mathrm{D}}(0,1)$, it is irreducible and θ belongs to S. ■

In the following theorems we will not suppose θ algebraic; we need only to show that either the function f or the series $\sum_{n\in\mathbf{N}} u_n X^n$ is rational. This implies that θ is algebraic and the proofs are completed as in Theorem 5.4.1.

On the other hand we note that Theorem 5.4.3 implies Theorem 5.4.2. Nevertheless we prove Theorem 5.4.2 for the intrinsic interest of the proof, in which figure properties of functions of bounded characteristicas well as properties of the set H^2 (cf. §1.2).

Theorem 5.4.2. *A real θ greater than 1 belongs to the set S if and only if there exists a non-zero real λ such that the series $\sum_{n\in\mathbf{N}} \|\lambda\theta^n\|^2$ converges.*

Proof. The series $\sum_{n\in\mathbf{N}} \varepsilon_n^2$ being convergent, the function ε belongs to the set H^2 and is thus of bounded characteristicin the disk $D(0,1)$. The same is true for the function f whose Taylor expansion has integer coefficients. Then by Theorem 1.2.1 the function f is rational. ∎

Remark. Sometimes instead of assuming that the series $\sum_{n\in\mathbf{N}} \|\lambda\theta^n\|^2$ is convergent, one assumes that the infinite product $\prod_{n\in\mathbf{N}} \cos(\pi\lambda\theta^n)$ or the series $\sum_{n\in\mathbf{N}} \sin^2(\pi\lambda\theta^n)$ converges.

In the proofs of Theorems 5.4.3 and 5.4.4 we will use, as in Theorem 5.1.2, the sequences (s_n) and (t_n) defined by

$$s_0 = \lambda_0 + \varepsilon_0, \qquad s_n = \varepsilon_n - \theta\varepsilon_{n-1} \qquad (n \geq 1),$$
$$t_0 = 1, \qquad t_n = 0 \qquad (n \geq 2),$$

the notations being those of Theorem 1.2.2.

Theorem 5.4.3. *A real θ greater than 1 belongs to S if and only if there exists a non-zero real λ such that*

$$\|\lambda\theta^n\| = o(n^{-1/2}).$$

Proof. This follows directly from the corollary of Theorem 1.2.2. The series $\sum_{n\in\mathbf{N}} t_n X^n$ is a polynomial and the sequence (s_n) satisfies $\sum_{m=n}^{2n-1} |s_m|^2 = o(1)$. Thus the series $\sum_{n\in\mathbf{N}} u_n X^n$ is rational. ∎

The interest of the following theorem comes from the fact that in the condition, which is an asymptotic one, an effective constant appears.

Theorem 5.4.4. *A real θ greater than 1 belongs to S if and only if there exist two reals λ and a with $\lambda > 0$ and $0 < a < 1/2\sqrt{2}(\theta+1)^2$ and an integer $n_0 \geq 1$ such that*

$$\|\lambda\theta^n\| \leq \frac{a}{\sqrt{n}} \qquad (\forall n \geq n_0).$$

Proof. Replacing λ by $\lambda\theta^{n_0}$ we can suppose that $|\varepsilon_n| \leq a/\sqrt{n+n_0}$ for $n \geq n_0$. The proof is based on an application of Theorem 1.1.1. It uses Lemmas 1.2.3, 1.2.5 and 1.2.6.

Let D_r, $(r \geq 1)$, denote the Kronecker determinant of the series $\sum_{n\in\mathbf{N}} u_n X^n$. In order to obtain an upper bound we proceed as in the proof of Lemma 1.2.6. We set

$$u_{0,0} = s_0, \qquad u_{m,n} = \sum_{i=0}^{n} t_i s_{m+n-i} = s_{m+n} - \theta s_{m+n-1} \;\; ((m,n) \neq (0,0));$$

$$v_{m,0} = 0, \qquad v_{m,n} = -\sum_{i=0}^{n-1} s_i t_{m+n-i} \;\; (n \geq 1).$$

Hence $v_{0,0} = 0$, $v_{0,n} = \theta s_{n-1}$ $(n \geq 1)$, $v_{m,n} = 0$ $(m \geq 1)$.

We deduce from Lemma 1.2.6 that for every integer $r \geq 1$, $D_r = \det(x_{m,n})$ with $x_{m,n} = u_{m,n} + v_{m,n}$, $(m = 0,\ldots,r;\; n = 0,\ldots,r)$.

We wish to determine an upper bound for $\sum_{m=0}^{r}\sum_{n=0}^{r} x_{m,n}^2$. Because of the inequalities $x_{m,n}^2 \leq u_{m,n}^2 + v_{m,n}^2 + 2|u_{m,n}v_{m,n}|$, $(m = 0,\ldots,r;\; n = 0,\ldots,r)$, we consider successively $\sum_{m=0}^{r}\sum_{n=0}^{r} v_{m,n}^2$, $\sum_{m=0}^{r}\sum_{n=0}^{r} u_{m,n}^2$ and $\sum_{m=0}^{r}\sum_{n=0}^{r} |u_{m,n}v_{m,n}|$.

The condition $|\varepsilon_n| \le a/\sqrt{n+n_0}$ $(\forall n \ge 0)$ implies $|s_n| \le a(\theta+1)/\sqrt{n+n_0-1}$ $(\forall n \ge 1)$. It follows that

$$\sum_{m=0}^{r}\sum_{n=0}^{r} v_{m,n}^2 = \sum_{n=0}^{r} v_{0,n}^2 = \theta^2 \sum_{n=1}^{r} s_{n-1}^2 \le \theta^2 s_0^2 + a^2\theta^2(\theta+1)^2 \sum_{n=1}^{r-1} \frac{1}{n+n_0-1}$$

hence $\sum\limits_{m=0}^{r}\sum\limits_{n=0}^{r} v_{m,n}^2 = O(\log r)$.

For $\sum\limits_{m=0}^{r}\sum\limits_{n=0}^{r} u_{m,n}^2$, we use the inequality obtained in the proof of Theorem 1.2.2: $\sum\limits_{m=0}^{r}\sum\limits_{n=0}^{r} u_{m,n}^2 < (\theta+1)^2 \sum\limits_{m=0}^{r}\sum\limits_{j=0}^{r} s_{m+j}^2$. The sequence (s_h^2) satisfies $\sum\limits_{h=j}^{2j-1} s_h^2 < \dfrac{ja^2(\theta+1)^2}{j+n_0-1} < a^2(\theta+1)^2$. We now apply Lemma 1.2.3. We set $\sigma_0 = \max(a^2(\theta+1)^2, s_0^2)$. We then have

$$\sum_{m=0}^{r}\sum_{n=0}^{r} u_{m,n}^2 \le 8[ra^2(\theta+1)^4 + \sigma_0(\theta+1)^2].$$

On the other hand we have

$$\sum_{m=0}^{r}\sum_{n=0}^{r} |u_{m,n}v_{m,n}| \le a^2(\theta+1)^3\theta \sum_{n=1}^{r} \frac{1}{n+n_0-1} = O(\log r).$$

Then the following inequality

$$\sum_{m=0}^{r}\sum_{n=0}^{r} x_{m,n}^2 \le 8a^2(\theta+1)^4 r + O(\log r)$$

holds.

By Lemma 1.2.5 we have for r large enough $|D_r| < 1$, hence since D_r is an integer, $D_r = 0$. Therefore the series $\sum\limits_{n\in\mathbf{N}} u_n X^n$ is rational. ∎

5.5 Salem numbers and distribution modulo 1 of certain sequences $(\lambda\tau^n)$

Let τ be a T-number. We have seen in §5.3 that the sequence (τ^n) is dense modulo 1 but not unformly distributed. The following theorem shows that one can find a real $\lambda \in \mathbf{Q}(\tau)$ such that the sequece $(\lambda\tau^n)$ is dense modulo 1 in an arbitrarely small interval around 0.

Theorem 5.5.1. *Let τ be a T-number. Suppose η is real with $0 < \eta < 1/2$; then there exists a non-zero real λ such that the sequence $(\lambda\tau^n)$ satisfies $\|\lambda\tau^n\| < \eta$ for every integer $n \geq n_0$. Moreover the sequence $(\lambda\tau^n)$ is dense modulo 1 in an interval whose center is 0 and which is contained in the interval $[-\eta, +\eta]$.*

Proof. The field $\mathbf{Q}(\gamma)$ is totally real, so let θ belong to $S \cap \mathbf{Q}(\gamma)$ and set $\lambda = \theta^{2h}$ $(h \in \mathbf{N}^\star)$. The conjugates $\lambda^{(j)}$ $(j = 2, \ldots, s)$ of λ are real and positive. The real $\lambda[\tau^n + \tau^{-n}] + 2\sum_{j=2}^{s} \lambda^{(j)} \cos(2n\pi\omega^{(j)})$ is an integer and let $\delta = \max_{j=2,\ldots,s} |\theta^{(j)}|$. We have then $0 < 2\sum_{j=2}^{s} \lambda^{(j)} \leq 2(s-1)\delta^{2h}$. Thus we can choose h, and then n_0 such that $2(s-1)\delta^{2h} < \eta/2$ and $\lambda\tau^{-n_0} < \eta/2$. We get $\|\lambda\tau^n\| \leq \eta$, $\forall n \geq n_0$. Assume for example $|\lambda^{(2)}| = \max_{j=2,\ldots,s} |\lambda^{(j)}|$, one shows as in the proof of Theorem 5.3.2 that for every ρ belonging to the interval $[-\lambda^{(2)}, \lambda^{(2)}]$, an arbitrarily large integer m can be found such that ε_m belongs to a given neighborhood of ρ. ∎

The following characterization does not directly concern properties of a sequence $(\lambda\tau^n)$ for a suitable λ but rather those of the associated function ε.

Theorem 5.5.2. *A real τ greater than 1 belongs to T if and only if there exists a non-zero real λ such that the corresponding function ε has a real part bounded above in the disk $D(0,1)$ but ε does not belong to the class H^2.*

Proof.

Sufficient condition. The hypothesis concerning $\mathrm{Re}(z)$ implies according to proposition 1.2.4 that the function ε is of bounded characteristic in the disk $D(0,1)$. As in the proof of Theorem 5.4.2, we deduce that f is a rational function and τ is a U-number.

Because the function ε does not belong to H^2, τ belongs to $U \setminus S = T$.

Necessary condition. The notations are the same as in Theorem 5.5.1. Here we seek $\lambda = \theta^{2h}$ with $\theta \in S \cap \mathbf{Q}(\gamma)$ such that $\varepsilon_n = \lambda\tau^{-n} + \sum_{j=2}^{s} \lambda^{(j)}[\tau^{(j)^n} + \tau^{(j)^{-n}}]$ $(\forall n \geq n_0)$. Then we have in the disk $D(0,1)$ the following inequalities

$$\begin{aligned}\varepsilon(z) &= \sum_{n=0}^{+\infty} (\lambda\tau^n - u_n)z^n + \pi(z)\\ &= -\left[\frac{\lambda}{1-\tau^{-1}z} + \sum_{j=2}^{s} \lambda^{(j)}\left(\frac{1}{1-\tau^{(j)}z} + \frac{1}{1-\overline{\tau}^{(j)}z}\right)\right] + \pi(z),\end{aligned}$$

with π a polynomial of degree $n_o - 1$. The inequalities $\mathrm{Re}(\frac{1}{1-\tau^{-1}z}) \geq \frac{\tau}{\tau+1}$, $\mathrm{Re}(\frac{1}{1-\tau^{(j)}z}) \geq \frac{1}{2}$ and $\mathrm{Re}(\frac{1}{1-\overline{\tau}^{(j)}z}) \geq \frac{1}{2}$, $(j = 2, \ldots, s)$ imply

$$\mathrm{Re}\left[\frac{1}{1-\tau^{-1}z} + \sum_{j=0}^{s} \lambda^{(j)} \left(\frac{1}{1-\tau^{(j)}z} + \frac{1}{1-\overline{\tau}^{(j)}z}\right)\right] \geq \frac{\lambda\tau}{\tau+1} + \sum_{j=2}^{s} \lambda^{(j)}$$
$$\geq \frac{\lambda\tau}{\tau+1} \qquad (\forall z \in D(0,1)).$$

The real part of $\pi(z)$ being bounded, then $\mathrm{Re}(\varepsilon(z))$ is bounded above in the disk $D(0,1)$. ∎

5.6 Other sequences $(\lambda\alpha^n)$ non-uniformly distributed modulo 1

We proved in the previous sections that if α is a U-number, then a real $\lambda \in S \cap \mathbf{Q}(\alpha)$ can be found such that the sequence $(\|\lambda\alpha^n\|)$ belongs to an interval whose center is 0 and whose length is arbitrarily small. The following theorem concerns the set of pairs of reals (λ, α) with $\lambda > 0$ and $\alpha > 1$ such that the sequence $(\|\lambda\alpha^n\|)$ satisfies a weaker condition of non-uniform distribution modulo 1. It produces a remarkable set of sequences of integers.

Theorem 5.6.1. *The set of pairs of reals (λ, α) with $\lambda > 0$ and $\alpha > 1$ such that*

$$\sup_{n \geq n_0} \|\lambda\alpha^n\| < \frac{1}{2(1+\alpha)^2}$$

for an integer n_0 is countable.

Then α is the limit of a sequence (u_{n+1}/u_n) derived from a sequence (u_n) that satisfies the recurrence relation $u_{n+2} = E(u_{n+1}^2/u_n)$ for n large enough.

Proof. The equality $\lambda\alpha^n = u_n + \varepsilon_n$ $(n \geq 0)$ implies that $\lim_{n\to+\infty} u_n = +\infty$. Accordingly we can write

$$u_{n+2} - \frac{u_{n+1}^2}{u_n} = -\frac{\lambda\alpha^n}{u_n}(\alpha^2\varepsilon_n - 2\alpha\varepsilon_{n+1} + \varepsilon_{n+2}) + \frac{\varepsilon_n\varepsilon_{n+2} - \varepsilon_{n+1}^2}{u_n}.$$

Hence setting $\eta = \sup_{n \geq n_0} |\varepsilon_n|$

$$\left|u_{n+2} - \frac{u_{n+1}^2}{u_n}\right| \leq \eta(1 + \frac{\eta}{u_n})(\alpha+1)^2 + \frac{2\eta^2}{u_n} \leq (\alpha+1)^2\eta + c\frac{\eta^2}{u_n} \qquad (n \geq n_0)$$

with c a real constant.

Thus the inequality $\left|u_{n+2} - \frac{u_{n+1}^2}{u_n}\right| \leq \frac{1}{2}$ holds for $n \geq n_1 \geq n_0$. It follows that $u_{n+2} = E\left(\frac{u_{n+1}^2}{u_n}\right)$, hence u_{n+2} can be determined from u_n and u_{n+1}. Then the sequence (u_n) is defined by the information provided by all its terms before u_{n_1+1}. Two different pairs (λ, α) cannot produce the same sequence (u_n), so the set of pairs (λ, α) of reals satisfying the hypothesis is countable.

The equations $\lambda\alpha^{n+1} = u_{n+1} + \varepsilon_{n+1} = \alpha u_n + \alpha\varepsilon_n$ imply $\left|\alpha - \frac{u_{n+1}}{u_n}\right| = \left|\frac{\varepsilon_{n+1} - \alpha\varepsilon_n}{u_n}\right|$; hence $\alpha = \lim_{n\to+\infty} \frac{u_{n+1}}{u_n}$ in the disk $D(0,1)$. ■

The sequence (u_n) defined above satisfies $u_{n+2} = E(\frac{u_{n+1}^2}{u_n})$ for $n \geq n_0$, and is a **Pisot sequence**. The set of Pisot sequences will be studied in Chapter 13. Pisot numbers, and Salem numbers too because of Theorem 5.5.1, are limits of sequences $\left(\frac{u_{n+1}}{u_n}\right)$ with (u_n) a Pisot sequence. Then if E denotes the set of limits of the sequences, the inclusion $U \subset E$ holds. We will see in Chapter 13 that the set E contains non-algebraic numbers.

We end this chapter by discussing sequences $(\lambda\alpha^n)$ with finitely many limit points. According to Theorem 4.1 one can find a non-zero integer h such that $\varepsilon(h\lambda\alpha^n)$ belongs, for n large enough, to an arbitrarily small interval whose center is 0. We note that if θ is a Pisot number and λ belongs to $\mathbf{Q}(\theta)$ there exists an integer h such that $h\lambda$ is an algebraic integer. By considering then the residues modulo 1 of $\lambda\theta^n$ and $h\lambda\theta^n$ we see that the real $h\varepsilon(\lambda\theta^n) - \varepsilon(h\lambda\theta^n)$ is an integer.

The equality $\lim_{n\to+\infty} \|h\lambda\theta^n\| = 0$ implies that the sequence $(\|\lambda\theta^n\|)$ has finitely many limit points, every one of which is rational. The following theorem, which is a generalization of Theorem 5.4.1, shows that for an algebraic number greater than 1 this property characterizes the set S.

Theorem 5.6.2. *An algebraic number θ greater than 1 belongs to S if and only if there exists a non-zero real λ such that the sequence $(\lambda\theta^n)$ has finitely many limit points modulo 1.*

Proof. As in Theorem 5.4.1 we suppose that θ is a zero of a polynomial $\sum_{i=0}^{s} q_i X^i$ with integer coefficients. Let k denote the number of irrational limit points of the sequence $(\varepsilon(\lambda\theta^n))$. Let q be an integer with $q > 2(\sum_{i=0}^{s} |q_i|)$; then according to Theorem 4.1 there exists an integer h with $0 < h \leq q^k$ such that $h\lambda\theta^n = v_n + \eta_n$ with $v_n \in \mathbf{Z}$ and $|\eta_n| \leq 2/q$ for $n \geq n_0$. It follows that $\sum_{i=0}^{s} q_i v_{n+i} = 0$ for $n \geq n_0$. Thus the series $\sum_{n \in \mathbf{N}} v_n X^n$ is rational. As in Theorem 5.4.1, one shows that θ belongs to U and cannot belong to T. ∎

As for Theorem 5.4.1, the condition θ is algebraic can be replaced by a condition concerning the rapidity of the convergence of the sequences extracted from $(\varepsilon(\lambda\theta^n))$.

Of course, the problem of the existence of pairs (λ, α), with α transcendental and greater than 1 such that the sequence $(\varepsilon(\lambda\alpha^n))$ has finitely many limit points, is unsolved. What we have shown is that there is no pair (λ, α), with α algebraic and greater than 1, such that the sequence $(\varepsilon(\lambda\alpha^n))$ has finitely many limit points of which at least one is irrational.

Notes

The results in this chapter were not presented in chronological order. Actually, apart from some very recent results, most of the theorems proved are anterior to 1950. We have attempted to make a synthesis of the classical results by focusing on a comparison between the sets of Pisot and Salem.

The oldest results go back to 1912 (Thue [12]) and 1919 (Hardy [5]). They are thus contemporaneous with Weyl's works on uniform distribution. In particular Thue proved, by using a method based on the box principle, that if α is real and greater than 1 such that $\|\alpha^n\| < c\rho^n$, then it is algebraic.

The set S was simultaneously and independently defined by Pisot [7] and Vijayaraghavan [13], [14]. Thus S-numbers are sometimes called P.V. numbers. Pisot improved Thue's condition by replacing it by the condition $\sum \|\lambda\theta^n\|^2 < +\infty$. He also established the relation between the set S and certain sequences of rational integers now called Pisot sequences, which will be studied in Chapter 13.

In 1945 Salem defined and characterized the set T [10], [11]. The following year Pisot gave a characterization of the union of S and T, designated here by U, by again using Thue's method. We endow this set with a letter because the classical point of view, according to which it is merely the union of Pisot and Salem sets, does not suit most of the generalizations.

In this chapter there is almost no result proved between 1947 and 1977. During that period research—and in particular that of Pisot and his students—focused on studying limit points of the sets S and T and then on various generalizations. Improving the condition $\sum \|\lambda\theta^n\|^2 < +\infty$ was made difficult by the insufficiency of the criteria for rationality. Then in 1977 a proof of a new criterion by Cantor restarted investigations in that direction. This led not only to the improvement of the condition $\sum \|\lambda\theta^n\|^2 < +\infty$ but also that of Theorem 5.1.1 (Cantor [2], Decomps-Guilloux and Grandet-Hugot [3]. For this last condition Boyd's result (mentioned at the end of §5.1) shows the present state of research.

We may add that there exists a result due to Guelfond [4] (1941), whose terms are nearly those of Theorem 5.4.4 with a constant $a = 1/\sqrt{2e}(\alpha+1)$. That is better than the given constant. Due to its date, the very small number of copies of the articles, and the Russian language in which is was written, Pisot and Salem were not aware of the existence of this result.

The result was rediscovered by Korneyei in 1984 [6].

Some results proved in this chapter are mentioned in Pisot's courses, and in particular in the Montréal course [9].

References

[1] D.W. Boyd, Transcendental numbers with badly distributed powers. *Proc. Amer. Math. Soc.*, 23, (1969), 424-427.

[2] D.G. Cantor, On power series with only finitely many coefficients mod 1: solution of a problem of Pisot and Salem. *Acta Arith.* 34, (1977), 43-55.

[3] A. Decomps-Guilloux and M. Grandet-Hugot, Nouvelles caractérisations des nombres de Pisot et de Salem., *Acta Arith* 50, (1987), 154-174.

[4] A.O. Guelfond, *Math. Sb.* (NS) 9.51, (1987), 721-725.

[5] G. Hardy, A problem of diophantine approximation. *Journ. of Indian Math. Soc.* 11, (1919), 205-243.

[6] I. Korneyei, On a theorem of Pisot. *Publ. Math. Debrecen*, n° 3.4, (1919), 169-179.

[7] C. Pisot, La répartiton modulo 1 et les nombres algébriques. *Ann.Scu. Norm. Sup. Pisa*, série 27, (1938), 205-208,

[8] C. Pisot, La répartiton modulo 1 des puissances succerssives des nombres réls. - *Comm. Math. Helv.* 19, (1946-1947), 153-160.

[9] C. Pisot, Quelques aspects de la théorie des entiers algébriques. *Sem. Math. Sup. Montréal*, (1963).

[10] R. Salem, Power series with integral coefficients. *Duke Math. Journ.* 12, (1945), 153-173.

[11] R. Salem, *Algebraic Numbers and Fourier Analysis*, Heath Math. Monographs, Boston, Mass., (1963).

[12] A. Thue, Über eine Eigenschaft die keine transcendete Grosse haben kann. *Skrifter Vidensk Kristinia,* (1912), 1-15.

[13] T. Vijayaraghavan, On the fractioanl part of the powers of a number. *Proc. Camb. Phil. Soc.* 37, (1941), 349-357.

[14] T. Vijayaraghavan, On the fractioanl part of the powers of a number. *London Math. Soc.* 17, (1942), 137-138.

CHAPTER 6

LIMIT POINTS OF PISOT AND SALEM SETS

The purpose of this chapter is to study the limit points of the sets S and T. In particular we will show that S is a closed set and that the closure $\overline{\mathrm{T}}$ of T contains U. Even more than in the previous chapter, we will notice that while a great deal is known about the set S, very little is known about the set T. Thus we still do not know if the only limit points of T are S-numbers.

6.0 Notation

We recall that if $A = \sum_{i=0}^{d} a_i X^i$ is a polynomial with $a_i \in \mathbf{C}$ $a_0 \neq 0$, its reciprocal polynomial $\sum_{i=0}^{d} a_{d-i} X^i$ is denoted by $A^\star$.

In this chapter if θ is a number in S its minimal polynomial will be written $P = X^s + q_{s-1}X^{s-1} + \cdots + q_0$, $(q_i \in \mathbf{Z}, i = 0, \ldots, s-1)$.

In general we will denote by Q the reciprocal polynomial $P^\star$ of P. We will also consider (especially in § 6.1, 6.2, 6.3) the polynomial $P^+ = \varepsilon P$ with $\varepsilon = \pm 1$ such that $P^+(0) = \varepsilon q_0 > 0$. As in the previous chapter, we will write $\theta^{(1)} = \theta, \theta^{(j)}$, $(j = 2, \ldots, s)$ for the conjugates of θ.

6.1 Closure of the set S

To prove that S is a closed set we use the compactness of the families $\mathcal{F}(1,1,\delta)$, which was proved in Chapter 2. We will write $\mathcal{F}(\delta)$ to designate these families.

The following lemma sets a link between S and $\mathcal{F}(\delta)$.

Lemma 6.1. *Let θ be an S-number. Then there exists at least one polynomial A with integer coefficients, different from Q, and such that*

(i) $\qquad A(0) \geq 1$

(ii) $\qquad |A(z)| \leq |Q(z)| \qquad (\forall z \in C(0,1)).$

Proof. If the polynomials P and Q are not identical, we take $A = P^+$. Otherwise θ is a quadratic unit. We then have $P^+ = Q = X^2 - q_1X + 1$ ($q_1 \geq 3$) and we take as polynomials $A_1 = 1$ and $A_2 = (1 - X)^2$. ∎

We remark that it is equivalent to assume that A is different from Q and that A is prime to Q. The polynomial Q is irreducible, thus if A is not prime to Q it is a multiple of Q. Then the equality $A = BQ$ in $\mathbf{Z}[X]$ with $|B(z)| \leq 1$, $\forall z \in C(0, 1)$, and $B(0) \geq 1$ implies $B = 1$.

According to the previous lemma we can associate to every number θ in S at least one function in $\mathcal{F}(\delta)$. For we set $f(z) = A(z)/Q(z)$, the function f belongs to $\mathcal{F}(\delta)$ with $0 < \delta < 1/\theta$ and the inequality $f(0) \geq 1$ holds. We will see later that θ is a limit point of S if and only ifthere exist several polynomials satisfying the conditions of Lemma 6.1. It is equivalent to say that we can associate to it several functions of $\mathcal{F}(\delta)$.

Theorem 6.1. *The set S is closed on the real line.*

Proof. Let ω be a number belonging to the closure $\overline{\mathrm{S}}$ of S; we remark that the Smyth inequality proved in §3.5 implies that all Pisot numbers are greater than 1.32, and that therefore ω cannot be equal to 1.

If ω is an isolated point it belongs to S, otherwise it is a limit point of S and there exists a sequence (θ_ν) of numbers of S such that $\lim_{\nu \to +\infty} \theta_\nu = \omega$.

It follows from Lemma 6.1 that we can associate to every number θ_ν a rational fraction A_ν/Q_ν that defines a function f_ν of $\mathcal{F}(\delta)$ with $\delta < \inf_{\nu \in \mathbf{N}} 1/\theta_\nu$ such that $f_\nu(0) \geq 1$.

Since the family $\mathcal{F}(\delta)$ is compact we can extract from the sequence (θ_ν) a subsequence still denoted (θ_ν) such that the sequence (f_ν) of the associated functions converges in $\mathcal{F}(\delta)$. Let f be the limit function. We must show that f has effectively a pole that lies in $D(0, 1)$.

We can write the Taylor expansion of the functions f_ν ($\nu \in \mathbf{N}$) and f

$$f_\nu(z) = \sum_{n=0}^{+\infty} u_{\nu,n} z^n \qquad \text{and} \qquad f(z) = \sum_{n=0}^{+\infty} u_n z^n,$$

where all coefficients $u_{\nu,n}$ and u_n are integers. Hence there exists for every $n \in \mathbf{N}$ an integer $\nu_0(n)$ such that $u_{\nu,n} = u_n$, ($\forall n \in \mathbf{N}, \forall \nu \geq \nu_0(n)$). For every

$\nu \in \mathbf{N}$ the function f_ν has a pole in the disk $D(0,1)$ and satisfies $u_{\nu,0} \geq 1$. Hence considering the function $z \mapsto (f_\nu(z) - u_{\nu,0})Q(z)$ we deduce according to Lemma 2.2 that $u_{\nu,1} \neq 0$. Then the conditions $u_0 \geq 1$ and $u_1 \neq 0$ are satisfied, the function f has one pole in the disk $D(0,1)$, otherwise it would be holomorphic in $D(0,1)$, so $u_0 = 1, f(z) = 1$ and $u_n = 0, \forall n \geq 1$. This pole is at $1/\omega$.

Then we write $f(z) = A(z)/Q(z)$ where A and Q are polynomials with integer coefficients. A and Q are relatively prime and satisfy the condition $|A(z)| \leq |Q(z)|$, $(\forall z \in C(0,1))$. Hence the polynomial Q cannot have a zero on the circle $C(0,1)$. It follows then from the equality $Q(0) = 1$ that ω belongs to S. ∎

Remarks.

a) The limit function f belongs to the set $\mathcal{F}'(1,1,\delta)$, which we designate by $\mathcal{F}'(\delta)$. Hence the equality $|f(z)| = 1$ holds for finitely many points of the circle $C(0,1)$ and the polynomial A is different from the polynomial P^+.

b) The set S is closed and bounded below, so it has a least element. We will see in Chapter 7 that this number, denoted θ_0, is a zero of the polynomial $X^3 - X - 1$, $(\theta_0 = 1.3247\ldots)$.

The set of limit points of S is called the derived set of S. We denote it S'.

6.2 The derived set S' of S

The purpose here is to characterize the set S' and to exhibit certain remarkable elements.

Theorem 6.2.1. *A number θ of S belongs to the set S' if and only ifthere exists a polynomial A with integer coefficients that is different from the polynomials P^+ and Q and satisfies*

(*i*) $$A(0) \geq 1$$

(*ii*) $$|A(z)| \leq |Q(z)| \quad (\forall z \in C(0,1)).$$

Proof. It follows from Remark a) following the proof of Theorem 6.1, that the condition is necessary.

Conversely let θ be an S-number. We assume that there exists a polynomial A different from P^+ and Q satisfying conditions (*i*) and (*ii*). We will produce a sequence (f_ν) of functions belonging to $\mathcal{F}(\delta)$ and associated to a sequence (θ_ν) of S-numbers, and we will prove that (θ_ν) converges to θ. We recall that the polynomials P and P^+ verify $P^+ = \varepsilon P$ with $\varepsilon = \pm 1$ so that $P^+(0) = \varepsilon P(0) > 0$. We write B for the polynomial $\varepsilon A^\star$ and denote by a the degree of A.

Let f and f_ν ($\nu \in \mathbf{N}$) be the rational functions respectively defined by the rational fractions A/Q and $(A + X^{\nu+a}P^+)/(Q + X^{\nu+s}B)$.

The Taylor expansion of f is written $\sum_{n=0}^{+\infty} u_n z^n$ with $u_n \in \mathbf{Z}$. That of f_ν begins by $u_0 + u_1 z$ for $\nu \in \mathbf{N}$.

It follows from condition (i) that $u_0 \geq 1$. The function f has a single pole in the disk $D(0,1)$, so we deduce as before that $u_1 \neq 0$. For every $\nu \in \mathbf{N}$ this implies that f_ν has at least one pole in the disk $D(0,1)$. Consider then the function $z \mapsto z^{s+1}B(z) + \lambda Q(z)$ with $\lambda > 1$, $\lambda \to 1$. Following Rouché's theorem and using condition (ii) we conclude as in the proof of Lemma 2.2 that f_ν has at most one pole in $D(0,1)$. We denote this pole $1/\theta_\nu$.

From the equality $|A(z) + z^{\nu+a}P^+(z)| = |Q(z) + z^{\nu+s}B(z)|$ $\forall z \in C(0,1)$ it follows that, for every $\nu \in \mathbf{N}$, f_ν cannot have a pole on the circle $C(0,1)$. Let D_ν be the greatest common divisor of the polynomials $A + X^{\nu+a}P^+$ and $Q + X^{\nu+s}B$. We set

$$(1) \qquad Q + X^{\nu+s}B = D_\nu Q_\nu \quad \text{and} \quad A + X^{\nu+a}P^+ = D_\nu A_\nu.$$

The polynomials A_ν and Q_ν are relatively prime, and the polynomial Q_ν has only one zero in the disk $D(0,1)$. It has no zero on the circle $C(0,1)$ and satisfies $Q_\nu(0) = 1$. Then Q_ν is irreducible and θ_ν belongs to S. Equations (1) imply $X^aP^+Q - X^sAB = D_\nu[X^aP^+Q_\nu - X^sBA_\nu]$. We conclude that the polynomial D_ν divides a fixed polynomial. Hence the degree of D_ν is bounded and that of Q_ν goes to infinity. Hence there are infinitely many different numbers θ_ν.

Equations (1) imply $D_\nu(1/\theta_\nu)Q_\nu(1/\theta_\nu) = 0 = Q(1/\theta_\nu) + \theta_\nu^{-\nu-s}B(1/\theta_\nu)$, hence $Q(1/\theta_\nu) = -\theta_\nu^{-\nu-s}B(1/\theta_\nu)$.

The sequence (θ_ν) cannot converge to 1, hence $\lim_{\nu\to+\infty} Q_\nu(1/\theta_\nu) = 0$ and $\lim_{\nu\to+\infty} \theta_\nu = \theta$. ■

In the same way one can show that the sequence of rational fractions $(A - X^{\nu+a}P^+/Q - X^{\nu+s}B)_\nu$ is associated to a sequence (θ'_ν) of numbers of S that converges to θ. The numbers $Q(1/\theta_\nu)$ and $Q(1/\theta'_\nu)$ have different signs, hence θ is a limit of Pisot numbers from both sides. The interest of Theorem

6.2.1 is also due to the fact that with its help we can produce some remarkable numbers of S'. This is the purpose of Theorems 6.2.2 and 6.2.3.

We first remark that rational integers at least equal to 2 belong to S', since the polynomial $A = 1$ satisfies conditions (i) and (ii).

Theorem 6.2.2. *Let θ be an S-number. Then θ^m belongs to the set S' for every integer $m \geq 2$.*

Proof. Let $1, \zeta, \ldots, \zeta^{m-1}$ denote the mth roots of unity. Then we have

$$Q(z)\,Q(\zeta z)\cdots Q(\zeta^{m-1}z) = \hat{Q}_m(z^m)$$

where $\hat{Q}_m$ is a polynomial with integer coefficients, which has $1/\theta^m, 1/\theta^{(j)^m}$, $(j = 2, \ldots, s)$ as zeros. Let $h \in \{0, \ldots, m-1\}$ and φ_h the rational function defined by

$$\varphi_h(z) = \frac{1}{m}\sum_{k=0}^{m-1} \zeta^{-hk}\frac{P(\zeta^k z)}{Q(\zeta^k z)}.$$

If $\sum\limits_{n=0}^{+\infty} u_n z^n$ is the Taylor expansion of the rational function $z \mapsto P(z)/Q(z)$ and that of φ_h is $\sum\limits_{n=0}^{+\infty} u_{h+mn} z^{h+mn}$, the integers u_{h+mn} are not equal to zero because of the equations $u_n = E(\lambda\theta^n)$. Then we have $\varphi_h(z) = z^h A_h(z^m)/\hat{Q}_m(z^m)$, where A_h is a polynomial with integer coefficients not identical to zero.

We then have

$$\sum_{h=0}^{m-1}\left|\frac{A_h(z^m)}{\hat{Q}_m(z^m)}\right|^2 = 1 \qquad \forall z \in C(0,1).$$

It follows that

$$\sum_{h=0}^{m-1} |A_h(z)|^2 = |\hat{Q}_m(z)|^2 \qquad \forall z \in C(0,1).$$

We have exhibited m polynomials $A_0, \ldots, A_{m-1}$, none of which is identically zero and that all satisfy the conditions of Theorem 6.2.1. ∎

Theorem 6.2.3. *Every totally real S-number belongs to S'.*

Proof. Let θ be an S-number all of whose conjugates are real.

For $z \in C(0,1)$ we have

$$\left|\frac{z-\theta^{(j)}}{z-1}\right| \geq \frac{1+\theta^{(j)}}{2} \qquad \text{if } \theta^{(j)} > 0$$

and

$$\left|\frac{z-\theta^{(j)}}{z+1}\right| \geq \frac{1-\theta^{(j)}}{2} \qquad \text{if } \theta^{(j)} < 0.$$

Hence if there are p $\theta^{(j)}$'s positive and q negative the above inequalities imply

$$|P(z)| \geq |z-1|^p |z+1|^q \prod_{j=1\dots,s} \left(\frac{1+|\theta^{(j)}|}{2}\right).$$

Then we have

$$\prod_{j=1\dots,s} \frac{1+|\theta^{(j)}|}{2} \geq \prod_{j=1\dots,s} |\theta^{(j)}|^{1/2} \geq 1.$$

It follows that the polynomial $A = (X-1)^p(X+1)^q$ satisfies

$$|A(z)| < |P(z)| = |Q(z)| \qquad (\forall z \in C(0,1)).$$

■

We can also show that the inequality $1 \leq |Q(z)|$ holds for every $z \in C(0,1)$. Contrary to the previous case the inequality is not strict.

We remark that there are finitely many totally real Pisot numbers in every compact interval of **R**. For let θ be a totally real S-number, then θ^2 belongs to S' and is totally positive. It follows from a result of Siegel (cf. notes [14]) that we have $\theta^2 + {\theta^{(2)}}^2 + \cdots + {\theta^{(s)}}^2 \geq \frac{3}{2}s$. We deduce from the inequality $\theta^2 \leq M^2$ that $s \leq 2(M-1)$. The degree of θ is bounded and the coefficients of the minimal polynomial of θ are also bounded. They are integers, so there is a finite number of them.

Theorem 6.2.3 also shows that all Pisot numbers of degree 2 belong to S'. We write their minimal polynomial $X^2 + q_1X + q_0$ with $q_1 + |1+q_0| < 0$. We can easily show that the smallest is $\frac{1+\sqrt{5}}{2}$, which is also the least element of S' (cf. Chapter 7). This property follows from a more general result, which shows that if a polynomial A different from Q can be associated to a number θ of S then $\theta \geq \dfrac{1+\sqrt{5}}{2}$. It follows also that $\theta^2 > \dfrac{1+\sqrt{5}}{2}$, $\forall \theta \in S$.

6.3 Successive derived sets of S

Let $S^{(k)}$ $(k \geq 1)$ denote the kth derived set of S, i.e., the set of limit points of $S^{(k-1)}$ with $S^{(0)} = S$.

As in the case of the set S', the characterization of these sets is based on the fact that some polynomials with integer coefficients, satisfying certain remarkable inequalities on the circle $C(0,1)$, can be associated to the elements of the sets. Because the proofs are rather technical we only give the one concerning S''.

Theorem 6.3.1. *A number θ of S belongs to the set S'' if and only ifthere exist three polynomials A, B, C with integer coefficients, different from the polynomials P^+ and Q, satisfying the following conditions:*

(i)	$A(0) \geq 1,$	$B^2(0) + C^2(0) \geq 1$	
(ii)	$\|A(z) \leq \|Q(z)\|,$	$\|B(z)\| \leq \|Q(z)\|,$	$\|C(z)\| \leq \|Q(z)\|$
and	$\|A(z) + z^m C(z)\| \leq \|Q(z) + z^m B(z)\|$		

for every integer $m \geq m_0$ and every $z \in C(0,1)$.

We remark that condition (i) implies that polynomials B and C are not both identically zero.

Proof.

The condition is necessary. As in Theorem 6.2.1 the proof is based on the compactness of the families $\mathcal{F}(\delta)$. All rational functions that intervene are associated to a bounded set of numbers of S. Hence they belong to the same family $\mathcal{F}(\delta)$. The definitions of the polynomials A, B, C are derived from limit functions of sequences in $\mathcal{F}(\delta)$ or $\mathcal{F}'(\delta)$.

Let θ be an S''-number and (θ_μ) a sequence of S'-numbers converging to θ. We associate to the sequence (θ_μ) three sequences of functions (f_μ), (g_μ) and (h_μ) in the following way. At least one polynomial A_μ different from $P_\mu^+ = \varepsilon_\mu P_\mu$

can be associated to every number θ_μ. Let a_μ and s_μ denote respectively the degree of A_μ and P_μ, and we write $B_\mu = \varepsilon_\mu A_\mu^\star$. We set then

$$f_\mu(z) = \frac{A_\mu(z)}{Q_\mu(z)}, \qquad g_\mu(z) = \frac{B_\mu(z)}{Q_\mu(z)}, \qquad h_\mu(z) = \frac{P_\mu^+(z)}{Q_\mu(z)}.$$

Let f, g, h be the limit functions of converging sequences extracted from the sequences (f_μ), (g_μ) and (h_μ). We set $f(z) = A(z)/Q(z)$, $g(z) = \tilde{B}(z)/Q(z)$, $h(z) = \tilde{C}(z)/Q(z)$, and the polynomials $A, \tilde{B}$ and $\tilde{C}$ satisfy $|A(z)| \leq |Q(z)|$, $|\tilde{B}(z)| \leq |Q(z)|$, $|\tilde{C}(z)| \leq |Q(z)|$, $\forall z \in C(0,1)$. Our purpose is now to define polynomials B and C such that the other conditions are satisfied.

For this we construct a new sequence (f_ν) of functions whose limit f belongs to $\mathcal{F}'(\delta)$.

Set $\hat{P}_{\mu,\nu}^+ = A_\mu + X^{\nu+a_\mu} P_\mu^+$, $\hat{Q}_{\mu,\nu} = Q_\mu + X^{\nu+s_\mu} B_\mu$ for $(\mu,\nu) \in \mathbf{N}^2$. The polynomial $\hat{P}_{\mu,\nu}^+$ has a zero $\theta_{\mu,\nu}$ that belongs to S. Let $\varphi_{\mu,\nu}$ be the rational function of $\mathcal{F}(\delta)$ defined by $\varphi_{\mu,\nu}(z) = \dfrac{\hat{P}_{\mu,\nu}^+(z)}{\hat{Q}_{\mu,\nu}(z)}$.

We can extract from the sequence $(\varphi_{\mu,\nu})_\mu$ a sequence converging to a function $\hat{f}_\nu \in \mathcal{F}'(\delta)$. We set $\hat{f}_\nu(z) = \dfrac{\hat{A}_\nu(z)}{\hat{Q}_\nu(z)}$, then the rational fraction $\hat{A}_\nu/\hat{Q}_\nu$ can be written in several different ways depending on whether the sequence $(a_\mu - s_\mu)$ is bounded or not.

If the sequence $(a_\mu - s_\mu)$ is bounded. Then there exists an integer d such that $a_\mu - s_\mu = d$ for infinitely many integers μ.

If $d > 0$, we put $\nu' = \nu + s_\mu$, hence we have $\dfrac{\hat{A}_\nu}{\hat{Q}_\nu} = \dfrac{A + X^{\nu'+d}\tilde{C}}{Q + X^{\nu'}\tilde{B}}$. Then we set $C = X^d\tilde{C}$ and $B = \tilde{B}$.

If $d < 0$, we put $\nu' = \nu + a_\mu$, hence we have $\dfrac{\hat{A}_\nu}{\hat{Q}_\nu} = \dfrac{A + X^{\nu'}\tilde{C}}{Q + X^{\nu'-d}\tilde{B}}$. Then we set $B = X^{-d}\tilde{B}$ and $C = \tilde{C}$.

If the sequence $(a_\mu - s_\mu)$ is not bounded above. We can extract from the sequence $(a_\mu - s_\mu)$ a subsequence converging to $+\infty$. We put $\nu' = \nu + s_\mu$ hence we have $\dfrac{\hat{A}_\nu}{\hat{Q}_\nu} = \dfrac{A}{Q + X^{\nu'}\tilde{B}}$. Then we set $B = \tilde{B}$ and $C = 0$.

If the sequence $(a_\mu - s_\mu)$ *is not bounded below.* We can extract from the sequence $(a_\mu - s_\mu)$ a subsequence converging to $-\infty$. We put $\nu' = \nu + a_\mu$ hence we have $\frac{\hat{A}_\nu}{\hat{Q}_\nu} = \frac{A + X^{\nu'}\tilde{C}}{Q}$ and we set $B = 0$ and $C = \tilde{C}$.

In all cases the polynomials A, B, C satisfy conditions (i) and (ii). We remark that one of the polynomials B or C can be zero.

The condition is sufficient. Let θ be an S-number, and assume that there exist polynomials A, B, C of respective degree a, b, c such that conditions (i) and (ii) hold. Let μ and ν be integers satisfying $\mu \geq 1$, $\mu + c > b$, $\nu \geq 1$, $\nu + b > s$, $\nu + c > a$. We set $Q_{\mu,\nu} = Q + X^\mu A + X^\nu B + X^{\mu+\nu} C$. The polynomial $Q_{\mu,\nu}$ is of degree $\mu+\nu+c$; in the disk $D(0,1)$ it has only one zero $1/\theta_{\mu,\nu}$ with $\theta_{\mu,\nu} \in S$. The polynomial $P^+_{\mu,\nu}$ equals $P^+_{\mu,\nu} = C^\star + X^{\mu+c-b}B^\star + X^{\nu+c-a}A^\star + X^{\mu+\nu+c-s}P^+$. Let $\varphi_{\mu,\nu}$ denote the function defined by the rational fraction $P^+_{\mu,\nu}/Q_{\mu,\nu}$. The sequence $(\varphi_{\mu,\nu})_\nu$ for a fixed μ converges to the function f_μ defined by the rational fraction $(C^\star + X^{\mu+c-b}B^\star)/(Q + X^\mu A)$.

The function f_μ has one only pole $1/\theta_\mu$ in the disk $D(0,1)$. The number θ_μ belongs to S and we have $\theta_\mu = \lim_{\nu \to +\infty} \theta_{\mu,\nu}$.

The sequence (f_μ) converges in $\mathcal{F}'(\delta)$ to the function f defined by the rational fraction $C^\star/Q$ and the sequence (θ_μ) converges to θ.

Hence the number θ belongs to the set S''. ■

For the sets $S^{(k)}$ $(k \geq 3)$ we only state the following theorem.

Theorem 6.3.2. *A number θ of S belongs to the set $S^{(k)}$ if and only ifthere exist $2k-1$ polynomials A, B_j and C_j $(j = 1, \ldots, k-1)$ with integer coefficients, different from the polynomials P^+ and Q and satisfying*

(i) $\quad A(0) \geq 1, \quad B_j^2(0) + C_j^2(0) \geq 1, \quad (j = 1, \ldots, k-1)$

(ii) $\quad |A(z)| \leq |Q(z)|, \quad |B_j(z) \leq |Q(z)|, \quad |C_j(z)| \leq |Q(z)|$

and $\quad |A(z) + \sum_{j=1}^{k-1} z^{m_j} C_j(z)| \leq |Q(z) + \sum_{j=1}^{k-1} z^{m_j} B_j(z)$

for every integer m_j such that $m_j \geq M > 0$, every $j = 1, \ldots, k-1$ and every $z \in C(0,1)$.

The proof proceeds by induction on k.

The following theorem gives only a sufficient condition for θ to belong to $S^{(k)}$. We use it to produce numbers of $S^{(k)}$.

Theorem 6.3.3. *A sufficient condition for a number θ of S to belong to $S^{(k)}$ $(k \geq 1)$ is that there exist k polynomials A_j, $(j = 1, \dots k)$, with integer coefficients such that the inequality*

$$(1) \qquad \sum_{j=1}^{k} |A_j(z)| \leq |Q(z)|$$

holds for every $z \in C(0,1)$.

Proof. Let $(m_j)_{1 \leq j \leq k}$ be a k-tuple of integers. We define the polynomial $Q_{m_1 \dots m_k}$ by

$$Q_{m_1 \dots m_k} = Q + \sum_{j=1}^{k} X^{m_j} A_j.$$

It follows from inequality (1) that the polynomial $Q_{m_1 \dots m_k}$ has only one zero in the disk $D(0,1)$. This zero is the inverse of a number $\theta_{m_1 \dots m_k}$ of S.

Suppose the integers m_j $(1 \leq j \leq k-1)$ are fixed. The sequence of polynomials $(Q_{m_1 \dots m_k})_{m_k}$ converges to a polynomial $Q_{m_1 \dots m_{k-1}}$ on every compact set of the disk $D(0,1)$. The polynomial $Q_{m_1 \dots m_{k-1}}$ has as zero the inverse of a number $\theta_{m_1 \dots m_{k-1}}$ belonging to S'. By successively repeating the procedure k times, we show that θ belongs to $S^{(k)}$. ■

It follows from this theorem that if θ is an S-number such that the polynomial Q satisfies $|Q(z)| \geq k$, $\forall z \in C(0,1)$, then θ belongs to $S^{(k)}$. Therefore the integer $k+1$ belongs to $S^{(k)}$, and this is true for the zero greater than 1 of every polynomial $X^2 - (k+2)X \pm 1$.

As the sets $S^{(k)}$ are closed, they all have a smallest element $\min S^{(k)}$, which then satisfies $\min S^{(k)} \leq k+1$. In fact we get $\limsup(\min S^{(k)}/k) < 1$, a sharper inequality.

We will discuss in Chapter 7 how to find the least member of $S^{(k)}$, $k = 0, 1, 2.$ No explicit value of $\min S^{(k)}$ is known for $k \geq 3$, but the inequality $\min S^{(k)} \geq k^{1/2}$ holds. Thus the set S has no derived set of transfinite order.

6.4 Limit points of the set T

Among the unsolved problems concerning Pisot and Salem numbers, one of the best known is that of the limit points of T. This has been an open question for a long time and is related to other unsolved problems, e.g., Lehmer's conjecture. Theorems 6.4.1 and 6.4.4 show what is known at present.

Theorem 6.4.1. *The set S is included in the closure $\overline{\mathrm{T}}$ of the set T.*

Proof. Let θ be an S-number. The general proof does not apply if P and Q are identical, i.e., if θ is a quadratic unit. This case will be dealt with separately later.

a) *θ is not a quadratic unit.*

Let (R_n) be a sequence of polynomials defined by $R_n = X^n P + Q$. By Rouché's theorem and as in Theorem 6.2.1, R_n has at most one zero in the disk $D(0,1)$. It is a reciprocal polynomial that satisfies $R_n(1) = 2P(1) < 0$. It has then a single zero greater than 1. Let τ_n be this zero; then $1/\tau_n$ is also zero of R_n, and the remaining zeros are all on the circle $C(0,1)$. We wish to show that the sequence (τ_n) converges to θ and that for n large enough the numbers τ_n all belong to T. We first remark that the equality $R_n(1) = 2P(1) \neq 0$ implies that the sequence (τ_n) cannot converge to 1.

We have on one hand $P(\tau_n) = (\tau_n - \theta) \prod_{j=2}^{s} (\tau_n - \theta^{(j)})$, and on the other hand, because τ_n is a zero of R_n, $\tau_n^n P(\tau_n) = -Q(\tau_n) = -\tau_n^s P(1/\tau_n)$. It follows then that $|\tau_n - \theta| \prod_{j=2}^{s} |\tau_n - \theta^{(j)}| = \tau_n^{-n} |1 - \theta\tau_n| \prod_{j=2}^{s} |1 - \theta^{(j)}\tau_n|$. Hence

$$|\tau_n - \theta| \leq \frac{2\theta\tau_n^{1-n}(1+\tau_n)^{s-1}}{\prod_{j=2}^{s}(1 - |\theta^{(j)}|)} = c\tau_n^{s-n},$$

with c a constant.

We have then $\lim_{n\to+\infty} \tau_n = \theta$.

If the polynomial R_n is irreducible then R_n is the minimal polynomial of τ_n and τ_n belongs to T for $n + s \geq 4$.

Otherwise R_n is the product of a cyclotomic polynomial by the minimal polynomial of τ_n. The minimal polynomial of τ_n is a monic reciprocal polynomial, whose degree is at least equal to 4 for n large enough because a bounded neighborhood of θ cannot include infinitely many quadratic units. The sequence (τ_n) consists of only Salem numbers for n sufficiently large.

We can produce another sequence (τ'_n) of Salem numbers converging to θ by considering the sequence (τ'_n) of zeros greater than 1 of the polynomials $\Pi_n = \dfrac{X^n P - Q}{X - 1}$. The proof that (τ'_n) is a sequence of Salem numbers converging to

θ is the same as before, except for the existence of a zero of Π_n greater than 1. Here the condition $\Pi_n(1) < 0$ is not trivial. Let α be a real with $1 < \alpha < \theta_0$ (where θ_0 is the smallest Pisot number). We have $\Pi_n(\alpha) = \dfrac{\alpha^n P(\alpha) - Q(\alpha)}{\alpha - 1}$ with $P(\alpha) < 0$. For n large enough we have $\Pi_n(\alpha) < 0$ and the polynomial Π_n has a zero greater than α.

We also remark that, since $Q(\theta) \neq 0$, the equalities $Q(\tau_n) = -\tau_n^n P(\tau_n) = \tau'^n_n P(\tau'_n)$ imply that $P(\tau_n)$ and $P(\tau'_n)$ are of constant sign for n large. These signs are different so the sequences approach θ from the both sides.

b) *θ is a quadratic unit.*

Then the number θ is a zero of a polynomial $X^2 - q_1X + 1$ ($q_1 \geq 3$) and belongs to S'. Thus θ is a limit point of the sequence $(\theta_\nu)_{\nu\geq 1}$ where θ_ν is, for every $\nu \in \mathbf{N}^\star$, an S-number, and zero of a polynomial $X^{\nu+2} - q_1X^{\nu+1} + X^\nu + 1$. It follows from part a that θ_ν is, for every $\nu \in \mathbf{N}^\star$, the limit of a sequence $(\tau_{\nu,n})_{n\geq 1}$ where $\tau_{\nu,n}$ is the zero greater than 1 of the polynomial $X^n(X^{\nu+2} - q_1X^{\nu+1} + X^\nu + 1) + X^{\nu+2} + X^2 - q_1X + 1$. The numbers $\tau_{\nu,n}$ belong to T for n large. We consider then the sequence of polynomials obtained for $\nu = n$, $X^{2n+2} - q_1X^{2n+1} + X^{2n} + X^{n+2} + X^n + X^2 - q_1X + 1$. We define thus a sequence (τ_n) of numbers that belong to T for n large enough, and we have $\lim_{n\to+\infty} (\tau_n^2 - q_1\tau_n + 1) = 0$. Hence the sequence (τ_n) converges to θ.

In the same way if we consider the sequence (θ'_ν) of S-numbers that are zeros of the polynomials $X^{\nu+2} - q_1X^{\nu+1} + X^\nu - 1$, we can produce a sequence (τ'_n) of numbers of T converging to θ (from the other side). ■

In the preceding proof we constructed infinitely many numbers of T from one S-number. Conversely the purpose of Theorem 6.4.3 is to associate a T-number to infinitely many numbers of S. This possibility is based on the following theorem concerning the localization of zeros of certain polynomials.

Theorem 6.4.2. *Let V be a polynomial of degree s with real coefficients, satisfying $V(0) \neq 0$. Let n be a positive integer and W be the polynomial $W = X^nV + V^\star$. We suppose that following conditions are satisfied:*

(i) The polynomial W has a single zero in the disk $D(0,1)$, which is simple.

(ii) Every zero α of W belonging to the circle $C(0,1)$ satisfies

$$\alpha^{1-s}W'(\alpha)V(\alpha) \in \mathbf{R}^\star_-.$$

Then the polynomial V has $s-1$ zeros inside the disk $D(0,1)$ and one outside $D(0,1)$.

Proof. According to condition (i) the polynomial W has one simple zero in the disk $D(0,1)$. As W is reciprocal it has one simple zero outside $D(0,1)$. The remaining zeros are on the circle $C(0,1)$ and are simple because of condition (ii). The polynomial V has no zero on the circle $C(0,1)$. For, if V has a zero of modulus 1, then it would be a zero of $V^\star$, and hence a zero of W. This is in contradiction to condition (ii).

For t belonging to $\mathbf{R}_+^\star$ we set

$$W_t = X^n V + tV^\star = W + (t-1)V^\star.$$

Let α be a zero of the polynomial W that lies on the circle $C(0,1)$. There exists a function g_α analytic in the neighborhood of $t=1$ such that $g_\alpha(1)=\alpha$, $g'_\alpha(1) \neq 0$ and $W_t[g_\alpha(t)] \equiv 0$.

By taking the derivative function of $t \mapsto W_t[g_\alpha(t)]$ we get $g'_\alpha(1)W'(\alpha)+V^\star(\alpha) = 0$, and hence $g'_\alpha(1) = \dfrac{V(\alpha)V^\star(\alpha)}{V(\alpha)W'(\alpha)} = -\dfrac{\alpha^s V(\alpha)V(1/\alpha)}{V(\alpha)W'(\alpha)}$.

Then condition (ii) implies $g'_\alpha(1) = \lambda\alpha$ with $\lambda \in \mathbf{R}_+^\star$.

We define $G_\alpha(t) = |g_\alpha(t)|^2 = g_\alpha(t)\overline{g}_\alpha(t)$. We then have

$$G'_\alpha(1) = 2\,\mathrm{Re}[\overline{g}_\alpha(1)g'_\alpha(1)] = 2\lambda|\alpha|^2 > 0.$$

The function G_α is increasing in the neighborhood of 1, so we have $|g_\alpha(t)| < 1$ for t near 1, $t<1$.

The polynomial W_t has $n+s-1$ zeros inside the disk $D(0,1)$ and one zero outside. By Rouché's theorem the polynomial X^nV has $n+s-1$ zeros inside the disk $D(0,1)$ and one zero outside. ∎

Theorem 6.4.3. *Let R be the minimal polynomial of a Salem number; then there exists a polynomial P that is the minimal polynomial of a Pisot number such that $(X^2+1)R = XP + P^\star$.*

Proof. Let $2s$ be the degree of the polynomial R. We set $W = (X^2+1)R = X^{2s+2} + d_1X^{2s+1} + \cdots + 2d_{s+1}X^{s+1} + \cdots + d_1X + 1$, with $d_i \in \mathbf{Z}$ $(i = 1, \ldots, s+1)$.

Suppose P is the minimal polynomial of an S-number such that $W = XP + P^\star$. Then P can be written

$$P = X^{2s+1} + (d_1 - c_1)X^{2s} + \cdots + d_{s+1}X^s + \cdots + c_2X + c_1$$

with $c_i \in \mathbf{Z}$ $(1 = 1 \ldots, s)$ and $c_1 \neq 0$.

It is sufficient to show that one can find an s-tuple $(c_1, \ldots, c_s) \in \mathbf{Z}^s$ with $c_1 \neq 0$ such that the polynomial P has a zero outside the disk $D(0,1)$ and $2s$ zeros inside. So we write

(1) $$\alpha^{-2s} P(\alpha) W'(\alpha) \in \mathbf{R}_-^\star$$

for every α zero of W, $\alpha \in C(0,1)$. We begin by showing that $\alpha^{-s}P(\alpha)$ and $\alpha^{-s}W'(\alpha)$ are purely imaginary. We have $\alpha^{-s}P(\alpha) = c_1(\alpha^{-s} - \alpha^s) + \cdots + c_s(\alpha^{-1} - \alpha) + d_{s+1} + d_s\alpha + \cdots + \alpha^{s+1}$. Then $W(\alpha) = 0$ implies $d_{s+1} + d_s\alpha + \cdots + \alpha^{s+1} + \alpha^{-(s+1)}[d_{s+1}\alpha^{s+1} + d_s\alpha^s + \cdots + d_1\alpha + 1] = 0$

This implies that the number $1 + \sum_{i=1}^{s+1} d_i\alpha^i$ is purely imaginary and the same is true for $\alpha^{-s}P(\alpha)$. In the same way $W(z) = z^{2s+2}W(1/z)$ implies

$$W'(z) = (2s+2)z^{2s+1}W(1/z) - \frac{1}{z^2}z^{2s+2}W'(1/z),$$

hence $\overline{\alpha}^s W'(\alpha) + \alpha^s W'(\overline{\alpha}) = 0$. Then we can set $\alpha^{-s}P(\alpha) = i\delta$, $\alpha^{-s}W'(\alpha) = i\rho$ and $\alpha = e^{i\omega}$, with δ, s and ω real. We can write condition (i) in the following way:

(2) $$\rho \left[\sum_{k=1}^{s} c_k \sin(s+1-k)\omega - \delta \right] < 0.$$

Let now $\alpha^{(j)}$ and $\overline{\alpha}^{(j)}$ $(j = 1, \ldots, s)$ denote the zeros of W on the circle $C(0,1)$. We know they satisfy condition (ii); thus the following inequalities hold:

(3) $$\begin{aligned} &\sum_{k=1}^{s} c_k \sin(s+1-k)\omega^{(j)} < \delta_j \qquad \text{if } \rho_j > 0 \\ &\sum_{k=1}^{s} c_k \sin(s+1-k)\omega^{(j)} > \delta_j \qquad \text{if } \rho_j < 0. \end{aligned}$$

Let A be the matrix defined by $A = (a_{j,k})$ with $a_{j,k} = \sin(s+1-k)\omega^{(j)}$ $(j = 1, \ldots, s,\ k = 1, \ldots s)$. A is a non-singular matrix, for otherwise $\det A = 0$ would imply the existence of coefficients $c_1, \ldots, c_s$ that are not all equal to 0 such that $\sum_{k=1}^{s} c_k({\alpha^{(j)}}^{s+1-k} - {\overline{\alpha}^{(j)}}^{s+1-k}) = 0$, $(j = 1, \ldots, s)$. The polynomial $c_1X^{2s} + \cdots + c_sX^{s+1} - c_sX^s - \cdots - c_1$ would then have as zeros the numbers $\alpha^{(j)}, \overline{\alpha}^{(j)}$ $(j = 1, \ldots, s)$ and 1, that is, $2s+1$ zeros, and that is impossible.

The region of $\mathbf{R}^s$ that is defined by the inequalities (3) is a polyhedral cone that contains infinitely many points $(c_i)_{1 \leq i \leq s} \in \mathbf{Z}^s$ with $c_1 \neq 0$. ■

In the same way the polynomial $(X-1)R$ can be written $XP - P^\star$.

We remark that an arbitrarely larger S-number can be associated to a T-number because of the inequality $|c_1| > \theta$.

Thus on the one hand Theorem 6.4.1 associates to an S-number whose minimal polynomial is P infinitely many numbers of T that are zeros of the polynomials $X^n P \pm P^\star$, and on the other Theorem 6.4.3 associates to a number of T whose minimal polynomial is R infinitely many numbers of S that are zeros of the polynomials P defined by $(X^2+1)R = XP + P^\star$ or $(X-1)R = XP - P^\star$.

Hence the following definition.

Definition 6.4. *Let τ be a T-number and θ an S-number whose minimal polynomial is P. The numbers τ and θ are said to be associated if there exists an integer n such that τ is a zero of the polynomial $X^n P \pm P^\star$.*

Theorem 6.4.4. *Let (τ_n) be a sequence of distinct T-numbers. Suppose the sequence (τ_n) converges to a real σ that does not belong to S, and that (θ_n) is a sequence of S-numbers such that θ_n and τ_n are associated for every integer n. We have $\lim_{n\to+\infty} \theta_n = +\infty$.*

Proof. The number τ_n is a zero of a polynomial R_n, $R_n = X^{m_n} P_n \pm Q_n$, where P_n, and Q_n denote respectively the minimal polynomial of θ_n and the reciprocal polynomial of P_n. We wish to show that if there exists a sequence extracted from the sequence (θ_n) that is bounded, then σ belongs to S. We take a subsequence still denoted (θ_n) that converges to a number θ and we suppose that it does not contain quadratic units. Let f_n and φ_n be the rational functions defined by the rational fractions P_n/Q_n and $(1-\theta_n X)P/Q$.

The sequences (f_n) and (φ_n) converge in $\mathcal{F}(\delta)$ to functions defined respectively by A/Q and $(1-\theta X)A/Q$. The number τ_n is a zero of the polynomial R_n and we then have $\varphi_n(1/\tau_n) = \pm(1-\frac{\theta_n}{\tau_n})\tau_n^{m_n}$ with $\lim_{n\to+\infty} \varphi_n(1/\tau_n) = \varphi(1/\sigma)$.

If the sequence (m_n) is not bounded we consider a subsequence converging to $+\infty$. We then have $\lim_{n\to+\infty}(1-\frac{\theta_n}{\tau_n}) = 0$, hence $\theta = \sigma$ and σ belongs to S.

If the sequence (m_n) is bounded, there exists a subsequence that is stationary for $n \geq n_0$. Then we have

$$(1-\frac{\theta_n}{\tau_n})\frac{P_n(1/\tau_n)}{Q_n(1/\tau_n)} = \pm(1-\frac{\theta_n}{\tau_n})\tau_n^m \qquad \text{with } m \text{ fixed.}$$

Let n tend to $+\infty$, then we see that the number $1/\sigma$ is a solution of an equation of the form $A(\frac{1}{\sigma}) \pm Q(\frac{1}{\sigma})\frac{1}{\sigma^m} = 0$. The conditions defining the polynomials A and Q imply that σ belongs to S. ■

Notes

The theorem about the closure of the set S by Salem in 1944 [11] was a great surprise for the mathematicians interested in the topic. Pisot in particular had thought that S was dense. Salem gave a second proof of his theorem in 1945 [12], [13]. Salem's original idea of associating a rational fraction to an S-number was particularly fruitful. It oriented the study of sets S and T toward an analytical and functional direction. One can follow the steps that led to the study of functions $\mathcal{F}(\delta)$ from that of the Pisot numbers through Salem's two proofs and later Dufresnoy and Pisot's [4], [5] in 1953 and 1955. The functional angle is particularly well adapted to generalizations.

The study of the derived set of S was undertaken by Dufresnoy and Pisot from 1953 onward [4], [5], [6],[7], [10]. They also began the investigation of the successive derived sets, which was terminated by Grandet-Hugot in the early 1960s [8]. Their results were extended by Boyd in 1979 [3].

We should also mention Amara's result [1], which more generally deals with the sets S_q, and Siegel's paper on totally real numbers [14]. The problem of the adherence of T turned out from the start to be difficult. In 1945 Salem proved the inclusion $S \subset \overline{\mathrm{T}}$ [12],[13]. The proof we give is different from Salem's, especially part b) (this is the first time this proof appears in print). Salem used Tchebycheff polynomial properties. There were no results concerning the adherence of T between 1947 and 1977, which is why Boyd's paper [2] is important even though it did not solve the problem. The title of the paper, *Small Salem Numbers*, makes obvious the relationship between questions related to the elements of T and Lehmer's conjecture. Theorem 6.4.4 makes a fresh start in the slow and difficult study of the set T. The proof we give is close to Boyd's, with a few modifications (Pathiaux [9]).

References

[1] M. AMARA, Sur le produit des conjugués, extérieurs au disque unité de certains nombres algébriques, *Acta. Arith.* XXXIV, (1979), 307-314.

[2] D.W. BOYD, Small Salem numbers, *Duke Math. Journal*, 44, n°2, (1977), 315-328.

[3] D.W. BOYD, On the successive derived sets of the Pisot numbers, *Proc. Amer. Math. Soc.*, 73, n°2, (1977), 154-156.

[4] J. DUFRESNOY AND C. PISOT, Sur un ensemble fermé d'entiers algébriques, *Ann. Sci. E.N.S.*, série 3, 70, (1953), 105-133.

[5] J. DUFRESNOY AND C. PISOT, Sur les dérivés successifs d'un ensemble fermé d'entiers algébriques, *Bull. Sci. Math.*, série 2, 77, (1953), 129-136.

[6] J. DUFRESNOY AND C. PISOT, Etude de certaines fonctions méromorphes bornées sur le cercle unité. Application à un ensemble fermé d'entiers algébriques, *Ann. Sci. E.N.S.*, série 3, 72, (1955), 69-72.

[7] J. DUFRESNOY AND C. PISOT, Sur les éléments d'accumulation d'un ensemble fermé d'entiers algébriques, *Bull. Sci. Math.*, série 2, 79, (1955), 54-64.

[8] M. GRANDET-HUGOT, Ensembles fermés d'entiers algébriques, *Ann. Sci. E.N.S.*, série 3, 82, (1965), 1-35.

[9] M. PATHIAUX, Résultats de Boyd sur les nombres de Pisot-Salem, *Séminaire Delange-Pisot-Poitou, Théorie des nombres*, 20ème année (1978–79).

[10] C. PISOT, Quelques aspects de la théorie des entiers algébriques, *Séminaire de Mathématiques supérieures, Université de Montréal* (1963).

[11] R. SALEM, A remarkable class of algebraic integers. Proof of a conjecture of Vijayaraghavan, *Duke Math. Journ.*, 11, (1944), 103-108.

[12] R. SALEM, Power series with integral coefficients, *Duke Math. Journ.*, 12, (1945), 153-172.

[13] R. SALEM, *Algebraic Numbers and Fourier Analysis*, Heath Math. Monographs, Boston, Mass., (1963).

[14] C.L. SIEGEL, The trace of totally positive and real algebraic integers, *Annals of Math.*, 46, n°2, (1945).

CHAPTER 7

SMALL PISOT NUMBERS

Using Schur's algorithm for generating all Pisot numbers less than or equal to $\widehat{\theta}_{15} \simeq 1.6183608\ldots$, we prove that Inf $S = \theta_0$, where $\theta_0 = 1.3247179572\ldots$ satisfies the equation $X^3 - X - 1 = 0$, and that Inf $S' = (\sqrt{5}+1)/2$.

The characterization of the second derived set S'', together with Schur's algorithm, enable us to show that Inf $S'' = 2$.

Some improvements of these results can be found in the appendix.

7.1 Schur's approximations for elements of $\mathcal{N}_1^*$

Definition 7.1. *A meromorphic function f is said to belong to $\mathcal{M}_1^*$ if $f(x)$ is real for x real, if α, $0 < \alpha < 1$, is the unique simple pole of f in $D(0,1)$, if $f(0) \geq 1$ and if $|f(z)| \leq 1$ for $|z| = 1$.*

Let $F = \sum_{n\geq 0} u_n z^n$, $u_n \in \mathbf{R}, n \geq 0$, be the expansion of an element f of $\mathcal{M}_1^*$ in the neighborhood of the origin. Using the notation of Chapter 3 for $(n+1)\times(n+1)$ matrices, we recall that Schur's determinants $\delta_n(F)$ satisfy

$$\delta_n(F) = \varepsilon_n^-(F)\varepsilon_n^+(F), \tag{1}$$

where $\varepsilon_n^-(F) = \det({}^tF_n + J_{n+1})$ and $\varepsilon_n^+(F) = \det({}^tF_n - J_{n+1})$.

Lemma 7.1. *Let A and B belong to $\mathbf{R}[X]$ and such that $|B(z)| \leq |A(z)|$ if $|z| = 1$, and denote ϕ the function $\phi(x) = B(x)f(x) - A(x)$, where $f \in \mathcal{M}_1^*$.*

If $\phi \not\equiv 0$, let $\Phi(z) = \sum_{n\geq k} a_n z^n$ be the expansion of ϕ in the neighborhood of the origin, with $a_k \neq 0$. Then the polynomial A has at least $k-1$ roots in $D(0,1^-)$.

Conversely, if A has exactly $k-1$ roots in $D(0,1^-)$, then, if $z \in D(0,1^-)$, $\phi(z) = 0$ only for $z = 0$ and $a_k \neq 0$; if moreover $A(1) \neq 0$, then $A(1)\phi(x)(x-\alpha) < 0$ for x real, $0 < x < 1$, and therefore $A(1)a_k > 0$.

Proof. If $\lambda > 1$, we know by Rouché's theorem that each analytic function ϕ_λ defined by $\phi_\lambda(z) = B(z)f(z)(z-\alpha) - \lambda A(z)(z-\alpha)$, has the same number of zeros as the polynomial $\lambda A(z)(z-\alpha)$.

Since ϕ_λ tends to $(z-\alpha)\phi$ as λ tends to 1, and $(z-\alpha)\phi$ has at least k zeros in $D(0,1^-)$, then by the continuity of the zeros of bounded analytic functions in a disk we deduce that ϕ_λ has at least k zeros in $D(0,1^-)$ and A at least $k-1$ zeros in $D(0,1^-)$.

If A has exactly $k-1$ zeros in $D(0,1^-)$, then k zeros of ϕ_λ tend to 0 as λ tends to 1. Therefore the origin is a zero of order k for ϕ; hence $a_k \neq 0$.

Taking x real, $\rho_\lambda < x < 1$, where ρ_λ is the maximum modulus of the zeros of ϕ_λ in $D(0,1^-)$, then $\phi_\lambda(x)\phi_\lambda(1) > 0$, and since $-\phi_\lambda(1)A(1) > 0$, we deduce that $\phi_\lambda(x)A(1) < 0$. But ρ_λ tends to zero as λ tends to 1, so for $0 < x < 1$ we have $\phi(x)A(1)(x-\alpha) < 0$ and $(\phi(x)/x^k)A(1)(x-\alpha) < 0$, that is $a_kA(1) > 0$ as x tends to 0. ∎

Proposition 7.1.1. *Let $f \in \mathcal{M}^{**}{}_1$ and $F = \sum_{n\geq 0} u_n z^n$ be its expansion in the neighborhood of the origin.*

Then $u_1 > 0$ and $z = 0$ is the unique simple zero of $f - u_0$ in $D(0,1^-)$.

Proof. The proof follows from Lemma 7.1 with $B = 1$ and $A = u_0$. ∎

Definition 7.1.2. *We call $\mathcal{N}_1^*$ the set of rational functions f of $\mathcal{M}_1^*$, $f = A/Q$, where A and Q are elements of $\mathbf{Z}[z]$ and $Q(0) = 1$.*

Proposition 7.1.2. *Let $f \in \mathcal{N}_1^*$, s be its finite or infinite rank and $F = \sum_{n\geq 0} u_n z^n$ be its expansion in the neighborhood of the origin.*

a) If $u_0 > 1$, then $\delta_n(F) < 0, \quad 0 \leq n \leq s-1$.

b) If $u_0 = 1$, then $\delta_0(F) = 1 - u_0^2 = 0$ and $\delta_n(F) < 0, \quad 1 \leq n \leq s-1$.

Proof. a) We recall the relations (cf. Chapter 3)

$$\delta_0(F) = 1 - u_0^2 \qquad \text{and} \qquad \delta_n(F) = c^n(1-u_0^2)\delta_{n-1}(F^1), n \geq 1,$$

where F^1 is the first Schur transform of F,

$F^1 = z(\dfrac{u_0F - 1}{F - u_0})$, and $c = (u_0^2 - 1)/\left|F^1(0)\right|$.

From Proposition 7.1.1., the function $f^1 = z(\dfrac{u_0 f - 1}{f - u_0})$ belongs to $\mathcal{M}$, and by Theorem 3.2.1. we have the inequalities

$$\delta_k(F^1) > 0, \ \ 0 \le k < s-1 \quad \text{and} \quad \delta_k(F^1) = 0, \quad k \ge s-1.$$

Now we deduce easily a), since $1 - u_0^2 < 0$.

b) We define the function g_2 by

$$g_2(z) = \frac{(z^2 + u_1 z - 1)f(z) - (z^2 - 1)}{(z^2 - u_1 z - 1) - (z^2 - 1)f(z)} \ .$$

In $D(0,1^-)$ $z = 0$ is the only zero of order 2 of the denominator of g_2 . As $z = 0$ is also a zero of order greater than 2 , g_2 is analytic in $D(0,1^-)$. Since $|\, g_2(z) \,| \le 1$ for $|\, z \,| = 1$, g_2 belongs to $\mathcal{M}$, and by Theorem 3.2.1

$$\delta_n(G_2) > 0, 0 \le n < s-1 \qquad \text{and} \qquad \delta_n(G_2) = 0, n \ge s-1,$$

where G_2 is the expansion of g_2 in $D(0,1^-)$.

Assertion b) follows immediately from the relation

$$\delta_n(F) = -\,|\, u_0 \,|^{2(n+1-2)} \,|\, u_1 \,|^4 \, \delta_{n-2}(G_2). \qquad \blacksquare$$

Theorem 7.1.1. *Let $f \in \mathcal{N}_1^*$ and $F = \sum_{n \ge 0} u_n z^n$ be its expansion in the neighborhood of the origin.*

a) If the rank of f is infinite, then there exists, for every $n > 0$, unique polynomials D_n and *E_n of degree n, with rational coefficients, such that $E_n(0) = 1$, $D_n = -E_n^*$ (E_n^* is the reciprocal polynomial of E_n) and the expansion of the rational function $\dfrac{D_n}{E_n}$ in the neighborhood of the origin is of the form*

$$\frac{D_n(z)}{E_n(z)} = u_0 + u_1 z + \cdots + u_{n-1} z^{n-1} + w_n z^n + \cdots .$$

If the rank of f is s, $f = U/V$ and *$f(1) = 1$, then the same conclusions hold for $n \le s+2$, i.e.,*

$$D_{s+1}(z) = (1-z)U(z), \qquad D_{s+2}(z) = (1-z^2)U(z),$$

$$E_{s+1}(z) = (1-z)V(z), \qquad E_{s+2}(z) = (1-z^2)V(z).$$

If however the rank of f is s, $f = U/V$ and $f(1) = -1$, the above conclusions hold only for $n \leq s+1$, and

$$D_s(z) = U(z) \qquad D_{s+1}(z) = (1+z)U(z)$$

$$E_s(z) = V(z) \qquad E_{s+1}(z) = (1-z)V(z).$$

b) Similarly if the rank of f is infinite, there exists, for every $n > 0$ if $u_0 \neq 1$, and for $n > 0, n \neq 2$, if $u_0 = 1$, unique polynomials D_n^+ and E_n^+ of degree n, with rational coefficients, such that $E_n^+(0) = 1, D_n^+ = (E_n^+)^*$, and the expansion of the rational function $\frac{D_n^+}{E_n^+}$ in the neighborhood of the origin, is of the form

$$\frac{D_n^+(z)}{E_n^+(z)} = u_0 + u_1 z + \cdots + u_{n-1}z^{n-1} + w_n^+ z^n + \cdots.$$

If the rank of f equals s, $f = U/V$ and $f(1) = 1$, the same conclusions hold only for $n \leq s+1$, and

$$D_s^+(z) = U(z), \qquad D_{s+1}^+(z) = (1+z)U(z),$$

$$E_s^+(z) = V(z), \qquad E_{s+1}^+(z) = (1+z)V(z).$$

But if the rank of f equals s, $f = U/V$ and *$f(1) = -1$, the conclusions hold for $n \leq s+2$ and*

$$D_{s+1}^+(z) = (1-z)U(z), \qquad D_{s+2}^+(z) = (1-z^2)U(z),$$

$$E_{s+1}^+(z) = (1-z)V(z), \qquad E_{s+2}^+(z) = (1-z^2)V(z).$$

Proof. Let n be a positive integer. We seek rationals $\mu_1, \mu_2, \ldots, \mu_n$ such that $E_n(z) = 1 + \mu_1 z + \cdots + \mu_n z^n$ produces the expansion

$$\frac{D_n(z)}{E_n(z)} = u_0 + u_1 z + \cdots + u_{n-1}z^{n-1} + w_n z^n + \cdots$$

in the neighborhood of the origin.

This leads to the n linear equations:

$$u_0 + \mu_n = 0$$

$$u_1 + u_0\mu_1 + \mu_{n-1} = 0$$

$$u_2 + u_1\mu_1 + u_0\mu_2 + \mu_{n-2} = 0$$

$$\vdots$$

$$\vdots$$

$$u_{n-1} + u_{n-2}\mu_1 + u_{n-3}\mu_2 + \cdots + u_0\mu_{n-1} + \mu_1 = 0.$$

It should be observed that the determinant of the homogeneous system formed by the last $n-1$ equations is precisely $\varepsilon_{n-2}^-(F)$. By (1) and Proposition 7.1.2, $\varepsilon_{n-2}^-(F) \neq 0$ for $n \geq 2$ if $u_0 > 1$. But if $u_0 > 1$, $\varepsilon_0^-(F) = u_0 + 1 \neq 0$; so we reach the same conclusion.

Thus the above system is Cramérian, and from this we deduce the existence and uniqueness of polynomials D_n and E_n.

Notice that D_n and E_n can be expressed as determinants.

For example,

$$D_n(z) = \frac{-1}{\varepsilon_{n-2}^-(F)} \begin{vmatrix} z^n & z^{n-1} & \cdots\cdots & z^2 & z & 1 \\ u_{n-1} & u_{n-2}+1 & \cdots\cdots & u_1 & u_0 & 0 \\ u_{n-2} & u_{n-3} & \cdots\cdots & u_0 & 0 & 0 \\ \vdots & \vdots & \vdots & \vdots & \vdots & \vdots \\ u_2 & u_1 & \cdots\cdots & 1 & 0 & 0 \\ u_1 & u_0 & \cdots\cdots & 0 & 1 & 0 \\ u_0 & 0 & \cdots\cdots & 0 & 0 & 1 \end{vmatrix}.$$

If the rank of f is s, $f = U/V$ and $f(1) = 1$, the above conclusions remain valid as long as we have a Cramérian system, i.e. $\varepsilon_{n-2}^-(F) \neq 0$. But by Lemma 3.4.5, since $f(1) = 1$, we have $f^s(z) = 1$; thus $\varepsilon_0^-(F^s) = 2$ and $\varepsilon_s^-(F) \neq 0$. The existence and uniqueness of the polynomials D_n and E_n, for $2 \leq n \leq s+2$, follow immediately.

Moreover,

$$D_{s+1}(z) = (1-z)U(z), \qquad D_{s+2}(z) = (1-z^2)U(z),$$

$$E_{s+1}(z) = (1-z)V(z), \qquad E_{s+2}(z) = (1-z^2)V(z).$$

If the rank of f is s, $f = U/V$, $f(1) = -1$, then the corresponding system is a Cramérian system only for $n \leq s+1$, and

$$D_s(z) = U(z), \qquad D_{s+1}(z) = (1+z)U(z),$$

$$E_s(z) = V(z), \qquad E_{s+1}(z) = (1+z)V(z).$$

b) Let $E_n^+(z) = 1 + \mu_1^+ z + \mu_2^+ z^2 + \cdots + \mu_n^+ z^n$ such that

$$\frac{D_n^+(z)}{E_n^+(z)} = u_0 + u_1 z + \cdots + u_{n-1} z^{n-1} + w_n z^n + \cdots ,$$

in the neighborhood of the origin. We obtain as before n linear equations:

$$\begin{aligned}
&u_0 - \mu_n^+ = 0\\
&u_1 + u_0\mu_1^+ - \mu_{n-1}^+ = 0\\
&u_2 + u_1\mu_1^+ + u_0\mu_2^+ - \mu_{n-2}^+ = 0\\
&\vdots\\
&\vdots\\
&u_{n-1} + u_{n-2}\mu_1^+ + \cdots + u_0\mu_{n-1}^+ - \mu_1^+ = 0.
\end{aligned}$$

The determinant of the homogeneous system formed by the last $n-1$ equations is $\pm\varepsilon_{n-2}^+(F)$. Then by (1) and Proposition 7.1.2, if the rank of f is infinite, then $\varepsilon_{n-2}^+(F) \neq 0$ if $u_0 > 1$ and $n - 2 \geq 0$; but if $u_0 = 1$, then $\varepsilon_0^+(F) = u_0 - 1 = 0$ and $\varepsilon_{n-2}^+(F) \neq 0$ for $n > 2$. Therefore we obtain unique polynomials D_n^+ and E_n^+ for $n > 0$, except for $n = 2$ if $u_0 = 1$.

Moreover if f is of finite rank s, $f = U/V$ and $f(1) = 1$ (resp. $f(1) = -1$), then as in case a) we obtain unique D_n^+ and E_n^+ for $n \leq s + 1$ (resp. $n \leq s + 2$), except perhaps for $n = 2$.

The other assertions of b) are easily shown and left to the reader. ■

Remarks 7.1. 1) If f is of finite rank s, then for $n \geq s+2$ there exist infinitely many polynomials D_n, E_n, D_n^+, E_n^+ (for example, $D_n = KU, E_n = KV, D_n^+ = HU, E_n^+ = HV$ with the polynomials K and H of degree $n - s$ and satisfying $K = -f(1)K^*, H = f(1)H^*, K(0) = 1, H(0) = 1$.)

2) We obtain by calculating:

$$D_1(z) = u_0 - z, \qquad D_2(z) = u_0 + \frac{u_1}{1+u_0} z - z^2,$$

$$E_1(z) = 1 - u_0 z, \qquad E_2(z) = 1 - \frac{u_1}{1+u_0} z - u_0 z^2,$$

$$D_1^+(z) = u_0 + z, \qquad E_1^+(z) = 1 + u_0 z,$$

and if $u_0 \neq 1$,

$$D_2^+(z) = u_0 + \frac{u_1}{1-u_0} z + z^2, \qquad E_2^+(z) = 1 + \frac{u_1}{1-u_0} z + u_0 z^2,$$

$$w_1 = u_0^2 - 1, \qquad\qquad w_2 = u_0^2 - 1 + \frac{u_1^2}{1+u_0},$$

$$w_1^+ = 1 - u_0^2, \qquad\qquad w_2^+ = 1 - u_0^2 - \frac{u_1^2}{1-u_0}, \text{ if } u_0 \neq 1.$$

From now on, whenever the polynomials D_n, D_n^+ , E_n , E_n^+ are mentioned it will be only when they exist and are unique.

Theorem 7.1.2. *The following relations are satisfied by the polynomials D_n , D_n^+ , E_n , E_n^+.*

$$D_n^+ E_n - D_n E_n^+ \equiv (w_n^+ - w_n) z^n \tag{2}$$

$$D_{n+1} E_n - D_n E_{n+1} \equiv (u_n - w_n) z^n (1-z) \tag{3}$$

$$D_{n+1} E_n^+ - D_n^+ E_{n+1} \equiv (u_n - w_n^+) z^n (1+z) \tag{4}$$

$$D_{n+1}^+ E_n - D_n E_{n+1}^+ \equiv (u_n - w_n) z^n (1+z) \tag{5}$$

$$D_{n+1}^+ E_n^+ - D_n^+ E_{n+1}^+ \equiv (u_n - w_n^+) z^n (1-z) \tag{6}$$

$$D_{n+2} E_n - D_n E_{n+2} \equiv (u_n - w_n) z^n (1-z^2) \tag{7}$$

$$D_{n+2}^+ E_n^+ - D_n^+ E_{n+2}^+ \equiv (u_n - w_n^+) z^n (1-z^2). \tag{8}$$

Moreover, if f is of rank s, not necessarily finite, for $1 \leq n \leq s$ if $u_0 \neq 0$ and for $3 \leq n \leq s$ if $u_0 = 1$, we have

$$D_{n+1} \equiv \frac{w_n^+ - u_n}{w_n^+ - w_n}(1+z) D_n + \frac{u_n - w_n}{w_n^+ - w_n}(1-z) D_n^+ \tag{9}$$

$$E_{n+1} \equiv \frac{w_n^+ - u_n}{w_n^+ - w_n}(1+z) E_n + \frac{u_n - w_n}{w_n^+ - w_n}(1-z) E_n^+$$

$$D_{n+1}^+ \equiv \frac{w_n^+ - u_n}{w_n^+ - w_n}(1-z) D_n + \frac{u_n - w_n}{w_n^+ - w_n}(1+z) D_n^+$$

$$E_{n+1}^+ \equiv \frac{w_n^+ - u_n}{w_n^+ - w_n}(1-z) E_n + \frac{u_n - w_n}{w_n^+ - w_n}(1+z) E_n^+.$$

Likewise, for $1 \leq n \leq s-1$, the following relations hold:

$$D_{n+2} \equiv (1+z) D_{n+1} - \frac{u_{n+1} - w_{n+1}}{u_n - w_n} z D_n \tag{10}$$

$$E_{n+2} \equiv (1+z)E_{n+1} - \frac{u_{n+1} - w_{n+1}}{u_n - w_n} zE_n.$$

Similarly for $1 \leq n \leq s-1$ *if* $u_0 \neq 1$ *and for* $3 \leq n \leq s-1$ *if* $u_0 = 1$

$$D_{n+2}^+ \equiv (1+z)D_{n+1}^+ - \frac{u_{n+1} - w_{n+1}^+}{u_n - w_n^+} zD_n^+ \tag{11}$$

$$E_{n+2}^+ \equiv (1+z)E_{n+1}^+ - \frac{u_{n+1} - w_{n+1}^+}{u_n - w_n^+} zE_n^+.$$

Finally, the relation

$$w_{n+1}^+ - w_{n+1} = 4\,\frac{(w_n^+ - w_n)(u_n - w_n)}{w_n^+ - w_n} \tag{12}$$

shows that the sequence $(w_n^+ - w_n)$ *is decreasing and the series*

$$\sum_{n\geq 0} (u_n - \tfrac{w_n + w_n^+}{2})^2$$

converges.

Proof. From Theorem 7.1.1 we deduce

$$\frac{D_n^+}{E_n^+} - \frac{D_n}{E_n} = (w_n^+ - w_n)z^n + \cdots,$$

i.e.,

$$D_n^+ E_n - D_n E_n^+ = (w_n^+ - w_n)z^n + \cdots.$$

Since the left-hand side of the above inequality is a polynomial in z invariant under the transformations z to $1/z$ and multiplication by z^{2n}, we have identity (2). So the polynomials D_n and E_n on the one hand and the polynomials D_n^+ and E_n^+ on the other are relatively prime.

The same argument implies identities (3) to (8).

From (2), (3) and (4), and then from (2), (5) and (6), we deduce the relations (9).

Similarly (10) and (11) follow respectively from (3) and (7) and from (6) and (8).

Relation (12) is obtained from (2) and (9); so

$$w_{n+1}^+ - w_{n+1} \leq w_n^+ - w_n.$$

Thus the sequence $(w_n^+ - w_n)$ is decreasing.

If f is of infinite rank, $(w_n^+ - w_n)$ is an infinite positive decreasing sequence which thus tends to a positive limit or to zero.

Writing now (12) in the following way:

$$4(u_n - \frac{w_n + w_n^+}{2})^2 = (w_n^+ - w_n)[(w_n^+ - w_n) - (w_{n+1}^+ - w_{n+1})],$$

it can be easily shown that the series $\sum_{n\geq 0} (u_n - \frac{w_n+w_n^+}{2})^2$ converges. ∎

Theorem 7.1.3. *Let $f \in \mathcal{N}_1^*$ and f^n be its Schur transform. If f is of rank s, s not necessarily finite, then for $2 \leq n \leq s+1$ if $u_0 \neq 1$ and for $3 \leq n \leq s+1$ if $u_0 = 1$, we have the relations*

$$f^n(z) = \frac{[E_n^+(z) + E_n(z)]f(z) - [D_n^+(z) + D_n(z)]}{[E_n^+(z) - E_n(z)\,]f(z) - [D_n^+(z) - D_n(z)]}. \tag{13}$$

But for $n = 1$ and $u_0 \neq 1$, we have

$$f^1(z) = \frac{[E_1^+(z) - E_1(z)]f(z) - [D_1^+(z) - D_1(z)]}{[E_1^+(z) + E_1(z)]f(z) - [D_1^+(z) + D_1(z)]}, \tag{14}$$

and, for $n = 2$ and $u_0 = 1$,

$$f^2(z) = \frac{[(u_1/2)z - E_2(z)]f(z) - [(u_1/2)z - D_2(z)]}{[(u_1/2)z + E_2(z)]f(z) - [(u_1/2)z + D_2(z)]}. \tag{15}$$

In any case, for $2 \leq n \leq s+1$ and $n = 1$ if $u_0 \neq 1$, we get the inequalities

$$|\, f^n(z) \,| \leq 1 \quad \text{for } z \in D(0,1). \tag{16}$$

Proof. Suppose first $u_0 > 1$.

The first Schur transform is defined by $f^1(z) = z\,\dfrac{u_0 f - 1}{f - u_0}$ and gives (14), thanks to the expressions obtained for E_1, E_1^+, D_1 and D_1^+ in Remark 7.1.

Since, by Lemma 7.1, f^1 is analytic in $D(0,1)$ and satisfies $|\, f^1(z) \,| \leq 1$ if $|\, z \,| = 1$, inequality (16) holds for $n = 1$.

Since $|\, f^1(0) \,| < 1$, the second Schur transform f^2, defined by

$f^2 = \frac{1}{z} \frac{f_1 - f_1(0)}{1 - f_1 f_1(0)}$, is an analytic function bounded by 1 in $D(0,1)$. From (14) and (9), we thus obtain (13) for $n = 2$.

Now, assuming relations (13) to (16) up to n, we obtain the case $n+1$ by the use of (9) and the definition of the $(n+1)^{\text{st}}$ Schur transform.

Suppose now $u_0 = 1$.

From Definition 3.3.1, we get

$$f^2(z) = \frac{[(u_1/2)z - E_2(z)]f(z) - [(u_1/2)z - D_2(z)]}{[(u_1/2)z + E_2(z)]f(z) - [(u_1/2)z + D_2(z)]}.$$

By Lemma 7.1, f^2 is analytic in $D(0,1)$, and since $|\, f^2(z)\,| \le 1$ if $|\, z\,| = 1$, inequality (16) holds for $n = 2$.

It is obvious then that $f^3 = \frac{1}{z} \frac{f^2 - f^2(0)}{1 - f^2 f^2(0)}$ is analytic in $D(0,1)$ and satisfies (16).

Now, using (10) for $n = 1$ and the relations

$$D_3^+(z) \equiv (1-z)D_2(z) - \frac{u_2 - w_2}{u_1} z(1+z)$$

$$E_3^+(z) \equiv (1-z)E_2(z) - \frac{u_2 - w_2}{u_1} z(1+z),$$

obtained from (5) and (8), it follows that

$$f^3 = \frac{[E_3^+ + E_3]f - [D_3^+ + D_3]}{[E_3^+ - E_3]f - [D_3^+ - D_3]}.$$

The proof is concluded by induction as in the first case. ∎

Theorem 7.1.4. *Let $f \in \mathcal{N}_1^*$ and $F = \sum_{n \ge 0} u_n z^n$ be its expansion in the neighborhood of the origin. Then, for $n \le s$with the possible exception of $n = 1$, the polynomials D_n (resp. D_n^+) have a unique root $\tau_n > 1$ (resp. $\tau_n^+ > 1$) outside the unit circle, all the other roots lying inside the unit circle.*

Moreover we have the following inequalities:

1) if f is of infinite rank and if $u_0 \ne 1$

$$w_1 < u_1 \qquad \text{and } \tau_1 < (1/\alpha),$$

$$w_n < u_n < w_n^+ \qquad \text{and } \tau_n < (1/\alpha) < \tau_n^+, n \geq 2,$$

whereas if $u_0 = 1$,

$$0 < u_1, \quad w_2 < u_2 \qquad \text{and } \tau_2 < (1/\alpha),$$

$$w_n < u_n < w_n^+ \qquad \text{and } \tau_n < (1/\alpha) < \tau_n^+, n \geq 3,$$

2) if the rank of f *equals* s*, the previous inequalities hold for* $n < s$*, and for* $n = s$ *become*

$$w_s = u_s < w_s^+, \qquad \tau_s = (1/\alpha) < \tau_s^+ \quad \text{if } f(1) = -1,$$

$$w_s < u_s = w_s^+, \qquad \tau_s < (1/\alpha) = \tau_s^+ \quad \text{if } f(1) = 1,$$

and for $n \geq s+1$,

$$w_n = u_n = w_n^+, \qquad \tau_{n+1} = (1/\alpha) = \tau_{n+1}^+.$$

Proof. Consider $\phi_n = E_n f - D_n, n \geq 1$. Then $\phi_n(z) = (u_n - w_n)z^n + \cdots$, and by Lemma 7.1, the polynomial D_n has $n-1$ zeros inside the unit circle if $\phi_n \not\equiv 0$, i.e., if f is of infinite rank or if the rank of f equals s, with $n < s$ if $f(1) = -1$ and $n < s+1$ if $f(1) = 1$.

Since $D_n(0) = u_0 \geq 1$ is the absolute value of the product of all the roots of D_n, there is one root τ_n of D_n outside the unit circle. Moreover τ_n is positive, otherwise $D_n(-z)/E_n(-z)$ would belong to $\mathcal{N}_1^*$ and have an expansion in the neighborhood of the origin of the form

$$\frac{D_n(-z)}{E_n(-z)} = u_0 - u_1 z + \cdots, \qquad -u_1 > 0$$

according to Lemma 7.1, which contradicts Proposition 7.1.1.

The same argument shows that with the exception of $\tau_n^+, \tau_n^+ > 1$, all the zeros of D_n^+ lie in $D(0, 1^-)$.

Since $D_n(x)$ tends to $-\infty$ as x tends to $+\infty$ and $D_n(\tau_n) = 0$ for $1 < \tau_n$, we deduce $D_n(1) > 0$.

Again by Lemma 7.1, except for $\phi_n \equiv 0$, we have

$$E_n(x)f(x)(x-\alpha) - D_n(x)(x-\alpha) < 0,$$

for x real, $0 < x < 1$. Hence $u_n - w_n > 0$ for $n \geq 1$ if $u_0 \neq 1$ and for $n \geq 2$ if $u_0 = 1$.

We prove in the same way that $w_n^+ - u_n > 0$ for $n \geq 2$ if $u_0 \neq 1$ and for $n \geq 3$ if $u_0 = 1$. Inequalities (16) are equivalent to $E_n(\alpha) > 0$ and $E_n^+(\alpha) < 0$ for $n \geq 2$ if $u_0 \neq 1$ and for $n \geq 3$ if $u_0 = 1$; that is, $\tau_n < (1/\alpha) < \tau_n^+$ for $n \geq 2$ and $u_0 \neq 1$ and for $n \geq 3$ if $u_0 = 1$.

For $n = 1$ and $u_0 \neq 1$, we get $\tau_1 < (1/\alpha)$ and for $n = 2$, $u_0 = 1$, we get $\tau_2 < (1/\alpha)$.

If f is of finite rank s, w_n and w_n^+ are defined for $2 < n \leq s+1$. More precisely, $w_s = u_s$ if $f(1) = -1$ and $w_s^+ = u_s$ if $f(1) = 1$. Finally, $w_{s+1} = w_{s+1}^+ = u_{s+1}$.

For $n \geq s+2$, it follows from Remark 7.1 1) that $f = \dfrac{D_n}{E_n} = \dfrac{D_n^+}{E_n^+}$; hence $w_n = w_n^+ = u_n$.

The inequalities concerning τ_n , τ_n^+, and α for $n \geq s$ are deduced from Theorem 7.1.4. ∎

Theorem 7.1.5. *Let $f \in \mathcal{N}_1^*$, of infinite rank. Then sequences (τ_n) and (τ_n^+) defined as in Theorem 7.1.4 both converge to $1/\alpha$ as n tends to infinity.*

Proof. The first equation in (9) gives

$$D_{n+1}(\tau_n) = \frac{u_n - w_n}{w_n^+ - w_n}(1 - \tau_n)D_n^+(\tau_n).$$

Since $\tau_n < \tau_n^+$, i.e., $D_n^+(\tau_n) < 0$, it follows that $D_{n+1}(\tau_n) > 0$, and therefore $\tau_n < \tau_{n+1}$, since $D_{n+1}(z)$ tends to $-\infty$ as z tends to $+\infty$.

The sequence (τ_n) is thus an increasing sequence, with upper bound $1/\alpha$ by Theorem 7.1.4, so it converges to a limit τ, $\tau \leq (1/\alpha)$.

In the same way, it can be shown that the sequence (τ_n^+) is a decreasing sequence, with lower bound $1/\alpha$, and thus converges to a limit τ^+, $\tau^+ \geq (1/\alpha)$.

We now prove that $\tau = \tau^+ = 1/\alpha$.

We define ϕ_n by

$$\phi_n(z) = [\frac{D_n(z)}{E_n(z)} - f(z)](z - \alpha)(z - (1/\tau_n)).$$

The analytic functions ϕ_n/z^n, which are uniformly bounded on the unit circle, have the same property in $D(0,1)$ by the maximum principle; hence $\lim_{n \longrightarrow +\infty} \phi_n(\alpha) = 0$.

Since $z = \alpha$ is the unique single pole of f in $D(0,1)$, we deduce that $\lim_{n \longrightarrow +\infty} 1/\tau_n = \alpha$. ■

Remark 7.2. If $|z| = 1$, it may be observed that

$$E_n(z) = -z^n \overline{D_n(z)} \quad \text{and} \quad E_n^+(z) = z^n \overline{D_n^+(z)}.$$

Identity (2) thus becomes

$$D_n^+(z)\overline{D_n(z)} + D_n(z)\overline{D_n^+(z)} = w_n - w_n^+.$$

Theorem 7.1.6. *Let $f \in \mathcal{N}_1^*$ of infinite rank, $f = A/Q$, A and Q relatively prime.*

Let $F = \sum_{n \geq 0} u_n z^n$ be the expansion of f in the neighborhood of the origin. We denote $s = \deg(Q)$ and $a = \deg(A)$. If g_n is a rational function, $g_n = A_n/Q_n$, A_n and Q_n in $\mathbf{Q}[X]$ and relatively prime, such that g_n is bounded by 1 on the unit circle, and has the following expansion in the neighborhood of the origin:

$$G_n = u_0 + u_1 z + \cdots + u_{n-1} z^{n-1} + v_n z^n + \cdots, \quad v_n \neq u_n,$$

then there exists a natural integer N and two polynomials U_n and V_n with rational coefficients, satisfying

$$g_n = \frac{AU_n + z^{N+a} Q^* V_n}{QU_n + z^{N+s} A^* V_n},$$

$|V_n(z)| \leq |U_n(z)|$ if $|z| = 1$, and the rational function V_n/U_n is of finite rank if g_n is of finite rank and V_n/U_n is of infinite rank if g_n is of infinite rank.

Common factors of U_n and V_n are factors of the polynomial Ω defined by

$$\Omega = z^{a-s} QQ^* - AA^*, \quad \text{if } a > s$$

$$\Omega = QQ^* - z^{s-a} AA^*, \quad \text{if } s > a,$$

$$\Omega = z^{-r}(QQ^* - AA^*), \quad \text{if } s = a,$$

where the integer r is defined by the condition $\Omega(0) \neq 0$, and Ω is a polynomial.

More precisely, we have

$$N = n - s, \quad \deg(V_n) + a = \deg(\Omega), \qquad \text{if } a > s,$$

$$N = n - a, \quad \deg(V_n) + s = \deg(\Omega), \qquad \text{if } s > a,$$

$$N = n - r - a \text{ if } n \geq r, \quad N = -a \text{ if } n \leq r, \quad \deg(V_n) + s = \deg(\Omega) + r \text{ if } s = a.$$

Proof. We write

$$G_n - F = \frac{A_n}{Q_n} - \frac{A}{Q} = (v_n - u_n)z^n + u'_{n+1}z^{n+1} + \cdots, \qquad v_n - u_n \neq 0;$$

from this we derive the identity

$$A_nQ - Q_nA \equiv z^nV_n, \quad V_n \in \mathbf{Q}[z],$$

which can be written, after multiplication by Ω,

$$Q[A_n\Omega - z^{n+a-s}V_nQ^*] - A[Q_n\Omega - z^nA^*V_n] = 0, \text{ if } a > s,$$

$$Q[A_n\Omega - z^nV_nQ^*] - A[Q_n\Omega - z^{n+s-a}A^*V_n] = 0, \text{ if } s > a,$$

$$Q[A_n\Omega - z^{n-r}V_nQ^*] - A[Q_n\Omega - z^{n-r}A^*V_n] = 0, \text{ if } s = a \text{ and } n \geq r,$$

$$Q[z^{r-n}A_n\Omega - V_nQ^*] - A[z^{r-n}Q_n\Omega - A^*V_n] = 0, \text{ if } s = a \text{ and } n \leq r.$$

From these equalities, and since A and Q are relatively prime, there exists a polynomial U_n with rational coefficients such that

if $a > s$,

$$A_n\Omega = AU_n + z^{n+a-s}Q^*V_n,$$

$$Q_n\Omega = QU_n + z^nA^*V_n,$$

if $s > a$,

$$A_n\Omega = U_n + z^nQ^*V_n,$$

$$Q_n\Omega = QU_n + z^{n+s-a}A^*V_n,$$

if $s = a$ and $n \geq r$,

$$A_n\Omega = AU_n + z^{n-r}Q^*V_n,$$

$$Q_n\Omega = QU_n + z^{n-r}A^*V_n,$$

if $s = a$ and $n \leq r$,

$$z^{r-n}A_n\Omega = AU_n + Q^*V_n,$$

$$z^{r-n}Q_n\Omega = QU_n + A^*V_n.$$

Now, writing

$$g_n = \frac{A_n}{Q_n} = \frac{AU_n + z^{N+a}Q^*V_n}{QU_n + z^{N+s}A^*V_n}$$

as

$$z^{N+a}\frac{V_n}{U_n} = \frac{Q}{Q^*}\,\frac{A_nQ - AQ_n}{Q_nQ^* - z^{s-a}A_nA^*},$$

and observing that

$$z^{s-a}\frac{A^*(z)}{Q^*(z)} = \frac{\overline{A(z)}}{\overline{Q(z)}}, |\,A(z)\,| \leq |\,Q(z)\,| \,\text{and}\, |\,A_n(z)\,| \leq |\,Q_n(z)\,| \;\text{ if }\; |\,z\,| = 1,$$

we deduce the inequality $|V_n(z)| \leq |U_n(z)|$ if $|z| = 1$ and notice that the rational function V_n/U_n is of finite or infinite rank according to whether A_n/Q_n is. ■

7.2 Small Pisot numbers

Schur's algorithm derives from the theorems of §7.1. It generates all Pisot numbers less than or equal to $\widehat{\theta}_{15} \simeq 1.6183608\ldots$.

Theorem 7.2.1. *Pisot numbers less than or equal to $\widehat{\theta}_{15}$ can be arranged in increasing order as*

$$\theta_2 = \theta'_1 < \theta'_2 < \theta'_3 < \theta_3 < \theta'_4 < \theta_4 < \theta'_5 < \theta'' < \theta_5 < \theta'_6 < \cdots < \theta'_n < \theta_n < \theta'_{n+1} \cdots < \frac{1+\sqrt{5}}{2} < \cdots < \widehat{\theta}_n < \widehat{\theta}'_{n+1} < \widehat{\theta}_{n-1} \cdots < \widehat{\theta}_{15}.$$

The algebraic integers θ_{2p} (resp. θ_{2p+1}) are roots of the polynomials

$$P_{2p}(z) = \frac{1 - z^{2p}(1+z-z^2)}{1-z} \quad (\textit{resp. } P_{2p}(z) = \frac{1 - z^{2p+1}(1+z-z^2)}{1-z^2}),\ p \geq 1.$$

The algebraic integers $\widehat{\theta}_{2p}$(resp. $\widehat{\theta}_{2p+1}$) are roots of the polynomials

$$\widehat{P}_{2p}(z) = \frac{1 + z^{2p}(1+z-z^2)}{1+z} \quad (\textit{resp. } \widehat{P}_{2p+1}(z) = 1 + z^{2p+1}(1+z-z^2)),\ p \geq 7.$$

The Pisot numbers θ'', θ'_n , $\widehat{\theta}'_n$ are zeros respectively of the polynomials

$$1-z+z^2-z^4+2z^5-z^6,\quad 1-z^2+z^n(1+z-z^2),\quad 1-z^2-z^n(1+z-z^2),\quad n \geq 17.$$

The golden section $\dfrac{1+\sqrt{5}}{2}$, root of the polynomial $1+z-z^2$, is the smallest element of S', the first derived set of the set S of Pisot numbers.

Proof. We now describe Schur's algorithm with the help of Theorems 7.1.1, 7.1.2, and 7.1.4. Using the algorithm, we will determine all Pisot numbers up to $\widehat{\theta}_{15}$.

The previous theorems provide us with a finite set of polynomials D_1 and D_2. By induction, we assume that the polynomials D_1 to D_n are known.

The condition $D_{n+1}(\widehat{\theta}_{15}) \leq 0$, that is

$$u_n - w_n \leq \frac{1+\widehat{\theta}_{15}}{\widehat{\theta}_{15}} \frac{D_n(\widehat{\theta}_{15})}{D_{n-1}(\widehat{\theta}_{15})}(u_{n-1} - w_{n-1}),$$

gives an upper bound for $u_n - w_n$, and the lower bound is 0.

Moreover, because of the relation

$$\frac{A(z)}{Q(z)} - \frac{D_n(z)}{E_n(z)} = (u_n - w_n)z^n + \cdots,$$

$Q \in \mathbf{Z}[z], Q(0) = 1, E_n \in \mathbf{Q}[z], E_n(0) = 1$, then if d_n is the least common multiple of the denominators of all coefficients of E_n, we must have $d_n(u_n - w_n) \in \mathbf{N}$. Thus there are only a finite number of possible $u_n - w_n$, which correspond to the finite number of possible D_{n+1}.

Knowing $u_n - w_n$ and $w_n^+ - w_n$, we deduce $w_{n+1}^+ - w_{n+1}$ by formula (12).

Finally, the algorithm stops when $d_n(w_n^+ - w_n) < 1$.

And now let us carry out the calculations.

Since $D_1(z) = u_0 - z, u_0 \in \mathbf{N}$ and $\tau_1 = u_0 \leq \theta \leq \widehat{\theta}_{15} < 2$, we get $u_0 = 1$, so $D_1(z) = 1 - z$ and $w_1 = 0$.

Since $D_2(z) = 1 + (u_1/2)z - z^2$, $u_1/2 = \tau_2 - 1/\tau_2$ and $\tau_2 < 2$, we get $u_1 < 3$. But $u_1 - w_1 > 0$, so $u_1 > 0$.

Thus we get only $u_1 = 1$ or $u_1 = 2$.

If $u_1 = 2$, then $w_2 = 2$ and D_3 is derived from (10),

$$D_3(z) = (1+z)(1+z-z^2) - z(u_2 - 2)(1-z)/2.$$

If $u_2 > w_2$, then $\tau_3 > \widehat{\theta}_{15}$. So we get only the case $u_2 = w_2$, which corresponds to $\theta_\infty = \dfrac{1+\sqrt{5}}{2}$, the zero greater than 1 of the polynomial D_2, $D_2(z) = 1 + z - z^2$.

Thus the Pisot numbers up to $\widehat{\theta}_15$, and different from θ_∞, correspond to

$$D_1(z) = 1 - z, \qquad D_2(z) = 1 + (z/2) - z^2,$$

$$u_0 = 1, \quad u_1, u_1 - w_1 = 1, \quad w_2 = 1/2.$$

We now have $u_2 - w_2 \geq 0$, that is, $u_2 \geq 1/2$; but if $u_2 = 3$, we get $D_3 = \widehat{P}_2$, $\tau_3 \geq \widehat{\theta}_2 > \widehat{\theta}_{15}$.

So there remain only the cases $u_2 = 1$ and $u_2 = 2$.

1) Assume $u_2 = 1$

These Pisot numbers correspond to rational functions whose expansion in the neighborhood of the origin begins by $1 + z + z^2$. We shall prove that these numbers are θ'_n and $\widehat{\theta}'_n$.

For this purpose, let us consider

$$g_2(z) = \frac{(z^2 + z - 1)A(z) - (z^2 - 1)Q(z)}{(z^2 - z - 1)Q(z) - (z^2 - 1)A(z)}.$$

By Rouché's theorem, since $|(z^2 - 1)A(z)| < |(z^2 - z - 1)Q(z)|$ if $|z| = 1$, the denominator of g_2 has exactly two zeros in the unit circle.

But writing

$$g_2 = \frac{(z^2 + z - 1)f(z) - (z^2 - 1)}{z^2 - z - 1 - (z^2 - 1)f(z)},$$

we observe that the expansion of the denominator of g_2 whose coefficients are rational integers begins by z^2, and the expansion of the numerator by cz^3, $c \in \mathbf{Z}$. Therefore g_2 is analytic for $|z| \leq 1$. Moreover g_2 is bounded by 1 on the unit circle and its expansion has integer coefficients. It is thus easy to deduce from Parseval's inequality that either $g_2(z) = \pm z^n$ or $g_2(z) = 0$, that is, either

$$f(z) = \frac{1 - z^2 \pm z^n(1 + z - z^2)}{1 - z - z^2 \pm z^n(1 - z^2)}$$

or

$$f(z) = \frac{1 - z^2}{1 - z - z^2}.$$

These rational functions correspond to θ'_n, $\widehat{\theta}'_n$ for $n \geq 1$ or to $\theta_\infty = \dfrac{1 + \sqrt{5}}{2}$.

2) Assume $u_2 = 2$

Recall that we have

$$u_0 = 1, u_1 = 1, u_2 = 2, w_1 = 0, w_2 = 1/2, w_3 = 2,$$

$$D_1(z) = 1 - z, \quad D_2(z) = 1 + (z/2) - z^2, \quad D_3(z) = 1 + z^2 - z^3.$$

From (5) and (8), we deduce $D_3^+(z)$:

$$D_3^+(z) = (1-z)D_2(z) - \frac{u_2 - w_2}{u_1} z(1+z)$$

and $w_3^+ - w_3 = -2D_3(1)D_3^+(1) = 6$.

As $D_4(\widehat{\theta}_{15}) \leq 0$, it follows that

$$u_3 - w_3 \leq \frac{1+\widehat{\theta}_{15}}{\widehat{\theta}_{15}} \frac{D_3(\widehat{\theta}_{15})}{D_2(\widehat{\theta}_{15})}(u_2 - w_2) \leq 1.85.$$

Thus, the only possibilities are $u_3 - w_3 = 0$ or $u_3 - w_3 = 1$.

If $u_3 - w_3 = 0$, we get $f = D_3/E_3$, $D_3 = P_3$ and $\theta = \theta_3$.

If $u_3 - w_3 = 1$, then $D_4 = \dfrac{P_5 + (4/3)zP_3}{1+z}$ and the only possible values for $u_4 - w_4$ are $1/3$, $4/3$, or $7/3$.

Now by induction, we assume up to n, $n > 2$, the following situation: either

i) $u_{2n-1} - w_{2n-1} = 0$; thus $D_{2n-1} = P_{2n-1}$ and $\theta = \theta_{2n-1}$.

or

ii) $u_{2n-1} - w_{2n-1} = 1$ with

$$D_{2n} = \frac{P_{2n+1} + \frac{n+2}{n+1} zP_{2n-1}}{1+z} \tag{17}$$

and the only possibilities are a), b) or c):

a) $u_{2n} - w_{2n} = \dfrac{1}{n+1}$, $w_n^+ - u_{2n} = \dfrac{2n-1}{n-1}$, $w_{2n+1}^+ - w_{2n+1} = \dfrac{2(2n-1)}{n^2+n-1}$,

$$D_{2n+1} = P_{2n+1} + zP_{2n-1} = \frac{1-(1+z-z^2)z^{2n}}{1-z}$$

b) $u_{2n} - w_{2n} = \dfrac{n+2}{n+1}$, $w_{2n}^+ - u_{2n} = \dfrac{n}{n-1}$, $w_{2n+1}^+ - w_{2n+1} = \dfrac{2n(n+2)}{n^2+n-1}$,

$$D_{2n+1} = P_{2n+1}.$$

c) $u_{2n} - w_{2n} = \dfrac{2n+3}{n+1}$, $w^+_{2n} - u_{2n} = \dfrac{1}{n-1}$, $w^+_{2n+1} - w_{2n+1} = \dfrac{2(2n+3)}{n^2+n-1}$,

$$D_{2n+1} = P_{2n+1} - zP_{2n-1} = \frac{1+(1+z-z^2)z^{2n}}{1+z}.$$

Before continuing, notice the relation

$$P_{2n+1} = z^2 P_{2n-1} + 1. \tag{18}$$

a) Assume $u_{2n} - w_{2n} = \dfrac{1}{n+1}$.

We have $w^+_{2n+1} - w_{2n+1} = \dfrac{2(2n-1)}{n^2+n-1} < 1$ for $n \geq 3$;

thus, the only possibility is $u_{2n+1} = w_{2n+1}$, that is $f = \dfrac{D_{2n+1}}{E_{n+1}}$ and $D_{2n+1} = P_{2n+1} + zP_{2n-1}$. We obtain the numbers θ_{2n}.

If $n = 1$, then $u_2 - w_2 = 1/2$. This is precisely case 1.

If $n = 2$, then $u_4 - w_4 = 1/3$, $w^+_5 - w_5 = 6/5, D_5(z) = 1 + z + z^2 + z^3 - z^5$; hence there are two possibilities: either $u_5 = w_5$, which gives θ_4, or $u_5 - w_5 = 1, w^+_6 - w_6 = 2/3, D_6(z) = 1 - z + z^2 - z^4 + 2z^5 - z^6$, which give only $u_6 = w_6$ and thus θ'', since $w^+_6 - w_6 < 1$.

b) Assume $u_{2n} - w_{2n} = \dfrac{n+2}{n+1}$.

Since $w^+_{2n+1} - w_{2n+1} < 3$, the only possibilities are $u_{2n+1} - w_{2n+1} = 0, 1$ or 2.

a) If $\mathbf{u_{2n+1} - w_{2n+1} = 0}$, we get $f = \dfrac{D_{2n+1}}{E_{2n+1}}$, $D_{2n+1} = P_{2n+1}$ and $\theta = \theta_{2n+1}$.

b) If $\mathbf{u_{2n+1} - w_{2n+1} = 2}$, then $w^+_{2n+1} - u_{2n+1} = \dfrac{2(n+1)}{n^2+n-1}$ and

$$w^+_{2n+2} - w_{2n+2} = \frac{8(n+1)}{n(n+2)}.$$

Hence by (10), (17) and (18)

$$D_{2n+2} = \frac{1}{1+z}\left[P_{2n+3} - P_{2n+1} + 1 + \frac{2z}{n+2}P_{2n+1}\right].$$

For numbers less than $\widehat{\theta}_{15}$, the only possibility is

$$u_{2n+2}-w_{2n+2}=\frac{4}{n+2},\quad \overset{+}{w}_{2n+2}-u_{2n+2}=\frac{4n+8}{n(n+2)},\quad \overset{+}{w}_{2n+3}-w_{2n+3}=\frac{8}{n+1}.$$

From (10), we deduce

$$D_{2n+3}=P_{2n+3}-P_{2n+1}+1=1+(1+z-z^2)z^{2n+1}.$$

For $n<7$, all numbers are greater than $\widehat{\theta}_{15}$.

For $n>7$, we get only $u_{2n+3}=w_{2n+3}$ since $\overset{+}{w}_{2n+3}-w_{2n+3}<1$, which gives $\widehat{\theta}_{2n+1}$ as the zero greater than 1 of D_{2n+3}.

For $n=7$, the only number less than or equal to $\widehat{\theta}_{15}$ is provided by $u_{17}=w_{17}$. We obtain $\widehat{\theta}_{15}$ as a root of D_{17}.

c) If $\mathbf{u_{2n+1}-w_{2n+1}=1}$, then

$$\overset{+}{w}_{2n+1}-u_{2n+1}=\frac{n^2+3n+1}{n^2+n-1}\quad\text{and}\quad \overset{+}{w}_{2n+2}-w_{2n+2}=2\,\frac{n^2+3n+1}{n(n+2)}.$$

From (10), (17) and (18) we deduce

$$D_{2n+2}=\frac{P_{2n+3}+\frac{n+3}{n+2}\,zP_{2n+1}}{1+z}.$$

Now, the only possible values for $u_{2n+2}-w_{2n+2}$ are $\dfrac{1}{n+2}$, $\dfrac{n+3}{n+2}$ or $\dfrac{2n+5}{n+2}$.

If $u_{2n+2}-w_{2n+2}=\dfrac{1}{n+2}$, then $\overset{+}{w}_{2n+2}-u_{2n+2}=\dfrac{2n+1}{n}$ and from (12)

$$\overset{+}{w}_{2n+3}-w_{2n+3}=\frac{2(2n+1)}{n^2+3n+1}.$$

From (10) we deduce

$$D_{2n+3}=P_{2n+3}+zP_{2n+1}.$$

If $u_{2n+2}-w_{2n+2}=\dfrac{n+3}{n+2}$, then $\overset{+}{w}_{2n+2}-u_{2n+2}=\dfrac{n+1}{n}$ and from (12)

$$\overset{+}{w}_{2n+3}-w_{2n+3}=\frac{2(n+3)(n+1)}{n^2+3n+1}.$$

And from (10) we deduce $D_{2n+3}=P_{2n+3}$.

If $u_{2n+2}-w_{2n+2}=\dfrac{2n+5}{n+2}$, then $\overset{+}{w}_{2n+2}-u_{2n+2}=\dfrac{1}{n}$ and from (10)

$$w^+_{2n+3} - w_{2n+3} = \frac{2(2n+5)}{n^2+3n+1}.$$

From (10) we deduce $D_{2n+3} = P_{2n+3} - zP_{2n+1}$.

c) Assume $\mathbf{u_{2n} - w_{2n} = \frac{2n+3}{n+1}}$.

We have $w^+_{2n+1} - w_{2n+1} = \dfrac{2(2n+3)}{n^2+n-1} < 1$ for $n \geq 5$; thus the only possibility is $u_{2n+1} = w_{2n+1}$, that is $f = \dfrac{D_{2n+1}}{E_{2n+1}}$ and $D_{2n+1} = P_{2n+1} - zP_{2n-1}$, which gives the numbers $\widehat{\theta}_{2n}$.

For $n < 5$, we get only numbers greater than $\widehat{\theta}_8 > \widehat{\theta}_{15}$.

These three cases correspond to cases a), b), and c) up to $n+1$, which completes the proof by induction. ∎

7.3 The smallest number of S''

In this section we prove that Inf $S'' = 2$, where S'' is the second derived set of S.

Proposition 7.3.1. *Let θ be a Pisot number belonging to S''. Then there corresponds to θ a rational function A/Q , where the polynomials A and Q belong to $\mathbf{Z}[z]$, $Q(0) = 1$, $A(0) > 0$, $A(1/\theta) \neq 0$, and $|A(z)| \leq |Q(z)|$ if $|z| = 1$, where $1/\theta$ is the unique root of Q in $D(0,1)$.*

Moreover, if $\sum_{n\geq 0} u_n z^n$ denotes the expansion of A/Q in the neighborhood of the origin, the rational numbers w_n and w^+_n defined in Theorem 7.1.1 satisfy the inequalities

$$w_n + 2 \leq u_n \leq w^+_n - 2,$$

for n large enough.

Proof. From Theorem 6.3.1, there corresponds to θ a rational function A/Q for which there exist polynomials B and C belonging to $\mathbf{Z}[z]$ such that

$$|B(z)| \leq |Q(z)|, \quad |C(z)| \leq |Q(z)|, \quad |A(z) + z^m C(z)| \leq |Q(z) + z^m B(z)|$$

if $|z| = 1$ and $m \geq m_0$, the equality holding for at most finitely many z.

Denoting

$$A_m = A + \varepsilon z^m C, \qquad Q_m = Q + \varepsilon z^m B, \qquad \varepsilon = \pm 1,$$

we deduce

$$\frac{A_m}{Q_m} - \frac{A}{Q} = \varepsilon z^m \frac{AB - QC}{QQ_m}. \tag{1}$$

Let $\sum_{n\geq 0} u_n^{(m)} z^n$ (resp. $\sum_{n\geq 0} u_n z^n$) be the expansion in the neighborhood of the origin of A_m/Q_m (resp. A/Q) and r_1 the positive integer such that polynomial $z^{-r_1}(AB - QC)$ does not vanish for $z = 0$.

From (1) we deduce that the integers $u_n^{(m)}$ and u_n satisfy

$$u_n^{(m)} = u_n, \qquad n < m + r_1,$$

and

$$\left| u_{m+r_1}^{(m)} - u_{m+r_1} \right| \geq 1, \tag{2}$$

since this quantity is a non-zero rational integer.

Now let us consider the following sequence

$$\frac{A_{m,\nu}}{Q_{m,\nu}} = \frac{A_m + z^\nu \widetilde{Q}_m}{Q_m + z^\nu \widetilde{A}_m}, \qquad \nu \geq \nu_0$$

where $\widetilde{Q}_m$ (resp. $\widetilde{A}_m$) is the reciprocal polynomial of Q_m (resp. A_m) multiplied by a suitable power of z for the polynomials $A_{m,\nu}$ and $Q_{m,\nu}$ to become reciprocal.

We have thus an infinite sequence of distinct rational functions, since the equality $|A_m(z)| = |Q_m(z)|$ holds for at most finitely many z of modulus 1.

Denoting by $\sum_{n\geq 0} u_n^{(m,\nu)} z^n$ the expansion in the neighborhood of the origin of $A_{m,\nu}/Q_{m,\nu}$, and using the previous argument, we find

$$u_n^{(m,\nu)} = u_n^{(m)}, \qquad n < \nu + r_2,$$

and

$$\left| u_{\nu+r_2}^{(m,\nu)} - u_{\nu+r_2}^{(m)} \right| \geq 1, \tag{3}$$

where r_2 is the degree of the monomial of least degree in $Q_m^* Q_m - A_m^* A_m$, i.e., in $AC^* - QB^*$.

If ν is any rational integer, $\nu \geq 1$, the equalities $u_n^{(m,\nu)} = u_n^{(m)}$ for $n < \nu + r_2$ imply the equalities

$$w_n^{(m,\nu))+} = (w_n^{(m)})^+$$
$$w_n^{(m,\nu)} = w_n^{(m)},$$

for $n \leq \nu + r_2$, which gives by Theorem 7.1.4

$$w_{\nu+r_2}^{(m)} \leq u_{\nu+r_2}^{(m)} \leq (w_{\nu+r_2}^{(m)})^+,$$

$$w_{\nu+r_2}^{(m)} = w_{\nu+r_2}^{(m,\nu)} \leq u_{\nu+r_2}^{(m,\nu)} \leq (w_{\nu+r_2}^{(m,\nu)})^+ = (w_{\nu+r_2}^{(m)})^+.$$

These inequalities together with (3) lead to

$$w_{\nu+r_2}^{(m)} + 1 \leq u_{\nu+r_2}^{(m)} \leq w_{\nu+r_2}^{(m)} - 1, \qquad \nu \geq 1;$$

that is, for $n > r_2$,

(4) $$w_n^{(m)} + 1 \leq u_n^{(m)} \leq (w_n^{(m)})^+ - 1.$$

Now, if m is an arbitrary rational integer, $m \geq 1$, the equality $u_n^{(m)} = u_n$ for $n < m + r$ implies

$$(w_n^{(m)})^+ = w_n^+ \quad \text{and} \quad w_n^{(m)} = w_n, \qquad \text{for } n \leq m + r_1;$$

and we deduce from (4), for $m + r_1 > r_2$

(5) $$w_{m+r_1} + 1 \leq u_{m+r_1}^{(m)} \leq w_{m+r_1}^+ - 1.$$

Finally, from (1) and (5) and for $m + r_1 > r_2$, it follows that

$$w_{m+r_1} + 2 \leq u_{m+r_1} \leq w_{m+r_1}^+ - 2,$$

that is, for $n > r = \max(r_1, r_2)$,

$$w_n + 2 \leq u_n \leq w_n^+ - 2.$$

Remark 7.3. We could obtain by induction and from Theorem 6.3.2 the following result:

If a Pisot number θ belongs to the r-derived set $S^{(r)}$, $r \geq 1$, then there corresponds to θ a rational function A/Q, whose expansion in the neighborhood of the origin $\sum_{n \geq 0} u_n z^n$ satisfies

$$w_n + r \leq u_n \leq w_n^+ - r,$$

for n large enough, where w_n and w_n^+ are defined as in Theorem 7.1.1.

Proposition 7.3.2. *The rational integer* 2 *belongs to* S'' *and the two rational functions* $\frac{A}{Q} = \frac{1-z}{1-2z}$ *and* $\frac{A}{Q} = \frac{1}{1-2z}$ *satisfy the conditions of Proposition 7.3.1.*

Proof. Either the polynomials

$$A = 1 - z, \qquad B = z, \qquad C = -(1-z),$$

or the polynomials

$$A = 1, \qquad B = z^2, \qquad C = -1,$$

satisfy the conditions of the theorem characterizing elements of S'', since for $m \geq 2$ and $|z| = 1$ we can verify that

$$|1 - 2z + z^{m+1}|^2 = |1 - z - z^m(1-z)|^2 + |1 - z^{m-1}|^2. \qquad \blacksquare$$

Proposition 7.3.3. *a) If* $\frac{A}{Q} = \frac{1}{1-2z}$, *then the corresponding polynomials* D_n *and* D_n^+ *defined as in Theorem 7.1.1 satisfy*

$$(1-z)D_n \equiv 1 - z^n(2-z)$$

$$(1-z)^2 D_n^+ \equiv (1 - \frac{n}{n-2} z) + z^n(z-2)(z - \frac{n}{n-2}).$$

b) If $\frac{A}{Q} = \frac{1-z}{1-2z}$, *then the corresponding polynomials* D_n *and* D_n^+ *satisfy*

$$(1 - 3z + z^2)D_n \equiv (1-z)(1-a_n z) + (2-z)(z-a_n)z^n$$

$$(1 - 3z + z^2)D_n^+ \equiv (1-z)(1-a_n^+ z) - (2-z)(z-a_n^+)z^n,$$

$$a_n = \frac{1+\alpha^n}{\alpha(1+\alpha^{n-2})}, \qquad a_n^+ = \frac{1-\alpha^n}{\alpha(1-\alpha^{n-2})},$$

where α *denotes any root of the equation* $x^2 - 3x + 1 = 0$.

Proof. The polynomials D_n and D_n^+ are obtained as in the proof of Theorem 7.1.6.

a) If $\frac{A}{Q} = \frac{1}{1-2z}$, then $A^* = 1$, $Q^* = z - 2$ and $\Omega = -2(1-z)^2$.

We first determine D_n.

We know by Theorem 7.1.6 that $\deg(V_n) = 1$ and $\frac{V_n}{U_n}$ is of finite rank, as is $\frac{D_n}{E_n}$.

Moreover, following the proof of Theorem 7.1.6, we can write

$$(6) \qquad -2(1-z)^2 D_n = U_n + z^n(z-2)V_n;$$

hence $U_n(0) = -2$.

If U_n and V_n are relatively prime , and since $\frac{V_n}{U_n}$ is of finite rank, then $V_n = \varepsilon U_n^*$, $\varepsilon = \pm 1$.

So, if $U_n = -2 + a_n z$, then $V_n = \varepsilon(a_n - 2z)$.

Identifying the terms of highest degree in (6), it follows that $\varepsilon = -1$ and $V_n = -a_n + 2z$.

Now, by deriving (6) and evaluating at $z = 1$, we see that $a_n = 2$ and $U_n = V_n$, which contradicts the assumption.

Thus U_n and V_n have a common factor, which is a factor of Ω by Theorem 7.1.6.

Since U_n and V_n are of degree 1 and $U_n(0) = 2$, we can write $U_n = -2(1-z)$, $V_n = -2\varepsilon(1-z)$.

Similarly, by identifying terms of highest degree in (6), we find $\varepsilon = 1$.

After simplification in relation (6), we finally get

$$(1-z)D_n = 1 + z^n(z-2).$$

We now determine D_n^+.

Proceeding in the same way, we get the relation

$$(6') \qquad -2(1-z)^2 D_n^+ = U_n + z^n(z-2)V_n \ ;$$

hence $U_n(0) = -2$.

If U_n and V_n have a common factor, we can write

$$U_n = -2(1-z)$$
$$V_n = -2\varepsilon(1-z).$$

By identifying terms of highest degree in (6′), it follows that $\varepsilon = -1$.

After simplification, we get

$$(1-z)D_n^+ = 1 - z^n(z-2),$$

a contradiction since the right-hand side does not vanish for $z = 1$.

Thus U_n and V_n are relatively prime and we can write

$$U_n = -2 + a_n z$$

$$V_n = \varepsilon(a_n - 2z).$$

By identifying terms of highest degree in (6′) we get $\varepsilon = 1$.

By deriving (6′) and evaluating at $z = 1$, we see that $a_n = \dfrac{2n}{n-2}$; thus

$$(1-z)^2 D_n^+ = (1 - \frac{n}{n-2}z) + z^n(z-2)(z - \frac{n}{n-2}).$$

b) If $\dfrac{A}{Q} = \dfrac{1-z}{1-2z}$, then $A^* = z-1$, $Q^* = z-2$, $\Omega = -1+3z-z^2$ and $r = 0$.

By Theorem 7.1.6, since Ω is irreducible, the polynomials U_n and V_n are relatively prime . Moreover their degree is 1 and we can write

$$(-1+3z-z^2)D_n = (1-z)U_n + z^n(z-2)V_n; \tag{7}$$

hence

$$-D_n(0) = -1 = U_n(0), \quad U_n = -1 + a_n z, \quad V_n = \varepsilon(a_n - z), \quad \varepsilon = \pm 1.$$

We denote by α a zero of the equation $1 - 3z + z^2 = 0$.

From (7), and since $1 - \alpha = \alpha(2-\alpha)$, we deduce

$$a_n = \frac{1+\alpha^n}{\alpha(1+\alpha^{n-2})}$$

and

$$(1-3z+z^2)D_n \equiv (1-z)(1-a_n z) + (2-z)(z-a_n)z^n.$$

Similarly, we also find

$$(1-3z+z^2)D_n^+ \equiv (1-z)(1-a_n^+ z) - (2-z)(z-a_n^+)z^n$$

with

$$a_n^+ = \frac{1-\alpha^n}{\alpha(1-a^{n-2})}.$$

■

Theorem 7.3. Inf $S'' = 2$.

Proof. We wish to find Pisot numbers θ, $\theta \leq 2$, belonging to S'' and corresponding to suitable rational functions A/Q as in Proposition 7.3.1. For this purpose we use Schur's algorithm , as in Theorem 7.2.1, and the inequality

$$(9) \qquad w_n + 2 \leq u_n \leq w_n^+ - 2, \qquad \text{for } n > r,$$

given by Proposition 7.3.1.

We prove first that $r = 2$.

By definition and since $0 < u_0 < \theta \leq 2$, we get

$$D_1 = 1 - z, \qquad w_1 = 0,$$

$$D_2 = 1 + \frac{u_1}{2} z - z^2.$$

Since $\theta \leq 2$, $D_2(2) \leq 0$, that is, $u_1 \leq 3$; but $u_1 - w_1 = u_1 > 0$, thus $u_1 = 1$ or $u_1 = 2$ (since $u_1 = 3$ corresponds to $\dfrac{D_2}{E_2} = \dfrac{z-2}{1-2z}$).

a) If $u_1 = 1$, then

$$D_2 = 1 + \frac{z}{2} - z^2, \qquad w_2 = \frac{1}{2}, \qquad D_2(2) = -2.$$

But $D_3(2) \leq 0$, that is,

$$u_2 - w_2 \leq \frac{3}{2} \frac{D_2(2)}{D_1(2)} (u_1 - w_1) = 3;$$

moreover, u_2 and $2(u_2 - w_2)$ are rational integers. Hence $u_2 - w_2 = 1/2$ or $3/2$ or $5/2$.

If $u_2 - w_2 = 1/2$, then $u_2 = 1$ and the expansion in the neighborhood of the origin of A/Q begins by $1 + z + z^2$.

By Theorem 7.2.1, we obtain only Pisot numbers belonging to S or S', but not to S''.

If $u_2 - w_2 = 5/2$, then $u_2 = 3$, $w_3 = 5$, $D_3^+(1) = -5$, $w_3^+ - w_3 = 5$ and $D_3 = 1 - z + 2z^2 - z^3$.

Thus $u_3 - w_3$ is a rational integer satisfying

$$u_3 - w_3 \leq \frac{3}{2} \frac{D_3(2)}{D_2(2)} (u_2 - w_2) = \frac{15}{8};$$

we get only $u_3 - w_3 = 1$; hence $w_4^+ - w_4 = 16/5 < 4$, which is in contradiction with (9), since the sequence $(w_n^+ - w_n)$ is decreasing by Theorem 7.1.2. Thus $u_2 - w_2 = 3/2$, that is $u_2 = 2$.

b) If $u_1 = 2$, then $D_2 = 1 + z - z^2, w_2 = 2, D_2(2) = -1$; since the rational integer $u_2 - w_2$ satisfies

$$0 < u_2 - w_2 < \frac{3}{2} \frac{D_2(2)}{D_1(2)} (u_1 - w_1) = 3,$$

we get only $u_2 = 3$ or $u_2 = 4$.

Therefore the expansion in the neighborhood of the origin of the rational functions A/Q corresponding to Pisot numbers less than 2 and belonging to S'' is of the following type:

$$1 + u_1 z + u_2 z^2 + \cdots \tag{10}$$

with either $u_1 = 1$ and $u_2 = 2$ or $u_1 = 2$ and $u_2 = 3$ or 4.

Now, from the proof of Proposition 7.3.1, we get (9) for $n > r = \max(r_1, r_2)$, where the integer r_1 is defined by the fact that the polynomial $z^{-r_1}(AB - QC)$ does not vanish for $z = 0$ and the integer r_2 is the degree of the monomial of least degree in $AC^* - QB^*$.

If the polynomials B and C satisfy $B(0) = C(0) = 0$, dividing by a suitable power of z we can assume that either $B(0) = 0$ or $C(0) = 0$.

If $B(0)C(0) < 0$ or if either $B(0) = 0$ or $C(0) = 0$, then $r_1 = 0$; and if $B^*(0)C^*(0) < 0$, then $r_2 = 0$.

If $B(0)C(0) > 0$, then the rational functions $\varepsilon \dfrac{B}{Q}$ and $\varepsilon \dfrac{C}{Q}$ ($\varepsilon = 1$ if $B(0) > 0$ and $C(0) > 0$; $\varepsilon = -1$ otherwise) satisfy (10), and we have

$$\varepsilon \frac{A}{Q} \frac{B}{Q} - \varepsilon \frac{C}{Q} = z^{r_1} \Sigma, \quad \Sigma(0) \neq 0. \tag{11}$$

If $B^*(0)C^*(0) > 0$, then the rational functions $\varepsilon \dfrac{B^*}{Q}$ and $\varepsilon \dfrac{C^*}{Q}$ ($\varepsilon = 1$ if $B^*(0) > 0$ and $C^*(0) > 0$, $\varepsilon = -1$ otherwise) satisfy (10) and we have

$$\varepsilon \frac{A}{Q} \frac{C^*}{Q} - \varepsilon \frac{B^*}{Q} = z^{r_2} \Theta, \quad \Theta(0) \neq 0. \tag{12}$$

If $B(0)C(0) > 0$, we write

$$\frac{A}{Q} = 1 + u_1 z + u_2 z^2 + \cdots ,$$

$$\varepsilon \frac{B}{Q} = 1 + u'_1 z + u'_2 z^2 + \cdots ,$$

and
$$\varepsilon \frac{A}{Q} \frac{B}{Q} = 1 + u''_1 z + u''_2 z^2 + \cdots ,$$

with $u''_1 = u_1 + u'_1$ and $u''_2 = u_2 + u_1 u'_1 + u'_2$. Then $u''_2 \geq 5$, since A/Q and $\varepsilon B/Q$ are of type (10). If moreover $r_1 > 2$, then $\varepsilon \frac{A}{Q} \frac{B}{Q}$ is of type (10), which contradicts $u''_2 \geq 5$. Therefore $r_1 \leq 2$.

In the same way, if $B^*(0)C^*(0) > 0$, then $\frac{A}{Q}$, $\varepsilon \frac{C^*}{Q}$, and, if $r_2 > 2, \varepsilon \frac{A}{Q} \frac{C^*}{Q}$ are of type (10), which leads to a contradiction by a similar argument. Therefore $r_2 \leq 2$.

In any case, we get $r = \max(r_1, r_2) \leq 2$ and inequalities (9) hold for $n \geq 3$.

Now, returning to (10), if $u_1 = 1$ and $u_2 = 2$, we deduce

$$D_2 = 1 + (z/2) - z^2, \qquad D_3 = 1 + z^2 - z^3,$$

$$u_3 - w_3 \leq \frac{3}{2} \frac{D_3(2)}{D_2(2)} \frac{3}{2} = \frac{27}{8} .$$

Since $u_3 - w_3 \geq 2$, we have either $u_3 - w_3 = 2$ or $u_3 - w_3 = 3$.

If $u_3 - w_3 = 3$, that is $u_3 = 5$, then $w_4^+ - w_4 = 6$,

$D_4(2) = -1, u_4 - w_4 \leq \frac{3}{2} \frac{D_4(2)}{D_3(2)} (u_3 - w_3) = 1.5$; and since $u_4 - w_4$ is a rational integer, $u_4 - w_4 = 1$, which contradicts (9).

Thus, we get only $u_3 - w_3 = 2$, that is, $u_3 = 4$; hence

$$D_4 = 1 - \frac{z}{3} + \frac{z^2}{3} + \frac{4}{3} z^3 - z^4, \quad w_4 = \frac{17}{3}, \quad w_4^+ - w_4 = \frac{16}{3}, \quad D_4(2) = -\frac{11}{3} ,$$

$$u_4 - w_4 \leq \frac{3}{2} \frac{D_4(2)}{D_3(2)} 2 = \frac{11}{3} .$$

Because of (9), the possible values $1/3$, $4/3$, $10/3$ of $u_4 - w_4$ do not fit. Thus we get $u_4 - w_4 = 7/3$, i.e., $u_4 = 8$.

Therefore, if $n \leq 5$, the polynomials D_n and D_n^+ and the quantities w_n and w_n^+ correspond to the rational function $\dfrac{1-z}{1-2z}$.

Now, by induction, we assume the theorem true for $n \leq k$ and prove it for $n = k+1$.

By Proposition 7.3.3 b), we have

$$w_k = 2^{k-1} - a_k, \qquad w_k^+ = 2^{k-1} + a_k^+.$$

Since $u_k - w_k = a_k$ corresponds to $\dfrac{1-z}{1-2z}$, the possible values of $u_k - w_k$ are $a_k + n$, $n \in \mathbf{Z}$.

But we must have $u_k - w_k \geq 2$ by (9) and $a_k - 1 = \dfrac{1+\alpha^k - \alpha - \alpha^{k-1}}{\alpha(1+\alpha^{k-2})} < 2$. So $u_k - w_k = a_k - n$ $(n > 0)$ is impossible.

In the same way, since $w_k^+ - u_k \geq 2$ by (9) and $a_k^+ - 1 = \dfrac{1-\alpha^k - \alpha + \alpha^{k-1}}{\alpha(1-\alpha^{k-2})} < 2$, it is impossible to have $u_k - w_k = a_k + n$ $(n > 0)$, i.e., $w_k^+ - u_k = a_k^+ - n$ $(n > 0)$. Thus only $u_k = 2^{k-1}$ satisfies (9) and we have the theorem for $k+1$.

Therefore the only rational function corresponding to a Pisot number θ in S'', $\theta \leq 2$, whose expansion in the neighborhood of the origin begins by $1+z$, is the rational function $\dfrac{1-z}{1-2z}$.

If $u_1 = 2$, then $D_2 = 1 + z - z^2$, $w_2 = 2$ and the only possible u_2 are 3 or 4.

If $u_2 = 3$, that is, $u_2 - w_2 = 1$, then $D_3 = 1 + \dfrac{3}{2}z + \dfrac{z^2}{2} - z^3$, $w_3 = \dfrac{9}{2}$. By 7.2 (5) and (8), we get

$$D_3^+ = (1-z)D_2 - \frac{u_2 - w_2}{u_1} z(1+z) = 1 - \frac{z}{2} - \frac{5}{2}z^2 + z^3$$

and by 7.2 (2), it follows that $w_3^+ - w_3 = 4$.

Since $u_3 - w_3 \leq \dfrac{3}{2}\dfrac{D_3(2)}{D_2(2)}(u_2 - w_2) = 3$, we have $4.5 < u_3 \leq 7.5$ and no rational integer u_3 satisfies (9).

Thus, if $u_1 = 2$, we get $u_2 = 4$, that is, $u_2 - w_2 = 2$, $D_3 = 1 + z + z^2 - z^3$, and $w_3 = 6$.

Since $u_3 - w_3 \leq \dfrac{3}{2}\dfrac{D_3(2)}{D_2(2)}(u_2 - w_2) = 3$, the inequalities $6 < u_3 \leq 9$ and (9) imply $u_3 = 8$.

Thus for $n \leq 4$, the polynomials D_n and D_n^+ and the quantities w_n and w_n^+ correspond to the rational function $\dfrac{1}{1-2z}$.

Assume now, by induction, that this is true for $n = k$; we shall prove it for $n = k+1$.

From 7.3.3 b) we get

$$w_k = 2^k - 2, \qquad w_k^+ = 2^k + \frac{2k}{k-2}.$$

Since $u_k - w_k = 2$ corresponds to $\dfrac{1}{1-2z}$, the possible values of $u_k - w_k$ are $2+n, n \in \mathbf{Z}$.

But, $u_k - w_k \leq \dfrac{3}{2}\dfrac{D_k(2)}{D_{k-1}(2)}(u_{k-1} - w_{k-1}) = 3$, since by induction $D_h = 1 + z + \cdots + z^{h-1} - z^h, h \leq k$ and $D_h(2) = -1$.

Thus, from (9), the only possible value is $u_k - w_k = 2$, which gives $D_{k+1} = 1 + z + \cdots + z^k - z^{k+1}$, $w_{k+1} = 2^{k+1} - 2$, and therefore corresponds to $\dfrac{1}{1-2z}$.

So we have proved that the only rational function corresponding to an element of S'' less than or equal to 2 whose expansion in the neighborhood of the origin begins by $1 + 2z$ is the rational function $\dfrac{1}{1-2z}$.

This completes the proof. ■

Notes

M. Amara [1] determined all Pisot numbers of S' less than 2.

D.W. Boyd established a list of Salem numbers less than 1.295675372...; these are at present the smallest known Salem numbers. Moreover, in [6] and [7], he gave all reciprocal polynomials of Mahler measure less than 1.3 and of degree less than 26. (If $P = c_0 z^n + c_1 z^{n-1} + \cdots + c_n, c_i \in \mathbf{Z}$, has roots $\theta_1, \ldots, \theta_n$, Mahler's measure of P is $M(P) = |\, c_0 \,| \prod_{i=1}^{n} \max(|\, \theta_i \,|, 1)$.)

All these results are related to D.H. Lehmer's question (1933): Does there exist a constant $\varepsilon_0 > 0$ such that $M(P) < 1+\varepsilon_0$ and P monic imply that $\theta_1, \theta_2, \ldots, \theta_n$ are roots of unity?

At the same time, Lehmer exhibited a polynomial P of degree 10 and measure 1.1762808183..., corresponding to the smallest known Salem number and giving the smallest known Mahler's measure.

In Chapter 3 we gave Smyth's partial answer to Lehmer's problem. Other partial results are due, for instance, to Dobrowolski (1979-80) [10], Cantor-Straus (1982) [8], and Langevin (1985-86) [14], [15] and depend on the degree or the distribution of the roots of the polynomial P.

We should point out that the limit of the sequence $w_n^+ - w_n$ is related to Mahler's measure of the polynomial Ω defined in Theorem 7.1.6:

$$\lim_{n \longrightarrow \infty} \frac{w_n^+ - w_n}{2} = |\, Q(0)\,|^{-2} M(\Omega)$$

(cf. Boyd [4])

Theorem 7.2.1 has been generalized by F. Lazami-Talmoudi [16], who proved the following result: for every δ, $0 < \delta < 1$, all Pisot numbers lying in $[1, 2-\delta[$, except perhaps for a finite number, are roots of polynomials of the type $A(z) + z^n P(z)$, with A/P^* corresponding to an element of $S' \cap [1,2]$.

Since Boyd [5] gave an algorithm determining all Pisot numbers in $S \cap [\alpha, \beta]$, if the set is finite, we are supposed to know all Pisot numbers of $S \cap [1, 2-\delta]$.

References

[1] M. AMARA, Ensembles fermés de nombres algébriques, *Ann. Scient. Ec. Norm.*, Sup. 3, 83, (1966), 215-270.

[2] M.J. BERTIN AND M. PATHIAUX-DELEFOSSE, Conjecture de Lehmer et petits nombres de Salem, *Queen's Papers in Pure and Applied Mathematics*, Kingston, Canada, 81, (1989).

[3] D.W. BOYD, Small Salem numbers, *Duke Math. J.*, 44, (1977), 315-328.

[4] D.W. BOYD, Schur's algorithm for bounded holomorphic functions, *Bull. London Math. Soc.*, 11, (1979), 145-150.

[5] D.W. BOYD, Pisot and Salem numbers in intervals of the real line, *Math. Comp.*, 32, (1978), 1244-1260.

[6] D.W. BOYD, Reciprocal polynomials having small measure, *Math. Comp.*, 35, (1980), 1361-1377.

[7] D.W. BOYD, Reciprocal polynomials having small measure II, *Math. Comp.*, 53, (1989), n° 187, 355-357.

[8] D.C. CANTOR AND E.G. STRAUS, On a conjecture of Lehmer, *Acta Arith.*, 42, (1982), 97-100.

[9] C. CHAMFY, Fonctions méromorphes dans le cercle unité et leurs séries de Taylor, *Ann. Inst. Fourier*, 8, (1958), 214-261.

[10] E. DOBROWOLSKI, On a question of Lehmer and the number of irreducible factors of a polynomial, *Acta Arith.*, 34, (1979), 391-401.

[11] J. DUFRESNOY AND CH. PISOT, Sur un ensemble fermé d'entiers algébriques, *Ann. Scient. Ec. Norm.*, Sup. 3, 70, (1953), 105-133.

[12] J. DUFRESNOY AND CH. PISOT, Etude de certaines fonctions méromorphes bornées sur le cercle unité. Application à un ensemble fermé d'entiers algébriques, *Ann. Sc. Ec. Norm.*, Sup. 3, 72, (1955), 69-92.

[13] M. GRANDET-HUGOT, Ensembles fermés d'entiers algébriques, *Ann. Sc. Ec. Norm.*, Sup. 3, 82, (1965), 1-35.

[14] M. LANGEVIN, Méthode de Fekete-Szegö et problème de Lehmer, *C.R.A.S. Paris*, 301, Série I, (1985), 463-466.

[15] M. LANGEVIN, Minorations de la maison et de la mesure de Mahler de certains entiers algébriques, *C.R.A.S. Paris*, 303, Série I, (1986), 523-526.

[16] F. LAZAMI, Sur les éléments de $S \supseteq [1,2[$, *Séminaire Delange-Pisot-Poitou*, 20 ième année, (1978-79), n° 3.

[17] D.H. LEHMER, Factorization of certain cyclotomic functions, *Ann. Math.*, 34, (1933), 461-479.

[18] F. TALMOUDI, Sur les nombres de $S \cap [1,2[$, *C.R.A.S. Paris*, 287, (1978).

CHAPTER 8

SOME PROPERTIES AND APPLICATIONS OF PISOT NUMBERS

This chapter describes some lesser-known properties of Pisot numbers; Salem numbers appear only in Theorem 8.1.1. By this choice we have sought to demonstrate Pisot numbers' important role in many questions (applications to harmonic analysis will be given in Chapter 15). Notation is the same as in Chapter 5 (cf. §5.0).

8.1 Some algebraic properties of Pisot and Salem numbers

We first prove that U-numbers are zeros of polynomials whose coefficients satisfy certain inequalities, and then we will consider some properties of the conjugates of S-numbers.

Theorem 8.1.1. *Let α denote a U-number. Then α is a zero of a polynomial with integer coefficients whose absolute values are less than or equal to α.*

Proof. As in the proof of Theorem 5.1.1, we associate with the U-number α a sequence of integers (u_n), and consider for $k \geq 1$, the $\mathbf{R}^{k+1}$-linear forms V_n defined by equalities $V_n(\mathbf{x}) = \sum_{i=0}^{k} u_{n+i}x_i$, $\mathbf{x} = (x_i)_{0 \leq i \leq k}$. We wish to prove the existence of integers k and m ($k \geq 1$ and $m \geq 0$) and of a $(k+1)$-tuple $\mathbf{a} \in \mathbf{Z}^{k+1} \setminus \{\mathbf{0}\}$ with $|a_i| \leq \alpha$, $(i = 0, \ldots, k)$, such that $V_m(\mathbf{a}) = 0$ and that $V_n(\mathbf{a}) = 0$ implies $V_{n+1}(\mathbf{a}) = 0$ for every $n \geq m$. Pisot and Salem numbers are considered separately: the sequence (u_n) is defined by $u_n = E(\theta^n)$ if $\alpha = \theta \in S$ and by $u_n = E(\lambda\tau^n)$ if $\alpha = \tau \in T$ (λ is a suitable real number) . In both cases, the integer A is defined by $1 \leq A \leq \alpha < A + 1$.

1. *The number θ belongs to S.* Let $\delta = \sup_{j=2,\ldots,s} |\theta^{(j)}|$. Then $|\epsilon_n| = |\epsilon(\theta^n)| \leq (s-1)\delta^n$. From the inequalities $|a_i| \leq A \quad (i = 0, \ldots, k)$ we deduce

$$|V_{n+1}(\mathbf{a}) - \theta V_n(\mathbf{a})| \leq A(k+1)(\theta+1)(s-1)\delta^n.$$

Therefore $V_n(\mathbf{a}) = 0$ and

$$A(k+1)(\theta+1)(s-1)\delta^n < 1 \tag{1}$$

imply $V_n(\mathbf{a}) = 0$ for all $n \geq m$.

The linear form W_m defined by $W_m(\mathbf{x}) = \sum_{i=0}^{k} |u_{m+i}|x_i$, takes $(A+1)^{k+1}$ values for all $\mathbf{x} \in \mathbf{Z}^{k+1}$, $0 \leq x_i \leq A (i = 0, \ldots, k)$, such that $0 \leq W_m(\mathbf{x}) \leq (A+1)(k+1)\theta^{m+k} - 1$. By the *pigeonhole or box principle*, if $(A+1)^{k+1} > (A+1)(k+1)\theta^{m+k}$, i.e.,

$$(A+1)^k > (k+1)\theta^{m+k} \tag{2}$$

there exist two different elements $\mathbf{b}$ and $\mathbf{b}'$ in $\mathbf{Z}^{k+1}$, with $0 \leq b_i \leq A,\ \ 0 \leq b_i' \leq A$, $(i = 0, \ldots, k)$, such that $W_m(\mathbf{b}) = W_m(\mathbf{b}')$.

Then $\mathbf{a} = \mathbf{b} - \mathbf{b}'$ belongs to $\mathbf{Z}^{k+1} \setminus \{\mathbf{0}\}$ and we have $W_m(\mathbf{a}) = 0$ with $|a_i| \leq A$ $(i = 0, 1, \ldots, k)$. We now seek integers m and k $(m \geq 0$ and $k \geq 1)$ satisfying (1) and (2), i.e.,

$$m \log \delta + \log[A(s-1)(\theta+1)(k+1)] < 0, \tag{1'}$$

$$m \log \theta + \log(k+1) < k \log \frac{A+1}{\theta}. \tag{2'}$$

If we define m by

$$m - 1 \leq \frac{\log[A(s-1)(\theta+1)(k+1)]}{\log(\delta^{-1})} < m,$$

(1′) follows from the inequality on the right while the one on the left gives

$$m \log \theta + \log(k+1) \leq c_1 \log(k+1) + c_2$$

with $$c_1 = 1 - \frac{\log \theta}{\log \delta}, \quad c_2 = \log \theta \Big(1 - \frac{\log[A(s-1)(\theta+1)]}{\log \delta}\Big).$$

Finally, we obtain (2′) by choosing k such that

$$c_1 \log(k+1) + c_2 < c_3 k \qquad (c_1,\ c_2,\ c_3 \text{ positive constants}).$$

2. *The number τ belongs to T.* Let Δ denote the discriminant of the field $\mathbf{Q}(\tau+\tau^{-1})$. By Theorem 5.2.2, for every η, $0<\eta<1$, there exists a real number $\lambda \in \mathbf{Q}(\tau+\tau^{-1}) \cap S$ with $|\lambda^{(j)}| \leq \eta$, $(j=2,\ldots,s)$ and $|\lambda| \leq \sqrt{\Delta}\eta^{1-s}$. Hence there is an integer n_0 such that for all $n \geq n_0$,

$$\epsilon_n = \epsilon(\lambda\tau^n) = -\left[\sum_{j=2}^{s} \lambda^{(j)}\left(\tau^{(j)^n} + \tau^{(j)^{-n}}\right) + \lambda\tau^{-n}\right].$$

Hence
$$|\epsilon_n| \leq 2(s-1)\eta + \sqrt{\Delta}\eta^{1-s}\tau^{n},$$

Theorem 8.1.2. *Let θ be an S-element. Suppose there exists a subset J of $\{1,2\dots,s\}$ and integers ℓ_j $(j \in J)$ such that $\prod_{j\in J} \theta^{(j)^{\ell_j}} = 1$. Then either $\ell_j = 0$, $(\forall j \in J)$ or $J = \{1,2\dots,s\}$ and $\ell_j = \ell_1$ $(\forall j \in J)$.*

Proof. Write $\ell_{j_1} = \max_{j\in J} |\ell_j|$ with $j_1 = \min_{j\in J} j$. Since the Galois group of an irreducible polynomial acts transitively on its roots, we may assume $j_1 = 1$. Then we deduce

$$|\theta^{(1)} \prod_{j\in J\setminus\{1\}} \theta^{(j)}| \leq \left[\theta^{(1)^{\ell_1}} \prod_{j\in J\setminus\{1\}} \theta^{(j)^{\ell_j}}\right]^{1/\ell_1} \leq 1,$$

which contradicts $|\prod_{j=1,2,\dots,s} \theta^{(j)}| \geq 1$ except $\ell_j = 0$ $(\forall j \in J)$ or $J = \{1,\dots,s\}$ and $\ell_j = \ell_1$ $(\forall j \in J)$. ∎

Corollary *Let θ an S-element and let*

$$\rho^{(j)}e^{i\varphi^{(j)}}, \rho^{(j)}e^{-i\varphi^{(j)}} \qquad (j \in J_\theta \subset \{2,\dots,s\})$$

denote the non-real conjugates of θ. Then the real numbers 2π and $\varphi^{(j)} (j \in J_\theta)$ are $\mathbf{Q}$-linearly independant.

Proof. Otherwise a relation $2\pi\ell_1 + \sum_{j\in J_\theta} \ell_j\varphi^{(j)} = 0$ would imply $\prod_{j\in J_\theta} e^{i\ell_j\varphi^{(j)}} = 1$ and thus that $\prod_{j\in J_\theta} \theta^{(j)^{\ell_j}}$ is real and $\prod_{j\in J_\theta} \theta^{(j)^{\ell_j}} = \prod_{j\in J_\theta} \overline{\theta}^{(j)^{\ell_j}}$.

As in the previous theorem we deduce $\ell_j = 0$ $(\forall j \in J_\theta)$. ∎

Theorem 8.1.3. *Let θ denote an S-number, and suppose that two conjugates of θ have the same modulus. Then the two conjugates are imaginary conjugates.*

Proof. Let $P_1 = P$ denote the minimal polynomial of θ and let P_2 and P_3 be the polynomials defined by $P_2 = \prod_{1\leq j\leq s} (X - \theta^{(j)^2})$, $P_3 = \prod_{1\leq j<k\leq s} (X - \theta^{(j)}\theta^{(k)})$. We note that P_2 and P_3 are irreducible. Suppose there exists j and k satisfying $1 \leq j < k \leq s$ such that $|\theta^{(j)}| = |\theta^{(k)}|$ and $\theta^{(k)} \neq \overline{\theta}^{(j)}$.

There are three cases.

1) *$\theta^{(j)}$ and $\theta^{(k)}$ are both real.* Then ${\theta^{(j)}}^2 = {\theta^{(k)}}^2$ implies that P_2 has a multiple zero and hence is not irreducible.

2) *$\theta^{(j)}$ is real and $\theta^{(k)}$ is not real.* Then $P_3(\theta^{(k)}\overline{\theta}^{(k)}) = 0 = P_3(|\theta^{(k)}|^2) = P_3({\theta^{(j)}}^2)$. Hence P_2 divides P_3 while all zeros of P_3 have a modulus smaller than θ^2, which is impossible.

3) *Both $\theta^{(j)}$ and $\theta^{(k)}$ are not real.* Then P_3 has a multiple zero $|\theta^{(j)}|^2 = |\theta^{(k)}|^2$. Let P_4 be the minimal polynomial of $|\theta^{(j)}|^2$; then P_4^2 divides P_3. Now P_4 has zeros with modulus greater than 1, and the only zeros of P_3 satisfying the condition are $\theta\theta^{(h)}$ $(h = 2, \ldots, s)$. It follows from the equation $P_4(\theta\theta^{(h)}) = 0$ that $\theta\theta^{(h)}$ is a multiple zero of P_3 and thus $\theta\theta^{(h)} = \theta\theta^{(\ell)}$ for some $\ell \neq h$. But this is impossible since P_1 is irreducible. ■

By applying these properties of Pisot number conjugates we prove the following theorem, which we will need in §8.4.

Theorem 8.1.4. *Let K be a real extension of $\mathbf{Q}$ of degree s $(s \geq 2)$; then if γ is an arbitrary positive number there exist infinitely many numbers of S that generate K, are units of K, and zeros of a polynomial $X^s - c_{s-1}X^{s-1} + c_{s-2}X^{s-2} + \cdots + (-1)^s$ whose coefficients satisfy*

$$c_{s-k} > \gamma c_{s-k-1} \quad (k = 1, \ldots, s-1) \quad \textit{and} \quad c_0 = 1. \tag{3}$$

Proof. Let θ be an S-number that is a unit of K of degree s. Suppose the real conjugates of θ are positive, otherwise we consider θ^2. We arrange the conjugates of θ such that $\theta > |\theta^{(2)}| \geq \cdots \geq |\theta^{(s)}|$ and using the notations of the corollary of Theorem 8.1.2, the reals 2π and $\varphi^{(j)}$ $(j \in J_\theta)$ are $\mathbf{Q}$-linearly independent. If follows from Theorem 4.6.3 that the sequence $((h\varphi^{(j)}/2\pi)_{j\in J_\theta})_h$ is u.d. mod 1 in $\mathbf{R}^{\sharp J_\theta}$. We deduce that there exist infinitely many integers h such that

$$\frac{1}{2} < \cos(h\varphi^{(j)}) < 1 \qquad (\forall j \in J_\theta). \tag{4}$$

Let h denote an integer satisfying this condition. We write $X^s - c_{s-1}^{(h)}X^{s-1} + \cdots + (-1)^s$ for the minimal polynomial of θ^h, and we will show that, if h is large enough, the coefficients $c_{s-h}^{(h)}$ satisfy the inequalities (3).

Let k be an integer with $1 \leq k \leq s-1$. We have

$$c_{s-k}^{(h)} = \sum_{1 \leq i_1 < \cdots < i_k \leq s} (\theta^{(i_1)} \ldots \theta^{(i_k)})^h.$$

Let $\theta^{(\ell)}$ and $\theta^{(\ell+1)}$ be two successive conjugates of θ. If $|\theta^{(\ell)}| > |\theta^{(\ell+1)}|$, then the coefficient $c_{s-\ell}^{(h)}$ is written $c_{s-\ell}^{(h)} = (\theta \cdots \theta^{\ell)})^h(1+\varepsilon(h))$ with $\lim_{h\to+\infty} \varepsilon(h) = 0$. If $|\theta^{(\ell)}| = |\theta^{(\ell+1)}|$, then it follows from Theorem 8.1.3 that $\theta^{(\ell+1)} = \overline{\theta}^{(\ell)}$, hence we have $c_{s-\ell}^{(h)} = 2(\theta \cdots \theta^{(\ell-1)})^h|\theta^{(\ell)}|^h \cos(h\varphi^{(j)})(1+\varepsilon(h))$ with $\lim_{h\to+\infty} \varepsilon(h) = 0$.

For every real positive number ε there exists h satisfying (4), and large enough so that we have $1-\varepsilon \leq \dfrac{c_{s-\ell}^{(h)}}{|\theta \cdots \theta^{(\ell)}|^h} \leq 2+\varepsilon$, $(\ell = 1, \dots, s-1)$ and $\max_j 2|\theta^{(j)}|^h = 2\theta^h < 1/\gamma$; then the conditions (3) are satisfied. ∎

8.2 An application of Pisot numbers to a problem of uniform distribution

Pisot numbers were discovered during research in the uniform distribution of sequences, and their role has always been important in the development of that theory. In this section we give one example of the many results concerning this question. Some of these results are mentioned in the notes at the end of this chapter. Exceptionally, in this paragraph we write the decomposition modulo 1 of a real x in the following way: $x = [x] + \{x\}$ with $[x] \in \mathbf{Z}$ and $[x] \leq x < [x]+1$.

Let α be a real number and q an integer with $1 < \alpha < q$. We are concerned with the sequence $(u_n(\alpha,q))_n$ where $u_n(\alpha,q) = (q-\alpha)\sum_{k=1}^{+\infty}\{\frac{n}{q^k}\}\alpha^k$.

Theorem 8.2. *Let α be a real number greater than 1. Then the following statements are equivalent.*

(i) α does not belong to S.

(ii) There exists an integer $q > \alpha$ such that the sequence $(u_n(\alpha,q))$ is u.d. mod 1.

(iii) For all integers $q > \alpha$, the sequence $(u_n(\alpha,q))$ is u.d. mod 1.

This characterization uses the following lemma.

Lemma *Let (a_k) be a sequence of real numbers, and q a positive integer such that the series $\sum_{k\geq 1} a_k q^{-k}$ converges. We set $\rho_k = \sum_{h=k+1}^{+\infty} a_h q^{-h}$, and consider* *the sequences* (v_n) *defined by* $v_n = \sum_{k=1}^{+\infty}\{\frac{n}{q^k}\}a_k$. *We have then*

$$\limsup_{x\to+\infty}\frac{1}{x}\left|\sum_{n\leq x}\exp 2i\pi v_n\right| \leq \prod_{k=0}^{+\infty}\left|\frac{\sin(\pi q^{k+1}\rho_k)}{q\sin(\pi q^k\rho_k)}\right|.$$

Proof. Let n be a positive integer and let $n = \sum_{k=0}^{+\infty} e_k(n)q^k$ with $e_k(n) \in \{0,\dots,q-1\}$ be the q-adic expansion of n. We will use the following result (proved in [10]; cf. notes.) Let (b_k) be a sequence with $b_k = o(q^k)$ and let (w_n) be the sequence defined by $w_n = \sum_{k=0}^{+\infty} e_k(n)b_k$. We have then

$$\limsup_{x\to+\infty}\frac{1}{x}\left|\sum_{n\leq x}\exp 2i\pi w_n\right| = \prod_{k=0}^{+\infty}\left|\frac{\sin(\pi q b_k)}{q\sin(\pi b_k)}\right|.$$

We assume that (b_k) is the sequence defined by $b_k = q^k\sum_{h=k+1}^{+\infty} a_h q^{-h} = q^k\rho_k$; (b_k) verifies $b_k = o(q^k)$, and we have $b_k = qb_{k-1} - a_k$ $(k \geq 1)$.

Then $v_n = \sum_{k=1}^{+\infty}\{\frac{n}{q^k}\}(qb_{k-1} - b_k)$. Observing that $e_k(n) = q\{\frac{n}{q^{k+1}}\} - \{\frac{n}{q^k}\}$, we deduce that $v_n = \sum_{k=0}^{+\infty} b_k e_k(n)$. The lemma is then established by replacing w_n by v_n and b_k by $q^k\rho_k$. ■

Proof of the theorem. Let $j \in \mathbf{Z}^\star$ and let (a_k) be a sequence defined by $a_k = j(q-\alpha)\alpha^k$. The series $\sum_{k\geq 1} a_k q^{-k}$ converges, and it follows from the lemma that

$$\limsup_{x\to+\infty}\frac{1}{x}\left|\sum_{n\leq x}\exp 2i\pi j u_n(\alpha,q)\right| = \prod_{k=0}^{+\infty}\left|\frac{\sin(\pi j q\alpha^{k+1})}{q\sin(\pi j\alpha^{k+1})}\right|.$$

Suppose α is not an S-number; then the series $\sum_{k\in\mathbf{N}} \|j\alpha^k\|^2$ diverges and consequently so does the infinite product. By Weyl's criterion the sequence $(u_n(\alpha,q))$

is u.d. mod 1. Suppose α is an S-number; then the series $\sum_{k\in\mathbf{N}} \|j\alpha^k\|^2$ converges. Unless one of the factors of the infinite product vanishes, the infinite product is non-zero, hence $(u_n(\alpha, q))$ is not u.d. mod 1. If one of the factors is zero then there exists an integer k_0 such that $j\alpha^{k_0} \in \{\frac{1}{q}, \ldots, \frac{q-1}{q}\}$; then α^{k_0} is rational and an integer since it belongs to S. Then α too is an integer and we have $u_n(\alpha, q) = n(q-\alpha)\sum_{k=1}^{+\infty}(\frac{\alpha}{q})^k = \alpha n$; and in that case the sequence $(u_n(\alpha, q))$ is not u.d. mod 1. ∎

8.3 Application of Pisot numbers to a problem of rational approximations of algebraic numbers

The purpose of this section is to show that it is possible to characterize algebraic numbers by the existence of rational approximations with certain distribution properties. This characterization uses Pisot numbers. To generalize Lagrange's theorem (which shows that irrational quadratic numbers are characterized by periodic continued fractions) is not simple. But we observe that, assuming α is an irrational number and (p_n/q_n) the sequence of convergents to α, the sequence (p_n/q_n) contains the set of the solutions of the inequality

$$|q\alpha - p| \leq \frac{1}{2q} \qquad (p, q) \in \mathbf{Z} \times \mathbf{N}^\star. \tag{1}$$

Then Theorem 8.3.1 follows from Lagrange's theorem. Theorem 8.3.1 will then be generalized for algebraic numbers.

Theorem 8.3.1. *A real α is an irrational quadratic number if and only ifthere exists a sequence of pairs $(r_n, s_n) \in \mathbf{Z} \times \mathbf{N}^\star$ and a real number $\theta > 1$ such that, for every integer n,*

(1) $$|s_n\alpha - r_n| \leq \frac{1}{2s_n} \qquad \textit{and}$$

(2) $$|s_{n+1} - \theta s_n| < \frac{K}{s_n},$$

where K is a constant that depends only on α.

The sequence (r_n/s_n) is extracted from the convergents to α and condition (2) means that the sequence (s_n) is *almost* a geometrical progression, and this is also true for the sequence (r_n).

Proof. Suppose that α is an irrational quadratic number. The continued fraction which represents α is written $\alpha = [a_0, \dots, a_i, \overline{b_0, \dots, b_j}]$. Let (r_n/s_n) be the sequence extracted from the convergents to α defined by

$$\frac{r_n}{s_n} = \Big[a_0, \dots, a_i, \overbrace{\overline{b_0, \dots, b_j}, \dots, \overline{b_0, \dots, b_j}}^{n \text{ times}}\Big] \qquad (n \geq 0).$$

If we consider the last two convergents to the rational number $[b_0, \dots, b_j]$, we can show that the sequence (s_n) is such that $s_{n+1} = ms_n + (-1)^j s_{n-1}$ for an integer m greater than 1. It follows that $s_n = \lambda\theta^n + \mu\theta^{(2)^n}$, where θ and $\theta^{(2)}$ are the roots of the equation $X^2 - mX + (-1)^{j+1} = 0$ and thus satisfy $\theta > 1$ and $-1 < \theta^{(2)} < 1$. Then θ is an S-quadratic number, and λ and μ belong to $\mathbf{Q}(\theta)$. We then write $|s_{n+1} - \theta s_n| = |\mu(\theta - \theta^{(2)})|/\theta^n$, and there exists a constant K such that the inequality (2) is satisfied; since the fractions r_n/s_n are convergents to α, they are solutions of the inequality (1).

Suppose now that there exists a sequence (r_n/s_n) of fractions that satisfy (1) and such that the sequence (s_n) satisfies (2); then the sequence (r_n) satisfies an inequality similar to (2) with the same number θ. For we have

$$|r_{n+1} - \theta r_n| \leq \frac{K\alpha}{s_n} + \frac{\theta + 1}{2s_n} < \frac{K'}{s_n} < \frac{K"}{r_n}. \tag{2'}$$

It follows from inequality (2) that $\left|\frac{s_m}{\theta^m} - \frac{s_n}{\theta^n}\right| \leq \frac{K}{s_n(\theta - 1)\theta^n}$; hence the sequence (s_n/θ^n) is a Cauchy sequence. Let λ be its limit; then we have for n large enough $\lambda\theta^n = s_n + \varepsilon_n$ with $|\varepsilon_n| < K/(\theta - 1)\theta^n$. The series $\sum \varepsilon_n^2$ converges, and the real number θ belongs to S and λ to $\mathbf{Q}(\theta)$. The radius of convergence of the power series $\sum_{n\in\mathbf{N}} \varepsilon_n z^n$ is at least equal to θ, so we have $|\theta^{(j)}| < 1/\theta$, $(j = 2, \dots, s)$. It follows that $1 \leq |\theta\theta^{(2)} \cdots \theta^{(s)}| \leq 1/\theta^{s-2}$. This last inequality is satisfied only if $s = 2$. Then θ is a quadratic unit belonging to S, and λ is a quadratic number. In the same way it follows from inequality (2′) that the sequence (r_n/θ^n) converges to a quadratic number μ. Then $\alpha = \lim_{n\to+\infty} r_n/s_n = \lambda/\mu$ is a quadratic number. ∎

We wish to characterize algebraic numbers through properties of their rational approximations. The notion of rational approximations that are *almost* a geometric progression will be replaced by that of rational approximations that are *regularly distributed.* This characterization is based on the existence of S-numbers in every algebraic number field.

Theorem 8.3.2. *A real α is an algebraic number if and only ifthere exists a sequence of pairs $(r_n, s_n) \in \mathbf{Z} \times \mathbf{N}^\star$, a real $\theta > 1$ and a real $\eta > 0$ such that for every integer n*

$$(3) \qquad |s_n\alpha - r_n| < \frac{K_1}{s_n^\eta} \qquad \text{and}$$

$$(4) \qquad |s_{n+1} - \theta s_n| < \frac{K_2}{s_n^\eta}$$

where K_1 and K_2 are constants and depend only on α. Then the degree of α is not greater than $1 + 1/\eta$.

Proof.

Necessary condition. Let α be an algebraic number of degree s, $\alpha = \lambda/\mu$ with λ and μ algebraic integers of $\mathbf{Q}(\alpha)$. Let θ be a number of degree s in $S \cap \mathbf{Q}(\alpha)$; we set $\delta = \max_{j=2,\dots,s} |\theta^{(j)}|$. We denote (r_n) and (s_n) the sequences of integers defined by $r_n = \lambda\theta^n + \sum_{j=2}^{s} \lambda^{(j)}\theta^{(j)^n}$, and $s_n = \mu\theta^n + \sum_{j=2}^{s} \mu^{(j)}\theta^{(j)^n}$. r_n and s_n satisfy $|s_n\alpha - r_n| < c_1\delta^n$ and $|s_{n+1} - \theta s_n| < c_2\delta^n$. We define then the real η by the inequality $\theta^\eta = 1/\delta$. Inequalities (3) and (4) are satisfied, and we also have $1 \le \theta \prod_{j=2,\dots,s} |\theta^{(j)}| \le \theta\delta^{s-1} = \theta^{1-\eta(s-1)}$. Hence $\eta \le 1/(s-1)$.

a) The equality $\eta = 1/(s-1)$ is impossible, but θ is a unit all of whose remaining conjugates have the same modulus; it follows from Theorem 8.1.3 that either $s = 2$ or $s = 3$.

b) If the inequality $\eta < 1/(s-1)$ is satisfied, then we can choose θ such that $\theta\delta^{s-1} \le \sqrt{|\Delta|}$, where Δ is the discriminant of $\mathbf{Q}(\alpha)$. We deduce that $\eta \ge \frac{1}{s-1}(1 - \frac{\log|\Delta|}{2\log\theta})$. By considering θ^h with h large enough we can take η arbitrarily close to $1/(s-1)$. More precisely, let $\varepsilon > 0$; then there exists $\theta \in S$ such that inequalities (3) and (4) are satisfied and $\frac{1}{s-1} - \varepsilon < \eta \le \frac{1}{s-1}$.

Sufficient condition. We remark that the sequences (s_n/θ^n) and (r_n/θ^n) are Cauchy sequences; let μ and λ be their limits. The equality $\mu\theta^n = s_n + \varepsilon_n$ with $|\varepsilon_n| < \frac{K}{s_n^\eta(\theta - 1)}$ implies that θ belongs to S and μ to $\mathbf{Q}(\theta)$. We also see that λ is in $\mathbf{Q}(\theta)$. We then have $\alpha = \lim_{n\to+\infty} \frac{r_n}{s_n} = \frac{\lambda}{\mu}$ and α belongs to $\mathbf{Q}(\theta)$. The conjugates of θ are of modulus not greater than $1/\theta^\eta = \delta$. It follows from $1 \le \theta^{1-\eta(s-1)}$ that $s \le 1 + 1/\eta$. ∎

We remark that the sequences (r_n) and (s_n) satisfy $|s_{n+1}-\theta s_n| < c_2\delta^n$, $|r_{n+1}-\theta r_n| < c_3\delta^n$. It follows that we have, for n large enough, $r_{n+1} = E(\frac{r_n^2}{r_{n-1}})$ and $s_{n+1} = E(\frac{s_n^2}{s_{n-1}})$; (r_n) and (s_n) are Pisot sequences.

What inequalities (3) and (4) say is that algebraic numbers have rational approximations that are *regularly distributed.*

8.4 Pisot numbers and the Jacobi-Perron algorithm

The Jacobi-Perron algorithm generalizes continued fractions to finite sets of real numbers and gives simultaneous rational approximations of these numbers. The S-numbers are related to certain types of expansions.

We first recall the definitions and elementary properties used in this paragraph.

Let $\alpha_1,\dots,\alpha_n$ be n real numbers such that $1,\alpha_1,\dots,\alpha_n$ are $\mathbf{Q}$-linearly independent. We set $\boldsymbol{\alpha} = (\alpha_i)_{1\le i\le n}$ and we associate to $\boldsymbol{\alpha}$ three sequences $(\boldsymbol{\alpha}_\nu)$, $(\mathbf{a}_\nu)$ and $(\mathbf{A}_\nu)$, respectively defined in $\mathbf{R}^n$, $\mathbf{Z}^n$ and $\mathbf{Z}^{n+1}$ by the following equalities.

$$\boldsymbol{\alpha}_\nu = (\alpha_{\nu,i})_{1\le i\le n},\quad \mathbf{a}_\nu = (a_{\nu,i})_{1\le i\le n},\quad \mathbf{A}_\nu = (A_{\nu,i})_{0\le i\le n}.$$

$$\begin{cases} \alpha_{0,i} = \alpha_i & & (i=1,\dots,n) \\ \alpha_{\nu,1} = a_{\nu,1} + \dfrac{1}{\alpha_{\nu+1,n}} & \alpha_{\nu+1,n} > 1 & \\ \alpha_{\nu,i} = a_{\nu,i} + \dfrac{\alpha_{\nu+1,i-1}}{\alpha_{\nu+1,n}} & 0 < \alpha_{\nu+1,i-1} < \alpha_{\nu+1,n} & (i=2,\dots,n). \end{cases} \tag{1}$$

$$\begin{cases} A_{i,i} = 1 & (i=0,1,\dots,n) \\ A_{j,i} = 0 & (i,j=0,1,\dots,n,\ i\neq j) \\ A_{\nu+n+1,i} = A_{\nu,i} + a_{\nu,1}A_{\nu+1,i} + \cdots a_{\nu,n}A_{\nu+n,i} & (i=0,1,\dots,n). \end{cases} \tag{2}$$

The sequences $(\boldsymbol{\alpha}_\nu)$, $(\mathbf{a}_\nu)$ and $(\mathbf{A}_\nu)$ are respectively called the complete quotients, the partial quotients and the convergents to $\boldsymbol{\alpha}$. The sequences are related, and we have in particular $a_{\nu,i} = [\alpha_{\nu,i}]$; it has also been shown that $\alpha_i = \lim_{\nu\to+\infty} \frac{A_{\nu,i}}{A_{\nu,0}}$ $(i=1,2,\dots,n)$.

The sequence $(\mathbf{a}_\nu)$ of partial quotients is called the Jacobi-Perron expansion of $\alpha_1,\dots,\alpha_n$. When the sequence $(\mathbf{a}_\nu)$ is periodic, i.e., when there exist ν_0 and k

such that for $\nu \geq \nu_0$, $a_{\nu+k,i} = a_{\nu,i}$, $(i = 1\dots,n)$, the expansion is said to be periodic and k is the length of the period; the expansion is called purely periodic if moreover $\nu_0 = 0$. The main question of how to characterize the periodicity is still open, the usual conjecture being that the expansion of any basis of a real number field through this algorithm is periodic.

Let $\boldsymbol{\alpha} = (\alpha_i)_{1\leq i\leq n}$ belong to $\mathbf{R}^n$; we suppose that the expansion of $\boldsymbol{\alpha}$ is purely periodic and of length k and such that for evey integer $\nu \in \mathbf{N}$

$$\begin{cases} a_{\nu,n} \geq 1, \quad a_{\nu,n} \geq a_{\nu,i} \geq 0 & (i = 1,\dots,n-1) \\ (a_{\nu,n},\dots,a_{\nu+i,n-i}) \geq (a_{\nu,i},\dots,a_{\nu+i-1,1}) & (i = 1,\dots,n-1) \end{cases} \tag{3}$$

where the symbol $\geq$ refers to the lexicographic order. We associate to $\boldsymbol{\alpha}$ the matrix M defined by

$$M = \begin{pmatrix} A_{k,0} - X & A_{k+1,0} & \cdots & A_{k+n,0} \\ \vdots & \vdots & \ddots & \vdots \\ A_{k,n} & A_{k+1,n} & \cdots & A_{k+n,n} - X \end{pmatrix}$$

The polynomial Π defined by $\Pi(x) = (-1)^{n+1} \det M$ is called the characteristic polynomial of the expansion. We can show that it has a greatest positive zero β which is simple and that we have, when Π is irreducible, $\mathbf{Q}(\alpha_1,\dots,\alpha_n) = \mathbf{Q}(\beta)$.

Conversely, let $(\mathbf{a}_\nu)$ be a purely periodic sequence that satisfies conditions (3), and let $(\alpha_i)_{1\leq i\leq n}$ denote the real numbers defined by

$$\alpha_i = \frac{\Delta_{j,i}(\beta)}{\Delta_{j,0}(\beta)} \quad (i = 1,\dots,n), \quad j \in \{0,1,\dots,n\}, \tag{4}$$

where $\Delta_{j,i}$ are the co-factors of $\det M$. Then $\boldsymbol{\alpha}$ has $(\mathbf{a}_\nu)$ as the sequence of partial quotients.

We now turn to certain remarkable properties possessed by the Jacobi-Perron expansion when β is in S. A previous result showed that β belongs to S if and only if $\lim_{\nu\to+\infty} A_{\nu,i} - \alpha_i A_{\nu,0} = 0$. Theorem 8.4.1 is a consequence of the following lemma.

Lemma 8.4.1. *Let $(a_i)_{1\leq i\leq n}$ be non-negative integers satisfying*

$$a_n > \max(a_1,\dots,a_{n-1}). \tag{5}$$

The Jacobi-Perron purely periodic expansion $\mathbf{I}_0, \mathbf{I}_1, \dots, \mathbf{I}_n$ *defined by*

$$\mathbf{I}_0 = \mathbf{I}_1 = \cdots = \mathbf{I}_{n-1} = (0, 0, \dots, 1), \quad \mathbf{I}_n = (a_1, \dots, a_n) \tag{6}$$

has as characteristic polynomial $X^{n+1} - c_n X^n + \cdots + (-1)^n c_1 X + (-1)^{n+1}$ *with*

$$c_n = \binom{n+1}{1} + \sum_{k=1}^{n} a_k, \quad c_i = \binom{n+1}{i} + \sum_{k=1}^{i} \binom{n-1-k}{n-1-i} a_k, \quad (i = 1, \dots, n-1). \tag{7}$$

We only state the principle of the proof. We define $\mathbf{A}_\nu$ $(\nu = 0, 1, \dots, 2n+1)$ as in equations (2) and (6). We set $s_j = \sum_{k=n-j+1}^{n} a_k$ $(1 \leq j \leq n)$. The characteristic polynomial Π is

$$\Pi(X) = (-1)^{n+1} \begin{vmatrix} 1-X & 0 & \dots & 0 & 1 \\ 1 & 1-X & \dots & 0 & 1 \\ \vdots & \vdots & \ddots & \vdots & \vdots \\ s_n & s_{n-1} & \dots & s_1 & 1 + s_n - X \end{vmatrix}.$$

We deduce this result by linearly combining lines and columns and by expanding the determinant in terms of the bottom row.

Theorem 8.4.1. *Let* $(c_i)_{1 \leq i \leq n}$ *be positive integers. The integers* $(a_i)_{1 \leq i \leq n}$ *defined by the conditions* (7) *are supposed non-negative and satisfy inequality* (5)*. Let* β *be the greatest positive zero of the polynomial* Π*. We can determine numbers* α_i $(i = 1, \dots, n)$ *in* $\mathbf{Q}(\beta)$ *with purely periodic expansions of length equal to* $n+1$*,* $\mathbf{I}_0, \dots, \mathbf{I}_n$*, where the partial quotients* I_j *are defined by* (6)*.*

Furthermore if the inequality

$$c_n > 1 + \sum_{i=1}^{n-1} c_i \tag{8}$$

is satisfied, then β *is a Pisot number, the polynomial* Π *is irreducible, and the reals* $1, \alpha_1, \dots, \alpha_n$ *are a basis of* $\mathbf{Q}(\beta)$*.*

Proof. Equations (7) define a bijection between the set of n-tuples $(c_i)_{1 \leq i \leq n}$ and the set of n-tuples $(a_i)_{1 \leq i \leq n}$. Since the purely periodic development $\mathbf{I}_0, \dots, \mathbf{I}_n$ is given, the reals $(\alpha_i)_{1 \leq i \leq n}$ are determined by (4). We know that inequality (8) implies that β is a Pisot number, so we have $\mathbf{Q}(\alpha_1, \dots, \alpha_n) = \mathbf{Q}(\beta)$ and the set $1, \alpha_1, \dots, \alpha_n$ is a basis of $\mathbf{Q}(\beta)$. ∎

We now show that there exist purely periodic expansions in every real field.

Theorem 8.4.2. *In every real number field of degree $n+1$ there exists an infinity of n-tuples $(\alpha_i)_{1\le i\le n}$ whose Jacobi-Perron expansion is purely periodic and such that the set $(1, \alpha_1, \ldots, \alpha_n)$ is a basis of the field.*

Proof. It follows from the inequalities (7) that $a_i = \sum_{k=0}^{i} \mu_k^{(i)} c_i$ with $c_0 = 1$ where the constants $\mu_k^{(i)}$ depend only on n. We apply Theorem 8.1.4 with $s = n+1$. We choose γ large enough such that the conditions $c_{i+1} > \gamma c_i$ $(i = 0, \ldots, n-1)$ imply $a_i \geq 0$ $(i = 0, \ldots, n-1)$ and $a_n > \max(a_1, \ldots, a_{n-1})$. This is always possible because by (7) the condition can be written $\gamma > \gamma_0(n)$. (For example, if $n = 2$, $\gamma = 3$ works because $a_1 = c_1 - 3$ and $a_2 = c_2 - c_1$, then $c_1 \geq 3$ and $c_2 \geq 2c_1 - 2$). We then apply Theorem 8.4.1 and the result follows. ■

Corollary *The continued fraction representing a unit $\alpha > 1$ of norm 1 belonging to a real quadratic field is periodic and of the form $(a+1, \overline{1, a})$.*

Proof. The characteristic polynomial of the expansion $(\overline{1, a})$ is of the form $X^2 - (a+2)X + 1$.

Let α be a unit of norm 1; then α is a zero of a polynomial $X^2 - c_1 X + 1$ with $c_1 = \alpha + \dfrac{1}{\alpha} > 1$. If $c_1 = 2$ the field is not quadratic, and we have $c_1 \geq 3$, so we apply Theorem 8.4.1. ■

We will note that Theorem 8.4.2 is effective if we have a Pisot number in K that is a unit. It yields numerical examples in some cubic fields and in particular totally real cubic fields [14].

Notes

This chapter deals with very different questions. Apart from the fact that Pisot numbers appear in all of them, they have little in common. Our purpose was to restate results, some old, some recent; some having led to further developments whereas others have stayed isolated. Among the isolated results, Theorem 8.1.1 was proved by Grandet-Hugot and Pisot in 1958 [8]. The other theorems in § 8.1 were obtained by Boyd (unpublished) and generalized by Mignotte [12] during the past decade.

The rest of the chapter deals with two main themes: distribution modulo 1 and rational approximations. Here Pisot numbers have always played an important role. With regards to distribution modulo 1, Coquet and Mendès-France's works should be mentioned; in particular Theorem 8.2 is due to the latter. The interested reader may consult articles [2], [3], [4], [9], [10] and [11].

Rational approximations have been studied in two distinct ways: rational approximations of algebraic numbers, and simultaneous approximations related to the Jacobi-Perron algorithm. Theorem 8.3.2 on rational approximations of algebraic numbers is due to Pisot (1938) whereas those on the Jacobi-Perron algorithm were proved by Dubois and Paysant-Le Roux in 1985 [5], [6] and [7]. These latter together with the Caen Number Theory team are working at present on the characterization of families of numbers with periodic expansions. For generalities on the Jacobi-Perron algorithm the reader is referred to [1], [15] and for more recent works to [5], [6], [7], [13] and [14].

References

[1] L. BERNSTEIN, *The Jacobi-Perron Algorithm. Its Theory and Applications, Lecture Notes Math* 207, (1971), Berlin, Heidelberg, New York: Springer-Verlag.

[2] J. COQUET, Remarques sur les nombres de Pisot-Vijayaraghavan. *Acta Arith.* 22, (1977), 79-87.

[3] J. COQUET, M. MENDES-FRANCE, Sur les mesures spectrales de certaines suites arithmétiques. *Bull. Soc. Math. France*,105, (1977), 389-394.

[4] J. COQUET, M. MENDES-FRANCE, Suites à spectre vide et suites pseudo-aléatoires. *Acta Arith.* 22, (1987), 99 - 106.

[5] E. DUBOIS, Approximations diophantiennes simultanées. *Thèse*, (1960), Paris.

[6] E. DUBOIS, R. PAYSANT-LE ROUX, Développements périodiques par l'algorithme de Jacobi-Perron et nombres de Pisot. *C.R.A.S.* 272, Paris (1971), 649-652.

[7] E. DUBOIS, R. PAYSANT-LE ROUX, Algorithmes de Jacobi-Perron dans les extensions cubiques. *C.R.A.S.* 280, Paris (1975), 183-185.

[8] M. HUGOT, C. PISOT, Sur certains entiers algébriques, *C.R.A.S.* 246, Paris, (1958), 2831-2833

[9] M. MENDES-FRANCE, Nombre normaux. Applications aux fonctions pseudo-aléatoires. *Journ. Anal. Math. 20, Jérusalem*, (1967), 1-56.

[10] M. MENDES-FRANCE, Les suites à spectre vide et la répartition modulo 1. *Journ. Numb. Theory*, (1973), 1-15.

[11] M. MENDES-FRANCE, A characterizations of Pisot numbers, *Mathematika*, 23, (1976), 32-34.

[12] M. MIGNOTTE, Une propriété des nombres de Pisot, *C.R.A.S.*, Paris.

[13] R. PAYSANT-LE ROUX, Périodicité de l'algoritme de Jacobi-Perron dans un corps de séries formelles et dans le corps des nombres réels. *Monatsheft. Math. 98, Springer Verlag*, (1984), 145-155.

[14] R. PAYSANT-LE ROUX, E. DUBOIS, Une applicartion des nombres de Pisot à l'algorithme de Jacobi-Perron. *Monatsheft. Math. 98 Springer Verlag* (1984), 145-155.

[15] O. PERRON, Grundlagen für die Theorie des Jacobischen Kettenbruch algorithmus. *Math. Ann.* 64, (1907), 1-79.

CHAPTER 9

ALGEBRAIC NUMBER SETS

In this chapter we extend some properties of Pisot numbers to algebraic number sets and to n-tuples of algebraic integers.

9.1 S_q sets

Definition 9.1. *Let q denote a natural integer; a real θ greater than 1 belongs to S_q if*

(i) θ is a zero of a polynomial $P \in \mathbf{Z}[X]$: $P(X) = qX^s + q_{s-1}X^{s-1} + \cdots + q_0$, all of whose remaining zeros lie in $D(0,1)$;

(ii) there exists a polynomial $A \in \mathbf{Z}[X]$ satisfying $A(0) \geq q$ such that A/Q belongs to the family $\mathcal{F}(q,1,\delta)$ (with $\delta < 1/\theta$).

Remarks

- As the set S is included in every set S_q, none of these sets is empty.
- The polynomial P is not pre-supposed irreducible or primitive.

Theorem 9.1.1. *Every set S_q is closed in* $\mathbf{R}$.

We do not give the proof of this theorem, as it is analogous to that already proved for the set S.

Remarks

- Every $\theta \in S_q$ satisfies the inequality: $\theta \geq 1 + \dfrac{1}{4q}$. It follows that the number 1 does not belong to the set S_q'.
- The condition $A(0) \geq q$ in the definition of sets S_q, may seem artificial, but is needed for the sets to be closed. We may ask if it is not satisfied by

any polynomial P satisfying (i), but in fact, this is not the case. Consider for instance the polynomial

$$P(X) = qX^4 + (q-1)X^3 - X^2 - qX - (q-1);$$

it has three zeros in $D(0,1)$, while the fourth lies in the interval $]1, 1+\frac{1}{4q}[$, and is therefore not an element of S_q.

We now wish to characterize the set S'_q in the same manner as we already have S'; we cannot however use the same method because the polynomial P is not necessarily irreducible.

Theorem 9.1.2. *An S_q-element θ belongs to S'_q if and only if it can be associated to a function A/Q belonging to the family $\mathcal{F}'(q,1,\delta)$ satisfying the condition $A(0) \geq q$.*

Proof. The necessary condition follows from the definition, so we only prove the sufficient condition.

Let A/Q be a function belonging to $\mathcal{F}'(q,1,\delta)$ with a pole $1/\theta$ in $D(0,1)$. As the polynomial Q is not necessarily irreducible, we have two cases:

(i) $A(\theta) \neq 0$

Then A is prime to the minimal polynomial of θ and we can proceed in the same way as for S.

(ii) $A(\theta) = 0$

We can write:

$$A(X) = Pm_\theta(X)A_1(X) \qquad \text{and} \qquad Q(X) = Qm_\theta(X)Q_1(X)$$

where Pm_θ is the minimal polynomial of θ and Qm_θ is its reciprocal. Then the function A_1/Q_1 is analytic and bounded by 1 in absolute value in $\hat{D}(0,1)$.

If the sequences (P_μ) and (Q_μ) are determined as in §6.2, we note that $Q_\mu(1/\theta) = 0$ for all μ, and thus the sequence (θ_μ) is stationary. This however is not sufficient to prove that θ belongs to S'_q. For this purpose, let

$$A^\star(X) = A_1(0)N(\theta^{-1})Pm_\theta(X) + (Q_1(0) - A_1(0))Qm_\theta(X)$$
$$Q^\star(X) = qQm_\theta(X).$$

It is now easy to see that $A^\star/Q^\star$ belongs to $\mathcal{F}'(q,1,\delta)$ and that

$$A^\star(\theta) \neq 0 \qquad \text{and} \qquad |A^\star(0)| \geq q.$$

This allows us to construct a sequence of elements of S_q converging to θ. ∎

Several corollaries can be deduced from this characterization.

Corollary 1. *Let q and q_1 be positive integers. If q divides q_1, then any element of S_{q_1} belongs to S'_q. In particular, for all q every Pisot number belongs to S'_q.*

Proof. Let $\theta \in S'_{q_1}$ and an associated function $A/Q \in \mathcal{F}'(q_1, 1; \delta)$. We define $\tilde{A}$ and $\tilde{Q}$:

$$\tilde{A}(x) = aA(X) + (\frac{q}{q_1} - a)Q(X) \qquad \text{with} \qquad 1 \leq a < \frac{q}{q_1}$$
$$\tilde{Q}(X) = \frac{q}{q_1}Q(X).$$

Then $\tilde{A}/\tilde{Q}$ belongs to $\mathcal{F}'(q_1, 1; \delta)$ and has the pole $1/\theta$. This implies that θ belongs to S'_q. ∎

Corollary 2. *If θ belongs to S_q and $|N(\theta)| < 1$, then θ belongs to S'_q.*

Proof. Let A/Q be a function belonging to $\mathcal{F}(q,1,\delta)$ associated to the number θ. It is sufficient to show that $A(\theta)$ is not 0. The equation $A(\theta) = 0$ implies

$$A(X) = Pm_\theta(X)A_1(X) \qquad \text{and} \qquad Q(X) = Qm_\theta(X)Q_1(X)$$

with A_1 and Q_1 belonging to $\mathbf{Z}[X]$ and such that $|A_1(0)N(\theta)| \geq q$ and $Q_1(0) = q$. Moreover the maximum principle implies $|A_1(0)| \leq Q_1(0)$ and this is not compatible with the hypothesis. ∎

Corollary 3. *Every totally real θ belonging to S_q belongs to S'_q.*

Proof. According to Corollary 2, it is sufficient to show this result for $N(\theta) > 1$. Then the proof is similar to that of the real case (Theorem 6.2.3), and is left to the reader. ∎

At present there exist no characterizations of S''_q, or of the derived sets of superior orders. However the next theorem gives a sufficient condition for θ to belong to a derived set.

Theorem 9.1.3. *Let s and k be two natural integers, $s \geq 2$. Consider the polynomial $P \in \mathbf{Z}[X]$ defined by $P(X) = qX^s + q_1X^{s-1} + \cdots + q_s$, with*

$$q_s \geq q \tag{1}$$

$$q_1 > (k+1)q + \sum_{n=2}^{s} |q_n|. \tag{2}$$

Then P admits a single zero θ inside $D(0,1)$ and θ belongs to $S_q^{(k)}$.

Proof. The first statement follows from Rouché's theorem, the second is proved by induction on k. The proposition is true for $k = 0$, and we assume it is true for $k \leq m-1$. Let

$$P_\nu(X) = q + X^\nu P(X), \qquad \nu \in \mathbf{N}.$$

The polynomial P_ν satisfies conditions (1) and (2) for rank $m-1$, therefore θ_ν belongs to $S_q^{(m-1)}$. Then it is sufficient to show that $\lim_{\nu\to\infty} \theta_\nu = \theta$. We set

$$Q_\nu(X) = X^{s+\nu} P_\nu(X^{-1}) = qX^{s+\nu} + Q(X);$$

then P_ν/Q_ν belongs to $\mathcal{F}'(q,1,\delta)$ and the sequence (P_ν/Q_ν) converges to q/Q. The result follows. ∎

The determination of the least elements of S_q and S'_q uses the same methods as for S, i.e., considering the Taylor expansions of the associated functions $f \in \mathcal{F}(q,1,\delta)$. The sequence (u_n) of the coefficients of the Taylor expansion defines a number $\theta \in S_q$, and can be determined with the Schur algorithm. (cf. Chapter 3). The sequences $(1, 1/q, 1/q^2, \ldots)$ lead to the four smallest elements of S_q, which are

$$\begin{array}{lll}
\theta_1 & \text{zero of} & qX^3 + (q-1)X^2 - qX - q, \\
\theta_2 & \text{zero of} & qX^4 - X^3 - q, \\
\theta_3 & \text{zero of} & qX^5 - X^4 - qX^3 + qX^2 - q, \\
\theta_4 & \text{zero of} & qX^3 + (q-2)X^2 - (q-1)X - q.
\end{array}$$

We remark that θ_1, θ_2 and θ_3 are associated to the rational fractions

$$\frac{q(1-X^2) + X^\nu(q + X - qX^2)}{q - X - qX^2 + qX^\nu(1-X^2)} \qquad \text{for} \qquad \nu = 1,2,3.$$

These come from a sequence converging to $q(1-X^2)/(q-X-qX^2)$, which is associated to an element in S'_q, but not to the least such. The sequence $(1, 1/q, 1/q^2, \ldots)$ gives the smallest element of S'_q, which is associated to the rational fraction

$$\frac{q+(q-1)X-(q-1)X^2-(q-1)X^3}{q-1+qX-(q-2)X^2-qX^3}.$$

9.2 n-tuples of algebraic numbers

Definition 9.2.1. *Let k be an integer. Let $\{\theta_1, \theta_2, \ldots, \theta_k\}$ be a k-tuple of algebraic integers, each having an absolute value greater than 1. Let $P \in \mathbf{Z}[z]$ be the monic polynomial of least degree, having $\theta_1, \theta_2, \ldots, \theta_k$ as zeros. If the remaining zeros of P belong to $D(0,1)$, then $\{\theta_1, \theta_2, \ldots, \theta_k\}$ is said to be a P.V. k-tuple and P the defining polynomial of $\{\theta_1, \theta_2, \ldots, \theta_k\}$.*

(We say that two P.V. k-tuples are equal if there is set equality.)

Definition 9.2.2. $\overline{\mathrm{S}}(2)$ *is the set of non-real algebraic integers θ, which have all their conjugates other than $\overline{\theta}$ in $D(0,1)$.*

Definition 9.2.3. *If $(t_n)_{n\in\mathbf{N}} = (\{\theta_{1,n}, \theta_{2,n}, \ldots, \theta_{k,n}\})_{n\in\mathbf{N}}$ is a sequence of P.V. k-tuples, we say that $\lim\limits_{n\to+\infty} t_n = \{\theta_1^\star, \theta_2^\star, \ldots, \theta_k^\star\}$ if we can order the elements of t_n so that*

$$\lim_{n\to+\infty} \theta_i^{(n)} = \theta_i^\star \qquad (\forall i \in \{1, 2, \ldots, k\}).$$

Lemma 9.2. *Let f be a non-constant rational function in $\mathcal{M}_\infty$ with $|f(0)| \geq 1$. Let $F = \sum\limits_{i\in\mathbf{N}} u_i z^i$ be the Taylor series of f at zero. Then*

(i) If there exists $i \geq 1$ such that $u_i \neq 0$, then f has at least one pole in $D(0,1)$.

(ii) If f has k poles in $D(0,1)$, then there exists $i \in \{1, 2, \ldots, k\}$ such that $u_i \neq 0$.

Proof.

(i) If f has no pole in $D(0,1)$, by the maximum principle $|u_0| = 1$ and $f \equiv u_0$. So $u_i = 0$, $\forall i \geq 1$.

(ii) Set $P(z) = \prod_{i=1}^{k} (z - \theta_i)$ where θ_i are the poles of f in $D(0,1)$. Then $g = fP$ is analytic on $D(0,r)$ where $r > 1$. Moreover $|g(z)| \leq |P(z)|$ if $|z| = 1$. Let ℓ be the smallest integer such that $u_\ell \neq 0$. Then $g(z) = P(z)(u_0 + u_\ell z^\ell + \cdots)$ and $\mathrm{ord}(g - u_0 P) = \ell$. As $|u_0| \geq 1$, we have $|g(z)| \leq |u_0||P(z)|$ if $|z| = 1$.

By Lemma 2.2.1, we obtain $k \geq \ell$. ■

Theorem 9.2.1. *Let $t_n = (\theta_1^{(n)}, \ldots, \theta_k^{(n)})$ be a sequence of P.V. k-tuples with limit $(\theta_1^\star, \ldots, \theta_k^\star)$ and with $|\theta_i^\star| > 1$, $\forall i \in \{1, \ldots, k\}$. Then there exists some non-void subset of $(\theta_1^\star, \ldots, \theta_k^\star)$ containing k' elements with $1 \leq k' \leq k$ that is a P.V. k'-tuple.*

Proof. Let P_n be the defining polynomial of t_n and $Q_n = P_n^\star$. Since there are only a finite number of reciprocal monic polynomials $Q \in \mathbf{Z}[z]$, of degree $\leq 2k$ and whose zeros lie in a given bounded region, we can suppose that P_n/Q_n has at least one pole in $D(0,1)$.

By using Theorem 2.2.1 and Lemma 9.2, as in the proof of Theorem 6.1, we obtain the result. ■

Corollary *$\overline{\mathrm{S}}(2) \cup S$ is closed.*

Proof. Let $(t_n)_{n\in\mathbf{N}} = (\theta_n, \overline{\theta}_n)_{n\in\mathbf{N}}$ be a sequence of distinct elements of $\overline{\mathrm{S}}(2)$.

By hypothesis, $\lim_{n\to+\infty} \theta_n = \theta$, and $\lim_{n\to+\infty} \overline{\theta}_n = \overline{\theta}$.

By Smyth's theorem we have $\theta_n \overline{\theta}_n \geq \theta_0$. So $|\theta| > 1$.

By the preceding theorem either $\theta \in S$ or $(\theta, \overline{\theta}) \in \overline{\mathrm{S}}(2)$. ■

Notes

The sets S_q were defined by Pisot in 1964 [8]. His main purpose was to obtain a generalization of the set U by constructing closed sets of algebraic numbers. In the same paper he defined sets $S_q(m)$ by adding a condition on the p-adic conjugates of θ; these sets are closed.

The condition $A(0) > q$ in the definition of the sets S_q and $S_q(m)$ intervenes in the demonstration of the closure of these sets, but it is not known if the condition is optimal. The examples given in §9.1 (Prenat [9]) show that suppressing the condition leads to another set.

For the characterization of S'_q, we followed in general the proof given by Bertin [2].

Theorem 9.1.2 is due to Amara [1] and Bertin [2], and Theorem 9.1.3 to Halter-Koch [4]. We have slightly modified the demonstration. The determination of the smallest elements of S_q is due to Pisot [8] and that of the smallest element of S'_q to Amara [1]. The latter also gives a list of small elements of S_q.

We should mention that the union of all S_q is the set of algebraic numbers greater than 1, all of whose conjugates except θ belong to the unit disk (Pathiaux [6]).

Theorem 9.2.1 is due to Cantor [3] and its corollary to Kelly [5]. We are unable to prove that the P.V. k-tuples are closed and unable to find a counterexample. Pathiaux [7] proved that certain families of k-tuples (not necessarily integers) are closed and moreover, as in the case $k = 1$, the union of these families is the set of k-tuples of algebraic numbers (not necessarily integers) satisfying Definition 9.2.1.

References

[1] M. AMARA, Ensembles fermés de nombres algébriques. *Ann. Scient. Ec. Norm. Sup.* 3e Série, 83, (1966), 215-270.

[2] M.J. BERTIN-DUMAS, Caractérisation de l'ensemble dérivé de l'ensemble S_q. *C.R. Acad. Sc. Paris*, 279, (1974), A251-A254.

[3] D.G. CANTOR, On sets of algebraic integers whose remaining conjugates lie in the unit circle. *Trans. Amer. Math. Soc.*, 105,(1962), 391-406.

[4] F. HALTER-KOCH, Abgeschlossenen Mengen algebraischer Zahlen. *Ab. Sem. Hamburg*, 39, (1972), 65-79.

[5] J.B. KELLY, A closed set of algebraic integers. *Amer. J. Math.*, 72, (1950), 565-572.

[6] M. PATHIAUX, Classification en familles fermées de l'ensemble des nombres algébriques de module supérieur à un, dont tous les conjugués sont de module inférieur à un. *C. R. Acad. Sc. Paris*, 284, série A (1977), 1319.

[7] M. Pathiaux, Familles fermées de n-uples de nombres algébriques. *Thèse Université Pierre et Marie Curie (Paris VI)*, (1980).

[8] C. Pisot, Familles compactes de fractions rationnelles et ensembles fermés de nombres algébriques, *Ann. Scient. Ec. Norm. Sup.* 3e Série, 81, (1964), 165-199.

[9] M. Prenat, Sur des ensembles fermés de nombres algébriques, contenant les ensembles S_q. *C.R. Acad. Sc. Paris*, 280, (1975), A487-A488.

CHAPTER 10

RATIONAL FUNCTIONS OVER RINGS OF ADELES

In Chapter 11 we will discuss various generalizations of Pisot and Salem numbers. We will define sets with properties analogous to those of S and T, such as distribution modulo 1, or topological properties such as the fact that S is closed. The first attempt at generalization was to consider not algebraic integers but algebraic numbers that are zeros of polynomials in $\mathbf{Z}[X]$ and whose leading coefficient is a fixed integer q. These sets were discussed in Chapter 9. In fact, generalizing the distribution modulo 1 leads us to consider a more general framework: not the ring of adeles of $\mathbf{Q}$, but certain subrings that provide an appropriate domain for such investigations through the Artin decomposition.

In §10.1 and 10.2, we recall the properties necessary for our investigations. Then in §10.3 we state certain rationality criteria, and introduce, in §10.4, certain compact families of rational functions similar to those studied in Chapter 2.

10.1 Adeles of Q

Let $\mathbf{P}$ be the set of absolute values of the field $\mathbf{Q}$ of rational numbers (the archimedean absolute value is noted $|.|_\infty$ and the normed p-adic absolute value $|.|_p$).

If $p \neq \infty$, the field of p-adic numbers is noted $\mathbf{Q}_p$. Let $\mathbf{C}_p$ be the completion of the algebraic closure of $\mathbf{Q}_p$, $\mathbf{Z}_p$ the valuation ring of $\mathbf{Q}_p$, and Γ_p the valuation group, $(\mathbf{Q}_\infty = \mathbf{R}, \mathbf{C}_\infty = \mathbf{C}, \Gamma_\infty = \mathbf{R})$. For $a \in \mathbf{C}_p$ and $r \in \Gamma_p$ we set

$$D_p(a,r) = \{x \in \mathbf{C}_p;\ |x-a|_p < r\}$$
$$\hat{D}_p(a,r) = \{x \in \mathbf{C}_p;\ |x-a|_p \leq r\}$$
$$\mathbf{C}_p(a,r) = \{x \in \mathbf{C}_p;\ |x-a|_p = r\}.$$

Definition 10.1.1. *Let V be a finite set of absolute values of $\mathbf{Q}$ including the archimedean absolute value. We set*

$$A'_V = \prod_{p\in V} \mathbf{Q}_p \prod_{p\notin V} \mathbf{Z}_p \quad \textit{and} \quad \mathbf{A} = \bigcup_{V\subset \mathbf{P}} A'_V.$$

An element of $\mathbf{A}$ *is called an* **adele of Q**.

Then $\mathbf{A}$ is a unitary commutative ring with zero divisors and locally compact for the product topology. Let I be a **finite** set of valuations of $\mathbf{Q}$. We set

$$A_I = \{\alpha \in \mathbf{A}/\alpha_p = 0 \quad \text{for} \quad p \notin I\}.$$

We call A_I the ***I*-adele ring of Q** and design by e_I its unit element. Morever we remark that A_I is isomorphic to the cartesian product $\prod\limits_{p\in I} \mathbf{Q}_p$ and contains the field $\mathbf{Q}.e_I$, which is isomorphic to $\mathbf{Q}$. We identify the two fields $\mathbf{Q}$ and $\mathbf{Q}.e_I$.

Finally we note

$$I^+ = I \cup \{\infty\}, \quad I^- = I \setminus \{\infty\}, \quad \text{and} \quad Q^I = \{x \in \mathbf{Q}/|x|_p \leq 1 \quad \text{for} \quad p \notin I^+\};$$

Q^I is a Dedekind ring.

Theorem 10.1.1.

(i) The field $\mathbf{Q}$ *is dense in* A_I.

(ii) Q^I *is a discrete subring of* A_I *and the quotient* A_I/Q^I *is locally compact.*

This theorem allows us to introduce the Artin decomposition in A_I, which plays the same role as the decomposition modulo 1 did for $\mathbf{R}$,and which uses the notion of fundamental domain. We define the set F_I as follows:

$$F_I = [-\frac{1}{2}, \frac{1}{2}[\times \prod_{p\in I^-} \mathbf{Z}_p \cong A_I/Q^I.$$

Theorem 10.1.2. *Every element* α *in* A_I *can be expressed in one and only one way as* $\alpha = E(\alpha) + \varepsilon(\alpha)$ *with* $E(\alpha) \in Q^I, \varepsilon(\alpha) \in F_I$.

In Chapter 14 we will use Artin decompositions with the fundamental domain modified slightly.

Notation

We set $\mathbf{C}_I = \prod\limits_{p\in I} \mathbf{C}_p$. Then $\mathbf{C}_I$ is a non-locally compact $\mathbf{Q}$-algebra. For $a = (a_p)_{p\in I} \in \mathbf{C}_I$ and $r = (r_p)_{p\in I}$ with $r_p \in \Gamma_p$ we define

$$\hat{D}_I(a,r) = \prod_{p\in I} \hat{D}_p(a_p, r_p) \quad \text{and} \quad D_I(a,r) = \prod_{p\in I} D_p(a_p, r_p).$$

Finally, the relations ($|x|_p > 1, \ \forall p \in I$) and ($|x|_p \geq 1, \ \forall p \in I$) are respectively denoted $\|x\| \gg 1$ and $\|x\| \ggg 1$.

We are now able give a few indications of the properties of the algebraic elements of A_I.

Let α be an algebraic element of A_I. Then the monic polynomial of minimal degree that vanishes at α is called the **minimal polynomial** of α and noted Pm_α^I. If Pm_α^I is not irreducible, then there exists a partition $(I_h)_{h=1,\dots,m}$ of I such that if we set $\alpha_h = \alpha.e_{I_h}$, the polynomials $Pm_{\alpha_h}^{I_h}$ are irreducible and we have $Pm_\alpha^I(X) = \prod_{h=1}^{n} Pm_{\alpha_h}^{I_h}(X)$. We will say that (I_h) is the **partition of I associated to α**. An algebraic element α of A_I is called an **algebraic integer** if Pm_α^I belongs to $Q^I[X]$. The extension of $\mathbf{Q}$ obtained by adding an algebraic element α of A_I is denoted $\mathbf{Q}_I[\alpha]$; this subring of A_I is a field if and only ifthe polynomial Pm_α^I is irreducible. One can now prove the following theorem.

Theorem 10.1.3. *In every ring $\mathbf{Q}_I[\alpha]$ there exists a base made up of algebraic integers, called a* **minimal base**.

As in the real case, we will use *Minkowski's theorem*, which can be stated in the following form.

Theorem 10.1.4. *Suppose that I includes the archimedean absolute value and consider a system of n linear forms $(L_i)_{1\leq i\leq n}$ with coefficients in A_I whose determinant is not zero. Let $(\delta_{i,p})_{1\leq i\leq n}, p \in I$ be real positive numbers satisfying the following inequality:*

$$\prod_{p\in I}\prod_{i=1}^{n}\delta_{i,p} \geq \prod_{p\in I}|\Delta|_p.$$

Then there exists a non-zero element $\mathbf{x}$ of $(Q^I)^n$ satisfying the inequalities

$$|L_i(\mathbf{x})|_p \leq \delta_{i,p} \quad \textit{for } 1 \leq i \leq n,\ p \in I.$$

It is possible to define several types of uniform distribution in A_I and we will restrict ourselves to *uniform distribution modulo Q^I*, which has a relationship with the elements of Pisot. In order to simplify the statement of the theorems, we will define this type of uniform distribution using *Weyl's criterion*, and simply state without proof the classical corollaries of Koksma's theorem. A more detailed treatment can be found in the papers cited in the references.

Definition 10.1.2. *A sequence (x_n) in A_I is said to be* **uniformly distributed modulo Q^I** *if, for all $a \in Q^I$,*

$$\lim_{N\to+\infty} \frac{1}{N}\sum_{n=1}^{N} \exp(2\pi i \varepsilon_\infty(ax_n)) = 0.$$

From this definition we immediately deduce

Theorem 10.1.5. *Let (x_n) be a sequence of elements of A_I uniformly distributed modulo Q^I. Then the sequence $(\varepsilon_\infty(x_n))$ is uniformly distributed modulo 1.*

Theorem 10.1.6. (Koksma's theorem)

(i) Let $\alpha \in A_I$, with $\|\alpha\| \gg 1$. Then the sequence $(x\alpha^n)$ is uniformly distributed modulo Q^I for almost every inversible element x in A_I.

(ii) Let λ be an inversible element of A_I. Then the sequence (λx^n) is uniformly distributed modulo Q^I for almost all $x \in A_I$ with $\|x\| \gg 1$.

The **exceptional set** of Koksma's theorem is the set of all $x \in A_I$ with $\|x\| \gg 1$, for which the sequence (x^n) is not uniformly distributed modulo Q^I.

10.2 Analytic functions in $\mathbf{C}_p$

The purpose of this section is to state certain properties of analytic p-adic functions, which will be used for studying functions with a variable in A_I or in $\mathbf{C}_I$.

All the functions considered are analytic in a neighborhood of zero in $\mathbf{C}_p$ and have values in $\mathbf{C}_p$. To each such function f we associate the formal power series $S(f)$ of its Taylor expansion at zero.

Theorem 10.2.1. *Let f be an analytic function on $\hat{D}_p(0,r), r \in \Gamma_p$. We set*

$$S(f) = \sum_{n\in\mathbf{N}} a_n X^n \quad \text{and} \quad M_f(r) = \sup_{|x|_p=r} |f(x)|_p.$$

We then have the following equalities:

$$M_f(r) = \sup_{n\in\mathbf{N}} |a_n|_p r^n \quad \textit{(Cauchy's inequalities)} \tag{1}$$

$$M_f(r) = \sup_{|x|_p=r} |f(x)|_p \quad \textit{(maximum modulus principle).} \tag{2}$$

Before stating an important result on the convergence of sequences of analytic functions, we recall the following definition.

Definition 10.2.1. *Let (S_ν) be a sequence of formal power series, $S_\nu = \sum_{n\in\mathbf{N}} a_{\nu,n}X^n \in \mathbf{C}_p[[X]]$; we say that this sequence is weakly convergent to $S = \sum_{n\in\mathbf{N}} a_nX^n$ if $a_n = \lim_{\nu\to+\infty} a_{\nu,n}$, $\forall n \in \mathbf{N}$.*

Let (f_ν) be a sequence of analytic functions in the disk $D_p(0,r)$. If the sequence $(S(f_\nu))$ is weakly convergent to a formal series S representing a function f analytic in a neighborhood of zero, then we can assume that the sequence (f_ν) is weakly convergent to f. In the following theorem we set $r = 1$.

Theorem 10.2.2. *Consider a sequence of functions $f_\nu : \hat{D}_p(0,1) \mapsto \mathbf{C}_p$.*

1) If the sequence (f_ν) is uniformly convergent to a function f on $\hat{D}_p(0,1)$, then

(i) The sequence (f_ν) is uniformly bounded and f is analytic on $\hat{D}_p(0,1)$.

(ii) The sequence $(S(f_\nu))$ converges weakly to $S(f)$.

2) If the sequence (f_ν) is uniformly bounded on $D_p(0,1)$ and if the sequence $(S(f_\nu))$ converges weakly to a formal series S, then

(i) The series S represents a function f analytic on $D_p(0,1)$.

(ii) The sequence (f_ν) converges to f uniformly on every circled disk in $D_p(0,1)$.

By analogy with the complex case, we introduce the notion of **function of bounded characteristic**:

Definition 10.2.2. *Let Δ be a subset of $\mathbf{C}_p$. A function $f : \Delta \mapsto \mathbf{C}_p$ is said to be of bounded characteristic if*

(i) f is analytic on Δ, except for isolated points.

(ii) f can be written as the quotient of two bounded analytic functions.

A computation analogous to those of Lemma 1.2.1 allows us to state

Theorem 10.2.3. *Let f be an analytic function in the neighborhood of zero and of bounded characteristic in $D_p(0,r)$. Then the Kronecker determinant of order n satisfies*

$$r^n|D_n(f)|_p^{1/n} = O(1).$$

10.3 Rationality criteria in $Q^I[[X]]$

In A_I as in $\mathbf{R}$, Pisot and Salem sets are associated to certain families of rational functions, so the importance of finding criteria for rationality is clear.

Because of its algebraic character, *Kronecker's criterion* (Theorem 1.1.1) can be applied without modification. *Cantor's algebraic criterion* (Theorem 1.1.2) is related to particular properties of $\mathbf{Z}$, so it has to be modified for applications (Theorem 10.3.1). Finally the generalization of the notion of a function with a bounded characteristic allows us to state an *analytic criterion* for rationality (Theorem 10.3.2).

Let (u_n) be a sequence of elements of Q^I. Then, as in §1.1, one can associate to (u_n) a family of matrices $A(L_n)$ in the following way: $\mathcal{L}_n$ denotes the set of increasing sequences of n natural integers $L_n = (0, \ell_1, \ldots, \ell_n)$, where $0 = \ell_0 < \ell_1 < \ldots < \ell_n$; for a sequence (t_n) of elements of A_I satisfying $t_0 = e_I$ we set

$$x_{m,n} = \sum_{i=0}^{m} \sum_{j=0}^{n} t_i t_j u_{m+n-(i+j)}, \quad (m,n) \in \mathbf{N}^2$$

and

$$A(L_n) = (x_{\ell_i,j}), \quad 0 \le i \le m; \quad 0 \le j \le n.$$

Theorem 10.3.1. *Let (u_n) be a sequence of elements of Q^I; the series $\sum_{n\in\mathbf{N}} u_n X^n$ is rational if and only ifthere exists a sequence (t_n) of elements of A_I satisfying $t_0 = e_I$, and a natural integer s such that*

$$\prod_{p\in I^+} |\det A(L_s)|_p < 1 \quad for \quad all \quad L_s \in \mathcal{L}_s.$$

The proof, which is similar to that of Theorem 1.1.1, is left to the reader.

Consider now a formal series $S = \sum_{n\in\mathbf{N}} u_n X^n \in Q^I[[X]]$. For $p \in \mathbf{P}$, we designate by ρ_p its radius of convergence in $\mathbf{C}_p$. Then

a) For $p \in I^+$ if $\rho_p > 1$ the series S represents an analytic function f_p on $D_p(0,1)$.

b) For $p \notin I^+$, $\rho_p \ge 1$ and the series S represents an analytic function f_p on $\hat{D}_p(0,1)$.

Set $f = (f_p)_{p\in I}$ and $\rho = (\rho_p)_{p\in I}$. Then $f : D_I \mapsto \mathbf{C}_I$ is analytic (its restriction to A_I is also noted f).

Let $\mathcal{H}_I$ be the set of functions $f = (f_p)_{p\in I} : \Delta \subset \mathbf{C}_I \to \mathbf{C}_I$ analytic in a neighborhood of zero and such that $S(f)$ belongs to $Q^I[[X]]$.

Definition 10.3.1. *A function $f \in \mathcal{H}_I$ is said to be of* **bounded characteristic** *on $D_I(0,r)$ if, for every $p \in I^+$, f_p is of bounded characteristic on $D_p(0,r_p)$ (where $r = (r_p)_{p\in I^+}$).*

Theorem 10.3.2. *If a function $f \in \mathcal{H}_I$ is of bounded characteristic in $D_I(0,r)$ with $\prod_{p\in I^+} r_p \geq 1$ then it is rational.*

Moreover, f can be written as the quotient of two coprime polynomials C and D, with coefficients in Q^I, such that D does not vanish in the disk $D_p(0,1)$ for $p \notin I^+$ and that $D(0) = 1$.

Proof. Lemma 1.1.1 and Theorem 10.2.3 allow us to write $r_\infty |D_n(f)|_\infty^{1/n} = o(1)$ and $r_p |D_n(f)|_p^{1/n} = O(1)$ for $p \in I^+$, hence $\prod_{p\in I^+} r_p |D_n(f)|_p^{1/n} = o(1)$. Then $D_n(f) = 0$ for n large enough. By Kronecker's criterion f is rational, and the last statement follows from Fatou's lemma. ∎

10.4 Compact families of rational functions

For $p \in I^+$ we denote by h_p a natural integer ($h_\infty = 0$ if $\infty \notin I$) and by (r_p, δ_p) a pair of elements of Γ_p satisfying $\delta_p \leq r_p$. We set

$$r = (r_p)_{p\in I^+}, \quad \delta = (\delta_p)_{p\in I^+}, \quad \text{and} \quad h = (h_p)_{p\in I^+}.$$

All these functions belong to $\mathcal{H}_I$ and satisfy, for every $p \in I^+$, one of the following conditions:

<u>Condition</u> A_p: There exists a real number $m_p(f)$ such that

$$|x|_p = r_p \Rightarrow |f(x)|_p \leq m_p(f).$$

<u>Condition</u> B_p: There exists $r'_p \in \Gamma_p$ and $m_p(f) \in \mathbf{R}$ such that

$$r'_p < |x|_p < r_p \Rightarrow |f(x)|_p \leq m_p(f).$$

Definition 10.4.1. *One designates by $\mathcal{F}_I(r,h,\delta)$ the family of rational functions $f \in \mathcal{H}_I$, analytic on $D_I(0,1)$ with the following properties:*

1) *For $p \in I^-$, f_p satisfies one of the two conditions A_p or B_p.*
2) *f_∞ satisfies A_∞.*
3) *For all $p \in I$, f_p admits at most h_p poles in $D_p(0,r_p)$, and the poles belong to $\mathbf{Q}_p$.*

Definition 10.4.2. *We call a set Φ of functions f a* **bounded family** *in $\mathcal{F}_I(r,h,\delta)$ when $(m_p(f))_{f\in\Phi}$ is bounded for all $p \in I^+$.*

Theorem 10.4.1. *If $\prod_{p\in I^+} r_p \geq 1$ then every bounded family in $\mathcal{F}_I(r,h,\delta)$ is compact for the topology of uniform convergence on every compact of $D_I(0,\delta)$.*

Proof. We will prove that from every bounded sequence of elements of $\mathcal{F}_I(r,h,\delta)$ one can extract a subsequence uniformly convergent to a function of $\mathcal{F}_I(r,h,\delta)$ on every circled disk in $D_I(0,\delta)$ (and *a fortiori* on every compact set). Let (f_ν) be a bounded sequence of elements of $\mathcal{F}_I(r,h,\delta)$. One can suppose that each function $f_{\nu,p}$ possesses exactly k_p poles in $D_p(0,r_p)$, $0 \leq k_p \leq h_p$, which we denote $\alpha_{\nu,p}^{(1)},\ldots,\alpha_{\nu,p}^{(k_p)}$. Set

$$m_p = \sup_\nu m_p(f_\nu) \text{ and } S(f_\nu) = \sum_{n\in\mathbf{N}} u_{\nu,n}X^n, \quad u_{\nu,n} \in Q^I.$$

Now we examine the different components of the functions f_ν.

a) $p = \infty$

We use the same method as in the proof of Theorem 2.2. We only indicate the results which will be of use later.

Given $\eta_\infty \in]0,\delta_\infty[$, there exists a constant M_∞ for which the following inequalities are satisfied

$$|u_{\nu,n}|_\infty \leq M_\infty \eta_\infty^{-n} \text{ for } n \in \mathbf{N}. \tag{1}$$

Moreover, the sequence $(f_{\nu,\infty})$ is uniformly convergent to a function f_∞ of bounded characteristic in $D_\infty(0,r_\infty)$on every compact set of $D_\infty(0,\delta_\infty)$.

b) $p \in I^-$

We set

$$\varphi_{\nu,p}(x) = \begin{cases} 1 & \text{if } k_p = 0, \\ \prod_{j=1}^{k_p}\left(1 - \frac{x}{\alpha_{\nu,p}^{(j)}}\right) & \text{if } k_p \neq 0; \end{cases}$$

the coefficients of these polynomials belong to $\mathbf{Q}_p$ and are bounded. Therefore one can extract from the sequence $(\varphi_{\nu,p})$ a sequence of polynomials with the same Newton polygon, and this new sequence is compact for the weak topology. Hence there exists a subsequence, uniformly convergent on every compact of $\mathbf{Q}_p$, and therefore on every circled disk of $\mathbf{C}_p$. This subsequence will be noted $\varphi_{\nu,p}$.

We define then a sequence $(F_{\nu,p})$ of analytic functions in $D_p(0,r_p)$ by $F_{\nu,p}(x) = \varphi_{\nu,p}(x)f_{\nu,p}(x)$.

Now we distinguish two cases according to whether condition A_p or condition B_p is satisfied for every $\nu \in \mathbf{N}$.

1. *Condition* A_p *is satisfied.* In this case the functions $f_{\nu,p}$ have no pole on $C_p(0,r_p)$ and hence the functions $F_{\nu,p}$ are analytic on $\hat{D}_p(0,r_p)$.

Let $\eta_p \in \Gamma_p,\ \ 0 < \eta_p < \delta_p$. According to Theorem 10.2.1 we can write

$$\sup_{|x|_p=\eta_p} |f_{\nu,p}(x)|_p = \sup_{|x|_p=\eta_p} |\varphi_{\nu,p}(x)f_{\nu,p}(x)|_p \le \sup_{|x|_p=r_p} |F_{\nu,p}(x)|_p;$$

and from the definition of $F_{\nu,p}$

$$\sup_{|x|_p=r_p} |F_{\nu,p}(x)|_p \le m_p\Big(\frac{r_p}{\delta_p}\Big)^{k_p} = M_p.$$

2. *Condition* B_p *is satisfied.* Then the functions $f_{\nu,p}$ may have poles on $C_p(0,r_p)$. Let η_p and r_p'' be two elements of Γ_p satisfying the inequalities $0 < \eta_p < \delta_p$ and $r_p' < r_p'' < r_p$. Theorem 10.2.1 allows us to write

$$\sup_{|x|_p=\eta_p} |f_{\nu,p}(x)|_p < \sup_{|x|_p=\eta_p} |F_{\nu,p}(x)|_p < \sup_{|x|_p=r_p''} |F_{\nu,p}(x)|_p;$$

and by the definition of $F_{\nu,p}$

$$\sup_{|x|_p=1} |F_{\nu,p}|_p \le m_p\Big(\frac{r_p}{\delta_p}\Big)^{k_p} = M_p.$$

In the above cases Theorem 10.2.1 allows us to write

$$|u_{\nu,p}|_p \le M_p\eta_p^{-n} \quad \text{for} \quad n \in \mathbf{N}. \tag{2}$$

From inequalities (1) and (2) we deduce

$$H(u_{\nu,p}) = \prod_{p\in I^+} \max(1,|u_{\nu,p}|_p) \le \prod_{p\in I^+} \max(1, M_p\eta_p^{-n}).$$

Therefore, for every index n, the sequence $(u_{\nu,n})_\nu$ is stationary from a certain rank on, hence the sequence $S(f_\nu)$ is weakly convergent to a formal power series $S = \sum_{n\in\mathbf{N}} u_nX^n \in Q^I[[X]]$.

For every $p \in I^-$, $(f_{\nu,p})$ is a uniformly bounded sequence of analytic functions on $\hat{D}_p(0,\eta_p)$, hence it converges uniformly to a function f_p analytic in $\hat{D}_p(0,\eta_p)$ on every circled disk contained in $D_p(0,\eta_p)$ (Theorem 10.2.2). This added to the uniform convergence of the sequence $(f_{\nu,\infty})$ on every compact set of $D_\infty(0,\delta_\infty)$ implies the uniform convergence of the sequence (f_ν) to $f = (f_p)_{p\in I}$ on every compact set of $D_I(0,\delta)$.

Morever according to Theorem 10.2.1, for $p \in I^-$ the sequence $(F_{\nu,p})$ is uniformly bounded on every circled disk in $D_p(0,r_p)$. Hence it converges uniformly on $D_p(0,r_p)$ to a function F_p and $F_p(x) = \varphi_p(x) f_p(x)$. This implies the inequality $\sup_{|x|_p<r_p} |F_p(x)|_p \leq M_p$.

The function f is then of bounded characteristic in $D_I(0,r)$, and hence is rational (Theorem 10.3.2). Examination of possible poles of f and Fatou's lemma allows us to finish the proof in the same way as in the complex case. ■

Remarks

1. It is easy to see that if for $p \in I^-$ the functions $f_{\nu,p}$ satisfy condition B_p, this also true for f_p, whereas if the functions $f_{\nu,p}$ satisfy condition A_p then the function f_p will satisfy either A_p or B_p.

2. One may ask whether a connection exists between the families $\mathcal{F}(q,h,\delta)$ studied in Chapter 2 and the real components of the functions studied in this section.

Take $r_p = 1$ for $p \in I^+$, and consider a convergent sequence of functions (f_ν) of $\mathcal{F}_I(r,h,\delta)$. The inequalities (2) show that there exists an integer q for which the functions $f_{\nu,\infty}$ belong to the family $\mathcal{F}(q,h_\infty,\delta_\infty)$.

Conversely, for $q \in \mathbf{N}^\star \backslash \{1\}$, set $I = \{p \in \mathbf{P}/p|q\} \cup \{\infty\}$. A function $\hat{f}$ belonging to a family $\mathcal{F}_I(r,h',\delta'')$ such that $\hat{f}_\infty = f, r = 1, h'_\infty = h$ and $\delta'_\infty = \delta$, can be associated to every function $f \in \mathcal{F}_I(q,h,\delta)$. But, given a convergent sequence (f_ν) of functions of $\mathcal{F}(r,h,\delta)$, one cannot ensure that $h'_{\nu,p}$ is bounded, as is shown by the following example. Let

$$f_\nu(x) = \frac{9 + x^\nu(15x-25)}{15 - 25x + 9x^{\nu+1}}, \quad \nu \in \mathbf{N}^\star.$$

These functions belong to a family $\mathcal{F}_I(15,1,\delta)$ and $I = \{3,5,\infty\}$, but the number of poles of f_ν in $D_5(0,1)$ is not bounded.

In Chapter 11 we will consider families of functions for which $r_p = 1$ for $p \in I^+$ and $h_p = 1$ for $p \in I$, which we note $\mathcal{F}_I(\delta)$. On can prove the following result:

Theorem 10.4.2. *The limit functions of a family $\mathcal{F}_I(\delta)$ possess the following property: the equality $|f(x)|_\infty = 1$ is satisfied in at most a finite number of points of $\mathbf{C}_\infty(0,1)$.*

The preceding example shows that the classical method is not sufficient for determining if the condition is sufficient.

Notes

In this chapter we indicated only the notions indispensable for understanding Chapters 11 and 15. The reader desiring further information is directed to the books of Goldstein [5] and Lang [12].

The algebraic elements of A_I were studied in 1965 by F. Bertrandias [2], and Minkowski's theorem for the adeles of an algebraic number field was also proved in 1965 by Cantor [3].

Uniform distribution modulo Q^I in A_I was defined by Grandet-Hugot [6], [7], and is based on the more general notion of uniform distribution in a locally compact group (cf. Kuipers and Niederreiter [11]).

For analytic functions in $\mathbf{C}_p$ one should consult the works of Amice [1] and Dwork [4]. The notion of a function of bounded characteristic was introduced by Grandet-Hugot [8], who also proved in 1987 the rationality criterion stated in Theorem 10.1.1 [10]; an other rationality criterion can be found in Vitiello [13].

A systematic study of compact families of rational functions is found in [8].

References

[1] Y. Amice, *Les nombres p-adiques*, P.U.F., Coll. Sup, (1975), 189 pages.

[2] F. Bertrandias, Ensembles remarquables d'adèles algébriques, *Bull. Soc. Math. France, Mémoire* 4, (1965).

[3] D.G. Cantor, On the elementary theory of diophantine approximation over the ring of adèles I, *Ill. J. Math.*, 9, (1965), 677-700

[4] B. Dwork, On the zeta function on a hypersurface, *Publ. Math. I.H.E.S.*, 12, (1962), 5-68.

[5] L.J. Goldstein, *Analytic Number Theory*, Prentice Hall Inc., (1972), 282 pages.

[6] M. Grandet-Hugot, Quelques résultats concernant l'équirépartition dans l'anneau des adèles d'un corps de nombres algébriques, *Bull. Sc. Math.* 99, (1975), 91-111 and 243-247.

[7] M. Grandet-Hugot, Etude des différents types d'équirépartition dans un anneau d'adèles, *Bull. Sc. Math.* 110, (1979), 349-360.

[8] M. Grandet-Hugot, P.V. eléments dans un corps de nombres algébriques, *Acta Arith.* 20, (1972), 204-214.

[9] M. Grandet-Hugot, Fonctions à caractéristique bornée et P.V. éléments, *Acta Arith.*, 34, (1979), 349-360.

[10] M. Grandet-Hugot, Une nouvelle caractérisation des éléments de Pisot dans l' anneau des adèles de **Q**, *Act Arith.*, 52, (1989), 27-37.

[11] L. Kuipers and H. Niederreiter, *Uniform Distribution of Sequences*, J. Wiley and Sons, (1974), 390 pages.

[12] S. Lang, *Algebraic Number Theory*, Addison Wesley; (1970), 354 pages.

[13] P. Vitiello, Etude de la fermeture de certains ensembles d'entiers algébriques, *Thèse de 3è cycle, Université de Paris* VII, (1977).

CHAPTER 11

GENERALIZATIONS OF PISOT AND SALEM NUMBERS TO ADELES

Most of the notation used in this chapter was introduced in Chapter 10.

As in the real case, the study of Pisot and Salem elements is connected to sequences $(\lambda\alpha^n)$, $(\alpha, \lambda) \in A_I \times A_I$ having remarkable distributional properties modulo Q^I.

Let (α, λ) be a pair of elements of A_I such that $\|\alpha\| \gg 1$ (if $\infty \notin I$, we set $\alpha_\infty = 0$ and $\lambda_\infty = 1$). Then it can be easily proved that there exists a natural integer q such that for n large enough, $q^{n+1}E(\lambda\alpha^n)$ belongs to $\mathbf{Z}$. In the following sections we set $u_n = E(\lambda\alpha^n)$. Moreover, as in the real case, two distinct sequences $(\lambda\alpha^n)$ and $(\lambda'\alpha'^n)$ cannot generate the same sequence (u_n).

11.1 Definition of the set U_I

This set plays a role analogous to that of the set U in the real case; in particular, U_I is included in the exceptional set of Koksma's Theorem. However, its study reveals noticeable differences with the set U.

Definition 11.1. *Let U_I be the set of algebraic elements α of A_I satisfying $\|\alpha\| \gg 1$, and such that the zeros of the minimal polynomial of α lie in $\hat{D}_p(0,1)$, with the exception of α_p if $p \in I$.*

We notice that U_I is never empty, since every element α of Q^I satisfying $\|\alpha\| \gg 1$ belongs to U_I. For $\alpha \in U_I$ we set $q = \prod_{p \in I^-} |\alpha|_p$ and

$$P(X) = qPm_\alpha^I(X) = qX^s + q_1X^{s-1} + \cdots + q_0$$
$$P^+(X) = \varepsilon P(X)$$

where $\varepsilon = \pm 1$ is chosen so that $P^+(0) > 0$.

By considering the Newton polygon of P in $\mathbf{Q}_p$, for $p \in I^-$ we notice that P is primitive and that $|q_{s-1}|_p = 1$ for $p \in I^-$.

The polynomial P is called **the associated polynomial to the element α of U_I**

If $Pm^I_\alpha(X)$ is not irreducible, we designate by $(I_h(\alpha))_{h=1,\ldots,m}$ the corresponding partition of I (when there is no ambiguity, we note more simply (I_h)), then $Pm^I_\alpha(X) = \prod_{h=1}^{m} Pm^{I_h}_{\alpha_{I_h}}(X)$ and $\alpha = \sum_{h=1}^{m} \alpha_{I_h}$, where $\alpha_{I_h} \in U_{I_h}$.

Conversely let $(I_h)_{h=1,\ldots,m}$ be a partition of I and assume that for $h = 1, \ldots, m$ there exists an element $\alpha_h \in U_{I_h}$ whose minimal polynomial is irreducible. Then the element α defined by $\alpha = \sum_{h=1}^{m} \alpha_h$ belongs to U_I, and the identity

$Pm^I_\alpha(X) = \prod_{h=1}^{m} Pm^{I_h}_{\alpha_h}(X)$ holds.

We can now state the following result.

Theorem 11.1. *Let (α, λ) be a pair of elements of A_I satisfying the inequalities $\|\alpha\| \gg 1$, $\|\lambda\| \gtrsim 1$, ($\alpha_\infty > 1$ and $\lambda_\infty \geq 1$ if $\infty \in I$). We set*

$$q = \prod_{p \in I^-} |\alpha|_p, \quad \ell = \prod_{p \in I} |\lambda|_p.$$

If the inequality (A) *holds for every integer $n \geq 0$, then α belongs to U_I and λ belongs to $\mathbf{Q}_I[\alpha]$.*

$$|\varepsilon(\lambda\alpha^n)|_\infty \leq \frac{1}{2eq^2(1+\alpha_\infty^2)(1+\log \ell)}. \tag{A}$$

This theorem is the analogue of Theorem 5.5.1 for A_I, and the principle of the proof is the same. The proof is left to the reader. Condition (A) can be replaced by one of the two conditions (B) and (C):

(B) *There exists a non empty subset J of I^+ such that, for every $n \in \mathbf{N}$*

$$\prod_{p \in J} |\varepsilon(\lambda\alpha^n)|_p \leq \frac{1}{2eq^2(1+\alpha_\infty^2)(1+\log \ell)}.$$

(C) *There exists an absolute value $p' \notin I$ such that, for every $n \in \mathbf{N}$*

$$|E(\lambda\alpha^n)|_{p'} \leq \frac{1}{2eq^2(1+\alpha_\infty^2)(1+\log \ell)}.$$

11.2 Subsets of U_I and characterizations

Throughout this section, some remarkable subsets of U_I will be defined.

Definition 11.2.1. *For $p' \in \mathbf{P}$ let $S_I^{p'}$ be the set of elements θ of U_I such that all zeros of the associated polynomial P in $\mathbf{C}_{p'}$ lie in the disk $D_{p'}(0,1)$ (except $\theta_{p'}$ if $p' \in I$).*

Then the polynomial P has no cyclotomic factor. Let J be a finite subset of $\mathbf{P}$. We denote $S_I^J = \bigcap_{p' \in J} S_I^{p'}$.

These sets possess properties close to those of classical Pisot numbers, and in particular the set S_I^∞, which will be called the **Pisot set of A_I.**

We have seen that the polynomial associated to an element α of U_I is not necessarily irreducible; hence the existence of a zero on the unit circle does not imply that P is reciprocal. Here, defining Salem elements will be more difficult than in $\mathbf{R}$.

Definition 11.2.2. *Let D_I be the set of elements σ of U_I associated to a polynomial P, where P is the product of two polynomials P_1 and P_2 such that*

- *the zeros of P_1 in $\mathbf{C}$ (except σ_∞ if $\infty \in I$) lie in the disk $D_\infty(0,1)$,*
- *the polynomial P_2 is a reciprocal polynomial of even degree and its complex zeros (except σ_∞ and $1/\sigma_\infty$ if $\infty \in I$) lie in $C_\infty(0,1)$.*

Let (I_1, I_2) be the partition associated to the decomposition of P as the product of P_1 and P_2. The definition implies that, if I_2 contains the archimedean absolute value, then the minimal polynomial of σ_{I_2} is of degree at least four.

Definition 11.2.3. *Let T_I be the set of elements τ of U_I associated to a reciprocal polynomial P of even degree and all of whose complex zeros, except possibly τ_∞ and $1/\tau_\infty$, lie on $C_\infty(0,1)$.*

This definition implies that if I contains the archimedean valuation, then the degree of P is at least four.

The set T_I is called the **Salem set of A_I**. Here the Pisot and Salem elements do not constitute a partition of U_I, as is shown in the next theorem.

Theorem 11.2.1. *The three sets S_I^∞, T_I and D_I constitute a partition of U_I.*

Proof. Consider an element α of U_I and the associated polynomial P. Two cases can arise.

- All complex zeros of P, except possibly α_∞, lie in $D_\infty(0,1)$. Then α belongs to S_I^∞.
- The polynomial P has at least one zero on $C_\infty(0,1)$, and, if it is not itself a reciprocal polynomial of even degree (at least equal to four if $\infty \in I$), it has a divisor of this type. In the first case α belongs to T_I, and in the second it is an element of D_I.

■

We will show now that every algebraic extension of $\mathbf{Q}$ included in A_I can be generated by a Pisot element, which generalizes a property of the set S.

Theorem 11.2.2. *Let I and J be two finite subsets of $\mathbf{P}$. Then every ring of algebraic elements in A_I can be generated by an element of S_I^J.*

Proof. Let $\mathbf{Q}_I[\alpha]$ be a ring of algebraic elements of A_I. We distinguish two cases: $\mathbf{Q}_I[\alpha]$ is a field and $\mathbf{Q}_I[\alpha]$ is not a field. In the first case the minimal polynomial of α is irreducible.

1. $\mathbf{Q}_I[\alpha]$ *is a field*

Denote by $\omega = (\omega_1, \dots, \omega_s)$ a base of $\mathbf{Q}_I[\alpha]$ composed of algebraic integers and let $\omega_{k,p}^{(j)}$, $(j = 1, \dots, s)$ be the zeros of $Pm_{\omega_k}^I$ in $\mathbf{C}_p$ for $p \in I$. Then, $|\omega_{k,p}^{(j)}|_p \leq 1$ for $p \notin I$ and $j = 1, \dots, s$.

We wish to determine the elements θ in A_I such that

$$\theta = \Theta(\omega; x) = \sum_{k=1}^{s} x_k \omega_k, \quad x_k \in Q^I.$$

We set

$$\theta_p^{(j)} = \Theta_p^j(\omega; x) = \sum_{k=1}^{s} x_k \omega_k^{(j)}, \begin{cases} j = 2, \dots, s & \text{for } p \in I, \\ j = 1, \dots, s & \text{for } p \notin I. \end{cases}$$

We should have

$$\begin{cases} |\theta|_p > 1 & \text{for } p \in I \\ |\theta^{(j)}|_p \leq 1 & \text{for } p \in I,\ j = 2, \ldots, s \\ |\theta^{(j)}|_p < 1 & \text{for } p \in J \begin{cases} j = 1, \ldots, s & \text{if } p \notin I \\ j = 2, \ldots, s & \text{if } p \in I. \end{cases} \end{cases}$$

In general these linear forms do not have their coefficients in A_I, so we will transform them in such a way that Minkowski's theorem applies.

(i) $p \in I^-$.

Every $\omega_{k,p}^{(j)}$ is of degree $s-1$ over $\mathbf{Q}_p$; therefore there exists a relation

$$\sum_{k=1}^{s} a_{k,p} \omega_{k,p}^{(j)} = 0, \quad a_{k,p} \in \mathbf{Q}_p, \quad j = 1, \ldots, s$$

where at least one of the coefficients is not zero. Suppose it is $a_{i,p}$. Then we have

$$\omega_{k,p}^{(j)} = \sum_{k=1,\ k \neq i}^{s} b_{k,p}^{j} \omega_{k,p}^{(j)}, \quad \text{with } b_{k,p}^{j} \in \mathbf{Q}_p$$

and
$$\Theta_p^j(\omega; x) = \sum_{k=1,\ k \neq i}^{s} (x_k + b_{k,p}^{j} x_i) \omega_{k,p}^{(j)}.$$

We set

$$L_{k,p}(x) = x_k + b_{k,p} x_i, \quad p \in I^-, \quad k = 2, \ldots, s, \quad \varpi_p = \max_{j,k} \left(|\omega_{k,p}^{(j)}|_p \right).$$

The inequalities

$$|L_{k,p}(x)|_p \leq \rho_p \varpi_p^{-1}, \quad p \in I^-, \quad k = 1, \ldots, s, \quad k \neq i \tag{1}$$

imply

$$|\Theta_p^j(\omega; x)|_p \leq \rho_p \quad \text{for } p \in I^-, \quad \text{and } j = 2, \ldots, s.$$

(ii) $p \notin I^+$.

For $|\omega_{k,p}^{(j)}|_p \leq 1$ for $p \in I^+$, $k = 1, \dots, s$ and $j = 1, \dots, s$ the inequalities

$$|x_k|_p \leq \rho_p, \quad k = 1, \dots, p \tag{2}$$

imply $|\Theta_p^j(\omega; x)|_p \leq \rho_p$, $j = 1, \dots, s$.

Then we set

$$L_{k,p}(x) = x_k, \quad p \notin I^+, \quad k = 1, \dots, s.$$

(iii) $p = \infty$.

We will keep the forms with real coefficients as they are, while to every complex coefficient form we associate its complex conjugate.

We can now apply Minkowski's theorem in $A_{I^+ \cup J}$ to the system of linear forms $L_1, \dots, L_s$ defined by the equations

$$\begin{aligned} L_{1,p}(x) &= \Theta_p(\omega; x), \quad p \in I, \\ L_{k,p}(x) &= x_k - b_{k,p}x_i, \quad p \in I^-, \quad k = 2, \dots, s, \\ L_{k,p}(x) &= x_k, \quad p \notin I^+, \quad k = 1, \dots, s, \\ L_{k,\infty}(x) &= \Theta_\infty^k(\omega; x) \quad \text{if } \infty \in I, \ k = 1, \dots, s. \end{aligned}$$

The determinant of this system is not zero, so one can find elements x_k in $Q^{I \cup J}$ such that the above equations hold. These equations imply

$$|\Theta_p^j(\omega; x)|_p \leq \rho_p < 1, \quad p \in I \cup J, \quad j = 1, \dots, s.$$

It suffices to consider the p-adic valuations of θ for $p \in I$. We have

$$\prod_{j=1}^s \Theta_p^j(\omega; x) = \prod_{j=1}^s \theta_p^{(j)} = q \in Q^I,$$

where q is not zero because $\mathbf{Q}_I[\alpha]$ is a field, hence

$$\prod_{p \in I^+} \prod_{j=1}^s |\Theta_p^j(\omega; x)|_p \geq 1$$

and, if p' belongs to I,

$$|\theta|_{p'} \geq \frac{1}{\prod_{j=2}^{s} |\Theta_p^j(\omega;x)|_{p'} \prod_{p\in I^+, p\neq p'} \prod_{j=2}^{s} |\Theta_p^j(\omega;x)|_p}. \tag{3}$$

If for $p \in I$ we have $|L_{1,p}(x)|_p < M_p$, i.e., $|\theta|_p < M_p$, the inequality (3) becomes

$$|\theta|_{p'} \geq \prod_{p\notin J+} \rho_p^{-(s-1)} \prod_{p\neq p'} M_p,$$

and the right member can be made strictly superior to 1 by a suitable choice of the numbers M_p.

The conditions (A_p) imply that every θ_p is of degree s over $\mathbf{Q}_p$, hence the same is true for θ.

2. $\mathbf{Q}_I[\alpha]$ *is not a field*

In that case, Pm_α^I is not irreducible over $\mathbf{Q}$. Let $(I_h)_{(h=1,\ldots,m)}$ be the associated partition. We designate the degree of the field $\mathbf{Q}_{I_h}[\alpha_h]$ by s_h.

Let θ_h be an element of $\mathbf{Q}_{I_h}[\alpha_h] \cap S_{I_h}^J$ of degree s_h. Then, for every positive integer n, θ_h^n are elements of $\mathbf{Q}_{I_h}[\alpha_h] \cap S_{I_h}^J$ of degree s_h, and their minimal polynomials are relatively prime.

Hence we can choose strictly positive integers $n_1, \ldots, n_m$ such that the minimal polynomials of the elements $\theta_h^{n_h}$ are relatively prime. Then the element θ of A_I defined by $\theta = \sum_{k=1}^{m} \theta_h^{n_h}$ belongs to $S_I^J \cap \mathbf{Q}_I[\alpha]$ and is of degree s, since $Pm_\alpha^I(X) = \prod_{h=1}^{m} P_h(X)$. ■

One can prove then that conditions (A), (B) and (C) of §11.1 characterize the set U_I. Then U_I is included in the exceptional set of Koksma's theorem. The proof uses the following lemma:

Lemma 11.2.1. *For every element α of U_I there exists an element λ in $\mathbf{Q}_I[\alpha] \cap S_I^\infty$ such that for every positive integer n*

$$E(\lambda\alpha^n) = \sum_{j=1}^{s} \lambda^{(j)} \alpha^{(j)^n}$$

$$\varepsilon_p(\lambda\alpha^n) = \begin{cases} -\sum_{j=1}^{s} \lambda_p^{(j)} \alpha_p^{(j)^n} & \text{for } p \in I, \\ -E(\lambda\alpha^n) & \text{for } p \notin I. \end{cases}$$

Proof. Let $\mu = (\mu_p)_{p\in I}$ be an element of degree s in $\mathbf{Q}_I[\alpha] \cap S_I^\infty$. We denote by $\mu_p^{(j)}$ $(j = 1, \ldots, s)$ the zeros of its minimal polynomial in $\mathbf{C}_p$.

Let ρ be a positive real number such that

$$|\mu^{(j)}|_\infty \leq \rho < 1 \quad \text{for} \begin{cases} j = 2, \ldots, s & \text{if } \infty \in I, \\ j = 1, \ldots, s & \text{if } \infty \notin I. \end{cases}$$

Then

$$\sum_{j=2}^{s} |\mu^{(j)} \alpha^{(j)^n}|_p \leq 1 \qquad \text{for } p \in I^-,$$

$$\sum_{j=1}^{s} |\mu^{(j)} \alpha^{(j)^n}|_\infty \leq s\rho \qquad \text{if } \infty \notin I,$$

$$\sum_{j=2}^{s} |\mu^{(j)} \alpha^{(j)^n}|_\infty \leq (s-1)\rho \quad \text{if } \infty \in I;$$

moreover $\sum_{j=1}^{s} \mu^{(j)} \alpha^{(j)^n}$ belongs to Q^I.

We now determine an integer ν such that the element $\lambda = \mu^\nu$ satisfies the assumptions of the lemma.

For every $\nu \in \mathbf{N}^\star$, $\lambda = \mu^\nu$ belongs to $\mathbf{Q}_I[\alpha] \cap S_I^\infty$, and we have

$$|\sum_{j=2}^{s} \lambda^{(j)} \alpha^{(j)^n}|_p \leq 1 \qquad \text{for } p \in I^-,$$

$$|\sum_{j=1}^{s} \lambda^{(j)} \alpha^{(j)^n}|_\infty \leq s\rho^\nu \qquad \text{if} \infty \notin I,$$

$$|\sum_{j=2}^{s} \lambda^{(j)} \alpha^{(j)^n}|_\infty \leq (s-1)\rho^\nu \quad \text{if } \infty \in I.$$

■

Moreover there exists an integer ν_0 such that we have for $\nu > \nu_0$: $s\rho^\nu < 1/2$, so the corresponding elements λ satisfy the assumptions.

The characterization resulting from the Theorem 11.1 can be improved with the help of Theorem 10.3.1. If I contains the archimedean absolute value, we assume $\theta_\infty > 1$ and $\lambda_\infty \geq 3/2$; these conditions imply $u_0 \geq 1$.

Theorem 11.2.3. *An element α in A_I satisfying $\|\alpha\| \gg 1$ ($\alpha_\infty > 1$ if $\infty \in I$) belongs to U_I if and only if there exists an element λ in A_I and real numbers $(\mu_p)_{p\in\mathbf{P}}$ satisfying $\|\lambda\| \gg 1$ and $\lambda_\infty \geq 3/2$ if $\infty \in I$, $o < \mu_p \leq 1$ with $\mu_p = 1$ for almost all p, and such that the following inequalities hold:*

$$|\varepsilon(\lambda\alpha^n)|_p \leq \mu_p \tag{1}$$

$$\max(\mu; q^2\mu_\infty) < \frac{1}{e(1+\alpha_\infty)^2(2+\sqrt{\log \ell})} \qquad \text{with} \qquad \mu = \prod_{p\in\mathbf{P}} \mu_p. \tag{2}$$

The proof of this theorem uses the following two lemmas. The first deals with the archimedean absolute value and the second with the p-adic ones. We will express this lemma in a general form, which will allow us to use it in §11.3.

Lemma 11.2.2. *Assume that there exist two real numbers ρ and σ satisfying $0 < \rho < 1$ and $0 < \sigma \leq 1/2$ such that*

$$(1+\alpha)^2 \sum_{i=m}^{m+n} |\varepsilon(\lambda\alpha^{i+1}) - \varepsilon(\lambda\alpha^i)|_\infty^2 \leq (\rho(n+1))^{2\sigma} \qquad \forall (m,n) \in \mathbf{N}^2,$$

then, for every $r \in \mathbf{N}$ and every $L_r \in \mathcal{L}_r$, the following inequalities hold:

$$|\det A(L_r)|_\infty \leq \sqrt{u_0^2 + \frac{(\rho(r+1))^{2\sigma}}{(\alpha_\infty+1)^3}} \quad \text{if} \quad \infty \in I, \tag{3}$$

$$|\det A(L_r)|_\infty \leq (\rho(r+1))^{\sigma(r+1)} \qquad \text{if} \quad \infty \notin I. \tag{4}$$

Inequality (3) is obtained with computations of the same type as those leading to relation (11) in the proof of Theorem 5.1.2, whereas inequality (4) follows immediately from Hadamard's inequality.

Lemma 11.2.3. *Assume that there exists $(\mu_p)_{p\in\mathbf{P}}$ and $(\gamma_p)_{p\in\mathbf{P}}$ satisfying $0 < \mu_\infty \leq 1$, $\gamma_p \geq 0$ where $\mu_p = 1$ and $\gamma_p = 0$ for almost all p, and such that the following inequalities hold:*

$$|\varepsilon(\lambda\alpha^n)|_p \leq \mu_p n^{-\gamma_p} \quad \forall n \in \mathbf{N}^\star \text{ and } |\varepsilon(\lambda)|_p \leq \mu_p \quad \forall p \in \mathbf{P}^-;$$

then, for every $L_r \in \mathcal{L}_r$ we obtain the following inequalities:

$$|\det A(L_r)|_p \leq |\lambda|_p[\mu_p|\alpha|_p^2]^r((r-1)!)^{-\gamma_p} \quad \text{if } p \in I^-, \tag{5}$$

$$|\det A(L_r)|_p \leq \mu_p^{r+1}((r-1)!)^{-\gamma_p} \qquad \text{if } p \notin I^+. \tag{6}$$

Proof. We associate to the series $\sum_{n\in\mathbf{N}} u_n X^n$ the sequence (t_n) defined by $t_0 = e_I$, $t_1 = -\alpha$, $t_n = 0$ for $n \geq 2$, and we set, as in §10.3.1,

$$x_{m,n} = \sum_{i=0}^{m}\sum_{j=0}^{n} t_i t_j u_{m+n-(i+j)}, \quad (m,n) \in \mathbf{N}^2.$$

a) $p \in I^-$

It is easy to obtain the following inequalities:

$$|x_{0,0}|_p = |\lambda_0|_p, \ |x_{0,1}|_p = |x_{1,0}|_p \leq \mu_p |\alpha|_p,$$
$$|x_{0,n}|_p = |x_{n,0}|_p \leq \mu_p |\alpha|_p (n-1)^{-\gamma_p} \qquad \text{for } n \geq 1,$$
$$|x_{m,n}|_p = |x_{n,m}|_p \leq \mu_p |\alpha|_p (m+n-2)^{-\gamma_p} \qquad \text{for } m \geq 1 \text{ and } n \geq 1;$$

then

$$|\det A(L_r)|_p \leq \max(|x_{0,i_0} \cdot x_{\ell_1,i_1} \cdots x_{\ell_r,i_r}|_p)$$

where $(i_0, \ldots, i_r)$ runs over the set of permutations of $(0, 1, \ldots, r)$.

Then one distinguishes two cases according to whether this maximum is attained for $i_0 = 0$ or not. We deduce inequality (5) from $|u_o|_p \geq 1$.

b) $p \notin I^+$

The following inequalities hold:

$|x_{0,0}|_p \leq \mu_p$ and $|x_{m,n}|_p = |x_{n,m}|_p = |\varepsilon(\lambda\alpha^n)|_p \leq \mu_p (m+n)^{-\gamma_p}$ for $m \geq 1$ and $n \geq 1$; hence (6) follows by the same procedure. ∎

Proof of the theorem. The proof is different depending on whether I includes the archimedean absolute value or not.

a) $\infty \in I$

Inequalities (3), (5) and (6) with $\gamma_p = 0$ and $\sigma = 1/2$ allow us to write for $r \in \mathbf{N}$ and $L_r \in \mathcal{L}_r$:

$$\prod_{p\in\mathbf{P}} |\det A(L_r)|_p \leq \ell^- \sqrt{u_0^2 + \frac{\rho(r+1)}{(1+\alpha_\infty)^3}} (k(r+1))^{r/2},$$

where $\rho = \mu_\infty^2 (1+\alpha_\infty)^4$; $\mu = \prod_{p\in\mathbf{P}} \mu_p$; $\ell^- = \prod_{p\in I^-} |\lambda|_p$ and $k = \mu^4 q^4 (1+\alpha_\infty)^4$.

Then we set $h = \max(k,\rho)$ and define the integer r with the inequality $\frac{1}{he} - 1 \leq r < \frac{1}{he}$. The proof is completed as in Theorem 5.1.3 with an application of Lemma 5.1.

b) $\infty \notin I$.

Inequalities (4), (5) and (6) with $\gamma_p = 0$ and $\rho = \mu_\infty^2$ allow us to write, for $r \in \mathbf{N}$ and $L_r \in \mathcal{L}_r$:

$$\prod_{p\in\mathbf{P}} |\det A(L_r)|_p \leq \ell\Big(q^2 \prod_{p\neq\infty} \mu_p\Big)^r (\mu_\infty^2(r+1))^{\frac{1}{2}(r+1)} \leq \ell(\mu q^2(r+1))^{\frac{r}{2}}(\mu_\infty^2(r+1)^{\frac{1}{2}}).$$

We set then $h = \max(\mu_\infty^2, \mu^2, q^4)$, and after having defined r as in a), we deduce $\mu_\infty^2(r+1) < 1$; hence

$$\prod_{p\in\mathbf{P}} |\det A(L_r)|_p \leq \ell(\mu q^2(r+1))^{r/2}$$

and we complete the proof as above. ∎

If $\mu_p = 1$ for $p \neq \infty$, we have the following corollary, which shows that this constitutes, at least for large enough values of ℓ, an improvement of Theorem 11.2.3.

Corollary *An element α of A_I satisfying $\|\alpha\| \gg 1$ and $\alpha_\infty > 1$ if $\infty \in I$ belongs to U_I if and only if there exists an element λ of A_I satisfying $\|\lambda\| \gg 1$ and $\lambda_\infty > 3/2$ if $\infty \in I$ such that, for every $n \in \mathbf{N}$,*

$$|\varepsilon(\lambda\alpha^n)|_\infty < \frac{1}{e(1+\alpha_\infty)^2(2+\sqrt{\log \ell})} \quad \textit{if } \infty \in I$$
$$|\varepsilon(\lambda\alpha^n)|_\infty < \frac{1}{eq^2(1+\sqrt{\log \ell})} \quad \textit{if } \infty \notin I.$$

Now we wish to characterize the subsets of U_I introduced in §11.2, especially those which occur in the partition of Theorem 11.2.1, through properties of distribution modulo Q^I.

11.3 The sets S_I^∞

The characterizations of S_I^∞ given in this section justify the name of **Pisot set of A_I** given to the set S_I^∞ in §11.2.

The two first theorems recall Theorems 5.4.1 and 5.4.2.

Theorem 11.3.1. *An algebraic element θ of A_I satisfying $\|\theta\| \gg 1$ belongs to S_I^∞ if and only if there exists an inversible element λ of A_I such that $\lim_{n\to+\infty} \varepsilon_\infty(\lambda\theta^n) = 0$. Then λ belongs to $\mathbf{Q}_I[\theta]$.*

Proof. The necessary condition follows from Lemma 11.2.1. Considering an element λ of $S_I^\infty \cap \mathbf{Q}_I[\theta]$ satisfying the assumptions of this lemma, one shows that $|\varepsilon(\lambda\alpha^n)|_\infty \leq s \max_j(|\lambda_\infty^{(j)}|_\infty \rho^n)$.

The proof of the sufficient condition is similar to that of the real case (Theorem 5.4.1). After having defined v_n in the same way, one shows that $v_n \prod_{p\in I} |v_n|_p < 1$; hence $v_n = 0$ for n large enough.

On the other hand, $(\varepsilon_\infty(\lambda\theta^n))$ converges to zero, and hence the function $x \mapsto \sum_{n=0}^{\infty} \varepsilon_\infty(\lambda\theta^n)x^n$ is analytic in $D_\infty(0,1)$ and has no pole on $C_\infty(0,1)$ (cf. Lemma 5.4). ■

Theorem 11.3.2. *An element θ in A_I, satisfying $\|\theta\| \gg 1$ belongs to S_I^∞ if and only if there exists an inversible element λ in A_I such that $\sum_{n=0}^{\infty} |\varepsilon(\lambda\theta^n)|_\infty^2 < +\infty$. Then λ belongs to $\mathbf{Q}_I[\theta]$.*

Proof. The necessary condition follows from the above proof, so we only show the sufficient condition.

Consider the functions f and ε defined in A_I by

$$f(x) = \sum_{n=0}^{\infty} E(\lambda\theta^n)x^n, \qquad \varepsilon(x) = \sum_{n=0}^{\infty} \varepsilon(\lambda\theta^n)x^n;$$

then $f(x) = \dfrac{\lambda}{1-\theta x} + \varepsilon(x)$. The assumptions imply that ε is of bounded characteristicon $D_I(0,1)$, hence (Theorem10.3.2) it is a rational function, and we complete the proof as in the preceding theorem. ■

One can improve this theorem in the following way.

Theorem 11.3.3. *An element θ in A_I satisfying $\|\theta\| \gg 1$ ($\theta_\infty > 1$ if $\infty \in I$) belongs to S_I^∞ if and only if there exists an inversible element λ in A_I and* *real numbers $(\mu_p)_{p\in I}$ with $0 < \mu_p \leq 1$ and $\mu_p = 1$ except for a finite set of subscripts, and such that we have, for n large enough*

$$|\varepsilon(\lambda\theta^n)|_\infty \leq \mu_\infty n^{-1/2}, \quad \textit{and} \quad |\varepsilon(\lambda\theta^n)|_p \leq \mu_p \quad \textit{for} \quad p \in I; \tag{1}$$

assuming the inequality

$$\mu < \frac{1}{2\sqrt{2}eq^2(1+\theta_\infty)^2} \quad \textit{where} \quad \mu = \prod_{p\in\mathbf{P}} \mu_p.$$

Proof. The necessary condition is included in the necessary condition of Theorem 11.3.1, so we only prove the sufficient condition.

By changing, if necessary, the value of λ and shifting the index, we can assume that the inequalities are satisfied for all n.

Lemma 11.2.2 allows us to write $\prod_{p\in\mathbf{P}^-} |D_r|_p \leq \ell^- q^2\mu$. If I contains the archimedean absolute value, a computation analogous to that of the real case (Theorem 5.4.4) leads to the inequality

$$\sum_{m=0}^{r}\sum_{n=0}^{r} |x_{m,n}|_\infty^2 \leq 8\mu_\infty^2(1+\theta_\infty)^4(r+1) + O(\log r).$$

Then by using Hadamard's upper bound line by line and the inequality between the arithmetic and geometric means, we have

$$|D_r|_\infty = |\det X_r|_\infty^2 \leq \Big(\frac{1}{r+1}\sum_{m=0}^{r}\sum_{n=0}^{r} |x_{m,n}|_\infty^2\Big)^{r+1}$$

hence $$|D_r|_\infty \leq \big(2\sqrt{2}\mu_\infty(1+\theta_\infty)^2 + o(1)\big)^{r+1}.$$

If I does not contain the archimedean absolute value, one proceeds in the same way and

$$\sum_{m=0}^{r}\sum_{n=0}^{r} |x_{m,n}|_\infty^2 = \sum_{m=0}^{r}\sum_{n=0}^{r} |u_{m,n}|_\infty^2 \leq 4\mu_\infty^2,$$

hence $$|D_r|_\infty \leq (2\mu_\infty)^{r+1}.$$

By a suitable combination of these inequalities, we can write in all cases

$$\prod_{p\in\mathbf{P}} |D_r|_p \le K\Big(2\sqrt{2}q^2(1+\theta_\infty)^2(\mu_\infty + o(1)) \prod_{p\in\mathbf{P}^-} \mu_p\Big)^r,$$

where K is a constant computable from the data. Therefore, if condition (1) is satisfied, we have $\lim_{r\to+\infty} \prod_{p\in\mathbf{P}} |D_r|_p = 0$. This implies that D_r is zero for r large enough, and hence the series $\sum_{n\in\mathbf{N}} u_n X^n$ is rational. Following Theorem 1.3.1, we can deduce that θ belongs to S_I^∞. ■

Corollary 1. *An element θ in A_I satisfying $\|\theta\| \gg 1$ belongs to S_I^∞ if and only if there exists an invertible element λ in A_I such that $\varepsilon_\infty(\lambda\theta^n) = o(n^{-1/2})$.*

Corollary 2. *Let θ be a real number strictly greater than 1. If there exists a non-zero real λ and a prime number p such that*

$$\|\lambda\theta^n\| = o(n^{-1/2}) \quad \textit{and} \quad \lim_{n\to+\infty} |\varepsilon_n|_p = 0$$

then θ is a Pisot number.

Remarks

1. Contrary to the condition in the previous section, which must be satisfied for every integer n, here the conditions are asymptotic.

2. It is not known if the condition $\lim_{n\to+\infty} \varepsilon_\infty(\lambda\theta^n) = 0$ implies the algebraicity of θ.

3. If θ belongs to S_I^∞ and λ is an integer of $\mathbf{Q}_I[\theta]$, then it can easily be seen that $\lim_{n\to+\infty} \varepsilon_\infty(\lambda\theta^n) = 0$. One can then show that if λ belongs to $\mathbf{Q}_I[\theta]$ then $(\varepsilon_\infty(\lambda\theta^n))$ can only have a finite number of limit points, all rational.

This remark suggests considering, as in the real case (cf. §5.6), the set of elements α in A_I satisfying $\|\alpha\| \gg 1$ and to which can be associated an invertible element $\lambda \in A_I$ such that the sequence $(\varepsilon_\infty(\lambda\alpha^n))$ has finitely many limit points. One shows that this set is countable, and then that S_I^∞ is the set of its algebraic elements. It follows from the general theorem that

Theorem 11.3.4. *The set of pairs (α, λ) of elements of A_I satisfying $\|\alpha\| \gg 1$ and λ invertible, for which $\sup_{n\ge 0} \prod_{p\in I^+} |\varepsilon(\lambda\alpha^n)|_p < \prod_{p\in I^+} (1+|\alpha|_p^2)^{-1}$, is countable.*

11.4 The sets T_I

The following characterization justifies the name of **Salem set** given to T_I.

Theorem 11.4.1. *An element $\tau \in A_I$ satisfying $\|\tau\| \gg 1$ belongs to T_I if and only if there exists an invertible element λ of A_I and a positive real number M such that*

(i) $\mathrm{Re}(\sum_{n=0}^{\infty} \varepsilon_\infty(\lambda\tau^n)x^n) \leq M$ *for* $x \in D_\infty(0,1)$.

(ii) For every subset J of I the sequence $(\varepsilon_\infty(\lambda_J\tau_J^n))$ has an infinite number of limit points.

Condition (ii) says that no τ_J belongs to S_J^∞, and that hence τ does not belong either to S_I^∞ or to D_I

Proof.

Necessary condition. Let $\tau \in T_I$. The associated polynomial P is reciprocal and of degree $2s$ $(s \geq 2)$; we denote its zeros in $\mathbf{C}_p$ by $\tau_p, \tau_P^{-1}, \alpha_p^{(j)}, \alpha_p^{(j)^{-1}}$, $j = 2, \ldots, s$ for $p \in I$, $\alpha_p^{(j)}, \alpha_p^{(j)^{-1}}$, $j = 1, \ldots, s$ for $p \notin I$.
We set $\gamma_p = \tau_p + \tau_p^{-1}$ for $p \in I$, $\gamma_\infty = 0$ if $\infty \notin I$.

Let θ be an element of degree s of the set $\mathbf{Q}_I[\gamma] \cap S_I^{I^+}$; one shows that there exists a positive integer h such that, if $\lambda = \theta^{2h}$, we have $E(\lambda\tau^n) = v_n$, where $v_n = (\tau^n + \tau^{-n}) + \sum_{j=2}^{s} \lambda^{(j)}(\alpha^{(j)^n} + \alpha^{(j)^{-n}})$, $\alpha^{(j)} \in \mathbf{C}_I$.

If I contains the archimedean absolute value, we complete the proof as in the real case (cf. Theorem 5.5.2).

If I does not contain the archimedean absolute value, the series $\sum_{n\in\mathbf{N}} v_n X^n$ is rational and represents a function whose complex component can be written

$$f_\infty(x) = \sum_{j=1}^{s} \lambda_\infty^{(j)}\Big(\frac{1}{1-\alpha_\infty^{(j)}x} + \frac{1}{1-\alpha_\infty^{(j)^{-1}}x}\Big).$$

From $|\alpha_\infty^{(j)}|_\infty = 1$ it follows by a classical result that for $x \in D_\infty(0,1)$, $\mathrm{Re}(\frac{1}{1-\alpha_\infty^{(j)}x}) \geq \frac{1}{2}$, and hence $\mathrm{Re}\, f(x) \geq \frac{1}{2}\sum_{j=1}^{s} |\lambda_\infty^{(j)}|_\infty$. The desired result follows.

Sufficient condition. The assumptions imply that the function ε is of bounded characteristicin $D_{I^+}(0,1)$. Therefore for f, which is a rational function, this implies that τ belongs to U_I, and from condition (ii) we deduce that τ belongs to T_I. ∎

Remark

If I contains the archimedean absolute value and if the associated polynomial P is irreducible, one can show that the sequence $(\varepsilon_\infty(\lambda\tau^n))$ is dense on $[-1/2, 1/2]$, but not uniformly distributed (Theorem 5.3.2), as in the real case.

We will not give an explicit characterization of the set D_I. Following the definition, we can associate to every element $\sigma \in D_I$ two subsets I_1 and I_2 of I such that the elements $\sigma_1 = \sigma \cdot e_{I_1}$ and $\sigma_2 = \sigma \cdot e_{I_2}$ belong respectively to $S_{I_1}^\infty$ and T_{I_2}. This property can be used for characterizing D_I.

11.5 The sets S_I^J

We obtain characterizations similar to those of the set S_I^∞.

Theorem 11.5.1. *An algebraic element $\theta \in A_I$, satisfying $\|\theta,\| \gg 1$ belongs to $S_I^{p'}$ if and only if there exists an inversible element $\lambda \in A_I$ such that $\lim_{n\to+\infty} \varepsilon_{p'}(\lambda\theta^n) = 0$. Then λ belongs to $\mathbf{Q}_I[\theta]$.*

The proof is the same as for Theorem 11.3.

Theorem 11.5.2. *Let J be a finite set of absolute values of $\mathbf{Q}$ not containing the archimedean absolute value.*

An element $\alpha \in A_I$ satisfying $\|\alpha\| \gg 1$ belongs to S_I^J if and only if there exists an element $\lambda \in A_I$ and real numbers $\rho, \sigma, (\gamma_p)_{p\in J}$ and $(\mu_p)_{p\in J}$ satisfying $\lambda \gg 1$, $\gamma_p > 0, 0 < \mu_p \leq 1, 0 < \sigma \leq 1/2, \rho > 0$, such that, for every $p \in J$:

$$|\varepsilon(\lambda\alpha^n)|_p \leq \mu_p n^{-\gamma_p} \qquad \text{for } n \in \mathbf{N}^\star,\ |\varepsilon(\lambda)|_p \leq \mu_p, \tag{1}$$

$$(1+\alpha_\infty)^2 \sum_{i=m}^{m+n} |\varepsilon(\lambda\alpha^{i+1}) - \alpha\varepsilon(\lambda\alpha^i)|_p < (\rho(n+1))^{2\sigma}, \tag{2}$$

assuming that one of the following conditions is satisfied:

(i) $\quad \sigma - \sum_{p\in J} \gamma_p < 0$

(ii) $\quad \sigma = \sum_{p\in J} \gamma_p$ *and* $\rho\mu < \dfrac{1}{e^2 q^2 (1+\alpha_\infty)}$ *where* $\mu = \sum_{p\in J} \mu_p$

Proof. When I contains the archimedean absolute value, computations analogous to those in the proof of Theorem 5.1.2 allow us to write

$$|D_r|_\infty^2 \leq \Big(u_0^2 + \frac{\rho(r+1)^{2\sigma}}{(\alpha+1)^3}\Big)(\rho(r+1))^{r\sigma}.$$

Hence $|D_r|_\infty = O(\rho(r+1))^{\sigma(r+1)}$; this inequality still holds if I does not contain the archimedean absolute value.

The inequalities of Lemma 10.2.2 allow us to write

$$\prod_{p\in\mathbf{P}} |D_r|_p = O\big((\rho^\sigma \mu q^2)^r((r-1)!)^{-\gamma}\big),$$

where we have set $\gamma = \sum_{p\in\mathbf{P}} \gamma_p$ and $\mu = \prod_{p\in\mathbf{P}} \mu_p$.

Stirling's formula gives $(r-1)! \geq (\frac{r-1}{e})^{r-1}\sqrt{2\pi(r-1)}$, but $\log(\frac{r+1}{r-1})^{r-1} = (r-1)\log(1+\frac{2}{r-1}) < 2$, hence $(r+1)^r < e^2(r+1)(r-1)^{r-1}$; thus

$$\prod_{p\in\mathbf{P}} |D_r|_p = O((r+1)^\sigma (r-1)^{\frac{\gamma}{2}+\sigma}(e\mu\rho^\sigma q^2)^r (r+1)^{r(\sigma-\gamma)}),$$

which was the desired result.

Proceeding the same way we get the result when I does not contain the archimedean absolute value. ■

By changing the value of λ if necessary and shifting the indexes, we obtain the following result.

Theorem 11.5.3. *An element $\alpha \in A_I$ satisfying $\|\alpha\| \gg 1$ ($\alpha_\infty > 1$ if $\infty \in I$) belongs to S_I^J if and only if there exists an invertible element $\lambda \in A_I$ and numbers $(\gamma_p)_{p\in J}$ such that $|\varepsilon(\lambda\alpha^n)|_p = O(n^{-\gamma_p})$ for $p \in J$, and one of the two following conditions is satisfied:*

(i) $\gamma = \sum_{p\in J} \gamma_p > \frac{1}{2}$

(i) $\gamma = \frac{1}{2}$ *and there exists $p' \in J$ such that $|\varepsilon(\lambda\alpha^n)|_{p'} = o(n^{-\gamma_{p'}})$.*

It should be remarked that in this characterization the archimedean absolute value does not play any particular role.

For $I = \{\infty\}$, this theorem gives **sufficient conditions** for a real number θ to be a Pisot number; for instance:

Theorem 11.5.4. *Let (λ, θ) be a pair of real numbers satisfying $\theta > 1$, $\lambda \neq 0$. If there exists a finite set J of ultrametric absolute values of $\mathbf{Q}$ and positive real numbers $(\gamma_p)_{p\in J}$ satisfying $\sum_{p\in J^+} \gamma_p \geq 1/2$, such that one of the two following conditions is satisfied, then θ belongs to S.*

(i) $|\varepsilon(\lambda\theta^n)|_\infty = o(n^{-\gamma_\infty})$ *and* $|u_n|_p = O(n^{-\gamma_p}), \qquad \forall p \in J$

(ii) $|\varepsilon(\lambda\theta^n)|_\infty = O(n^{-\gamma_\infty}), \gamma_\infty > 0,$ *and* $|u_n|_p = O(n^{-\gamma_p}), \qquad \forall p \in J.$

Moreover there exists an element $p' \in J$ for which $|u_n|_{p'} = o(n^{-\gamma_{p'}})$ in the case $\sum_{p\in J^+} \gamma_p = 1/2$.

Rational approximation of algebraic elements of A_I.

We wish to use the elements of S_I^J to characterize the algebraic elements of A_I with the help of rational approximations, as in the real case (§8.3). This is done in the following theorem.

Theorem 11.5.6. *An element $\alpha \in A_I$ is algebraic if and only if there exist two sequences (u_n) and (v_n) of elements of Q^I such that*

$$|v_n\alpha - u_n|_p < c\rho^n,$$
$$|v_{n+1} - \theta v_n|_p < c\rho^n \quad \text{for} \quad n \in \mathbf{N}^\star \quad \text{and} \quad p \in I^+,$$

where c and ρ are positive real numbers, $0 < \rho < 1$ and $\theta \in A_I$ satisfies $\|\theta\| \gg 1$. Moreover the degree s of α satisfies the inequality $s - 1 < \eta^{-\operatorname{card} I^+}$ where $\eta = -\log\rho/\log q$ if $\infty \notin I$ and $\eta = \log\rho/\log(q|\theta|_\infty)$ if $\infty \in I$.

Proof.

(i) Necessary condition. Let $\alpha \in A_I$ be an algebraic element of degree s; then α may be written $\alpha = \lambda/\mu$, where λ and μ are two integers of $\mathbf{Q}_I[\alpha]$. Let $\theta \in \mathbf{Q}_I[\alpha] \cap S_I^{I^+}$, and $u_n = E(\lambda\theta^n)$, $v_n = E(\mu\theta^n)$. It is easy to see that there exist two constants c and ρ such that

$$|\varepsilon(\lambda\theta^n)|_p < c\rho^n \text{ and } |\varepsilon(\mu\theta^n)|_p < c\rho^n \text{ for } p \in I^+;$$

hence

$$|v_{n+1} - \theta v_n|_p < c\rho^n \text{ and } |v_n\alpha - u_n|_p = |\varepsilon(\lambda\theta^n) - \alpha\varepsilon(\mu\theta^n)|_p < c\rho^n \text{ for } p \in I.$$

(ii) Sufficient condition. Suppose there exist two sequences (u_n) and (v_n) satisfying the stated conditions, then it is easy to see that the following limits exist for every $p \in I$: $\lim_{n\to+\infty}(u_n\theta_p^{-n}) = \lambda_p$, $\lim_{n\to+\infty}(v_n\theta_p^{-n}) = \mu_p$. Let us set $\alpha_p = \lambda_p/\mu_p$. If I does not contain the archimedean absolute value, we take here $\lambda_\infty = 0$ and $\mu_\infty = 1$. Then by setting $\lambda = (\lambda_p)_{p\in I^+}$ and $\mu = (\mu_p)_{p\in I^+}$ we have $|\varepsilon(\lambda\theta^n)|_p < c\rho^n$ and $|\varepsilon(\mu\theta^n)|_p < c\rho^n$ for $p \in I^+$. Hence θ belongs to $S_I^{I^+}$; λ and μ belong to $\mathbf{Q}_I[\theta]$, and α is algebraic.

The inequality for the degree of α is obtained by applying the product formula to the rational $N(\theta)$. ■

The study of limit points of the subsets of U_I introduced in §11.2 is more difficult here than in the real case. In particular, to obtain closed sets we have to add more or less artificial supplementary conditions. As in the previous sections the proofs use functions of bounded characteristicin the unit ball of A_I, and in particular rational functions.

11.6 The sets B_I

These are sets of adeles that are not necessarily algebraic; their interest for us is that they demonstrate the difficulties just mentioned.

Definition 11.6. *We denote by $B_I^{p'}$ the set of elements α of A_I, satisfying $\|\alpha\| \gg 1$ for which there exists a subset J of I such that the element $\alpha_J = \alpha \cdot e_J$ belongs to $S_J^{p'}$.*

Let α be an element of $B_I^{p'}$. Then there may exist several subsets J of I such that α_j belongs to $S_J^{p'}$. Among these, there exists at least one, say J_0, such that $Pm_{\alpha_{J_0}}^{J_0}$ is irreducible. Establishing the properties characterizing $B_I^{p'}$ is easy, since we know how this is done for $S_I^{p'}$. We will therefore omit this step and content ourselves with the following theorem.

Theorem 11.6. *The set $B_I^{p'}$ is closed in A_I.*

The proof uses the following lemma.

Lemma 11.6. *Let θ be an element of S_I^∞ whose minimal polynomial is irreducible, and of degree strictly greater than two. There exists an invertible element $\lambda \in A_I$ such that the following conditions are satisfied:*

$$(A) \quad \sum_{n=0}^{\infty}(\varepsilon_\infty(\lambda\theta^n))^2 \le cq^2 \quad \text{where} \quad q = \prod_{p\in I^-} |\theta|_p \quad \text{and} \quad c = \begin{cases} 1 & \text{if} \quad \infty \in I \\ 9 & \text{if} \quad \infty \notin I \end{cases}$$

$$(B) \quad |\lambda|_p \le 1 \quad \text{for} \quad p \in I^- \quad \text{and} \quad \lambda_\infty \le c\theta_\infty \quad \text{if} \quad \infty \in I$$

$$(C) \quad \max_{p\in I}(|\lambda\theta|_{p'}, \frac{|\lambda|_\infty}{q}) \ge 1.$$

Proof. From the assumptions the polynomial P associated to θ is irreducible and primitive. We set $Q(X) = X^s P(1/X)$; then P and Q are relatively prime, and the rational fraction P/Q is associated to the formal power series $\sum_{n\in\mathbf{N}} u_n X^n$ and the numbers $q^{n+1}u_n$ are rational integers. Moreover, $1/\theta$ is the one and only pole of the rational function $f = P/Q$ in $D_I(0,1)$.

For $p \in I$ there exists an element $\mu_p \in \mathbf{Q}_p$ such that the function $\varphi_p \mapsto f(x) - \dfrac{\mu_p}{1-\theta_p x}$ is analytic on $\hat{D}_p(0,1)$, and then

$$\mu_p = \lim_{x\to 1/\theta_p}((1-\theta_p x)\frac{P(x)}{Q(x)}) = -\theta_p \frac{P(1/\theta_p)}{Q'(1/\theta_p)} \tag{1}$$

and

$$|\mu|_p \le |\theta|_p \quad \text{for} \quad p \in I^-, \; |\mu|_\infty \le \left|\theta_\infty - \frac{1}{\theta_\infty}\right|_\infty < |\theta|_\infty, \quad \text{if} \quad \infty \in I. \tag{2}$$

By applying Cauchy's inequality to the expansion of φ_p, we obtain

$$|\mu_p\theta_n - u_n|_p \le |\theta|_p \quad \text{for} \quad p \in I. \tag{3}$$

Now we set $\mu = (\mu_p)_{p\in I}$. From (3) we have $|q\mu\theta^n - qu_n|_p \le 1$ for $p \in I^-$, hence

$$qu_n \equiv E(q\mu\theta^n) \bmod 1$$
$$q\mu_\infty\theta_\infty^n - qu_n \equiv \varepsilon_\infty(q\mu\theta^n) \bmod 1.$$

Moreover, if x belongs to $C_\infty(0,1)$ we have $|f(x)|_\infty = 1$, and hence

$$\sum_{n=0}^{+\infty}(\varepsilon_\infty(q\mu\theta^n))^2 \leq c \text{ where } c = \begin{cases} 1 & \textit{if } \infty \in I \\ 9 & \textit{if } \infty \notin I. \end{cases}$$

It follows that

$$\sum_{n=0}^{\infty}(\varepsilon_\infty(q\mu\theta^n))^2 = \sum_{n=0}^{\infty}(qu_n - q\mu_\infty\theta_\infty^n)^2 \leq cq^2.$$

Inequalities (1) and (2) imply $|q\mu|_p \leq 1$ for $p \in I^-$ and $|q\mu|_\infty \leq q|\theta|_\infty$ if $\infty \in I$. Hence, for every $p \in I$, there exists a natural integer h_p such that

$$|\theta|_p^{-h_p-1} \leq |q\mu|_p \leq |\theta|_p^{-h_p} \text{ for } p \in I^-$$
$$|\theta|_\infty^{-h_\infty-1} < \left|\frac{\mu}{\theta}\right|_\infty \leq |\theta|_\infty^{-h_\infty} \text{ if } \infty \in I.$$

Setting $h = \inf_{p\in I} h_p$ and $\lambda = q\mu\theta^h$ we can easily see that conditions (A), (B) and (C) are satisfied. ∎

Proof of the theorem. Consider a convergent sequence (α_ν) of elements of B_I^∞. We denote its limit α. We associate to each α_ν a subset J_ν of I such that $\alpha_\nu \cdot e_{J_\nu}$ belongs to $S_{J_\nu}^\infty$ and that $Pm_{\alpha_\nu \cdot e_{J_\nu}}^{J_\nu}$ is irreducible of degree strictly greater than two (this condition is not restrictive because in a convergent sequence there are only a finite number of elements of degree two). From this sequence one can extract a subsequence for which the sets J_ν are equal to a constant set J_0. We suppose $J_0 \neq \{\infty\}$ (then $S_{J_0}^\infty = S$). From a certain rank on we have $|\alpha_\nu|_p = |\alpha|_p$ for $p \in J_0^-$ and $q_\nu = \prod_{p\in J_0^-} |\alpha_\nu|_p = \prod_{p\in J_0^-} |\alpha|_p = q$; by suppressing if necessary a finite number of indices, we can suppose that these inequalities are satisified for every $\nu \in \mathbf{N}$.

We associate an element λ_ν satisfying the assumptions of Lemma 11.6 to every α_ν. Then by using conditions (A), (B) and (C) we can extract from the sequence (λ_ν) a subsequence converging to a limit λ. Then the following inequalities are satisfied:

$$|\lambda|_p \leq 1 \text{ for } p \in I^-, \quad |\lambda|_\infty \leq q \sup_\nu |\alpha_\nu|_\infty \text{ and } \sup_{p\in J}(|\lambda\alpha|_p, \frac{|\lambda|_\infty}{q}) \geq 1.$$

Hence λ is not zero. Moreover condition (A) implies $\sum_{n=0}^{\infty} \varepsilon_{\infty}^2(\lambda\alpha^n) < +\infty$. Hence there exists a non-empty subset J' of J such that $\alpha_{J'}$ belongs to $S_{J'}^{\infty}$, and α belongs to B_I^{∞} which is then closed.

In the particular case where $J_0 = \{\infty\}$, α_{J_0} belongs to S which is closed, hence the result. ∎

Remarks

Condition (C) of the lemma implies that at least one component of λ is not zero, but it says nothing about the other components. Hence this method cannot be used to determine if the sets S_I^{∞} are closed or not when I contains more than one element.

If I contains a single element, then the unique component of λ is not zero. One deduces from this that S_I is closed. For $I = \{\infty\}$ we obtain the set S of Pisot numbers, and for $I = \{p\}$ where $p \neq \infty$ we obtain the set of Chabauty numbers, which we will study in the next section.

11.7 Closed subsets of S_I^{∞}

The only elements of U_I to which rational functions of the family $\mathcal{F}_I(\delta)$ can be associated are the elements of S_I^{∞}. It is easy to see that every function $f \in \mathcal{F}_I(\delta)$ possessing exactly one pole in $D_I(0,\delta)$ can be associated to an element $\theta \in S_I^{\infty}$ and can be written $f(x) = \dfrac{B(x)}{q'Q(x)}$, where q' is an integer whose decomposition contains only elements of I. More precisely:

Lemma 11.7. *An element $\theta \in A_I$ satisfying $\|\theta\| \gg 1$ belongs to S_I^{∞} if and only if there exists a rational function f of a family $\mathcal{F}_I(\delta)$ associated with it and having $1/\theta$ as a unique pole in $D_I(0,1)$.*

Proof. It is sufficient to show that every element of S_I^{∞} can be associated to a rational function belonging to a family $\mathcal{F}_I(\delta)$.

First we assume that the polynomial P associated to θ is irreducible. Then if P and Q are relatively prime, i.e., if P is not a reciprocal polynomial of degree two, it is easy to see that if we choose δ such that $\delta_p < |\theta^{-1}|_p$ for $p \in I$, then the rational function P^+/Q belongs to the family $\mathcal{F}_I(\delta)$.

If P is a reciprocal polynomial of degree two, then I necessarily contains the archimedean absolute value and $P(X) = qX^2 - q_1X + q$, where $|q_1|_p = 1$ for

$p \in I^-$, $|q_1|_\infty > 2q$. A simple computation shows then that A/Q belongs to $\mathcal{F}_I(\delta)$ in the following cases:

$$A(X) = 2q - |q_1|_\infty$$
$$A(X) = qX^2 - aX + q, \quad \text{with } aq_1 > 0 \text{ and } 2q \le |a|_\infty < |q_1|_\infty.$$

Assume now that P is not irreducible. Let $(I_h)_{h=1,\dots,r}$ be the associated partition; then $P(X) = \prod_{h=1}^{r} P_h(X)$ where every P_h associated to the element $\theta_h = \theta \cdot e_{I_h} \in S_{I_h}^\infty$ is irreducible. Hence there exists a rational function A_h/Q_h belonging to the family $\mathcal{F}_{I_h}(\delta)$ which can be associated to θ_h. If we define the rational fraction A/Q by the equality

$$\frac{A(X)}{Q(X)} = \prod_{h=1}^{r} \frac{A_h(X)}{Q_h(X)},$$

we can easily see that it belongs to the family $\mathcal{F}_I(\delta)$.

We denote by $K(\theta)$ (resp. $L(\theta)$) the set of valuations $p \in I^-$ for which f has no pole on $C_p(0,1)$ (resp. has at least a pole on $C_p(0,1)$); and we distinguish two cases:

(1) f has no pole on $C_p(0,1)$, ($p \in I^-$).

In this case we have necessarily $|\theta^{(j)}|_p < 1$ for $j = 2, \dots, s$. Then

$$Q(X) = q(1 - \theta_p X) \prod_{j=2}^{s} (1 - \theta_p^{(j)} X);$$

hence we deduce $|Q(x)|_p = |x|_p$ for $|x|_p \le 1$ and $\left|\frac{A(x)}{Q(x)}\right|_p \le 1$ for $|x|_p = 1$. The condition (A_p) is then satisfied with $m_p = 1$ for every $p \in I^-$.

(2) f has at least one pole on $C_p(0,1)$.

Consider the annulus $\gamma_p = \{x \in \mathbf{C}_p; |\theta|_p^{-1} < |x|_p < 1\}$; then for every $x \in \gamma_p$ we have $|Q(x)|_p = |x|_p > |1/\theta|_p$. Hence $\sup_{x \in \gamma_p} \left|\frac{A(x)}{Q(x)}\right|_p < |\theta|_p$ and (B_p) is satisfied with $m_p = |\theta|_p$. Let (θ_ν) be a sequence of elements of S_I^∞ converging to an element $\theta \in A_I$. We set $q_\nu = \prod_{p \in I+} |\theta_\nu|_p$. The sequence $(|\theta_\nu|_p)$ is stationary from a rank ν_0 on, for every $p \in I^-$, and q_ν is constant. We set $q = q_\nu = \prod_{p\in I^-} |\theta_\nu|_p = \prod_{p\in I^-} |\theta|_p$ for $\nu \geq \nu_0$.

Now we consider this subsequence. To the sequence (θ_ν) we associate a bounded sequence of functions (A_ν/Q_ν) belonging to the family $\mathcal{F}_I(\delta)$ where $|\delta|_p < 1/|\theta|_p$, and from this sequence we extract a convergent subsequence whose limit we denote A/Q. ■

In general this argument does not necessarily imply that there exists a pole of A/Q in $D_I(0,1)$, so supplementary conditions must be introduced when I contains more than one element.

Definition 11.7.1. *Let $S_I^\star$ be the set of elements $\theta \in S_I^\infty$ for which there exists a function $f \in \mathcal{F}_I(\delta)$ satisfying the following properties:*

(i) $|f(0)|_p \geq 1$ *for* $p \in K(\theta)$,

(ii) $|f(0)|_p \geq |\theta|_p$ *for* $p \in L(\theta)$.

Theorem 11.7.1. *The set $S_I^\star$ is closed in A_I.*

Proof. Let (θ_ν) be a convergent sequence of elements of $S_I^\star$. We suppose that the set $K(\theta_\nu)$ does not depend of the index ν. We may associate to (θ_ν) a sequence of functions (f_ν) of the family $\mathcal{F}_I(\delta)$ satisfying conditions (i) and (ii) and extract a convergent subsequence. The limit also satisfies the conditions (i) and (ii); then the maximum principle shows that f possesses at least one pole in $D_I(0,1)$ and that the inverse of this pole belongs to $S_I^\star$. Therefore $S_I^\star$ is closed. ■

We can define other closed sets; for instance:

Definition 11.7.2. *Let I and J be two finite sets of absolute values of* **Q**. *We set $S_I^{\star J} = S_I^J \cap S_I^\star$.*

Theorem 11.7.2. *The sets $S_I^{\star J}$ are closed in A_I.*

The proof is similar to the one above.

It seems natural to try to characterize the derived set of $S_I^\star$ as in the case of S (§6.2), but here new difficulties appear; they come in particular from the fact that in general the polynomials P are not irreducible. However, one can state the following result.

Theorem 11.7.3. *Let $\theta \in S_I^\infty$, if a function $A/Q \in \mathcal{F}_I(\delta)$ with $A \neq \pm P$ can be associated to θ, then θ belongs to the derived set of S_I^∞.*

This condition is equivalent to: *the equality $|A(x)|_\infty = |Q(x)|_\infty$ is satisfied for at most finitely many points on the unit circle.*

Proof. As in the complex case, we consider the sequence (φ_ν) of rational functions defined by

$$\varphi_\nu(x) = \frac{A(x) + x^{\nu+a}P(x)}{Q(x) + x^{\nu+s}B(x)}, \quad \nu \in \mathbf{N}.$$

We then show that the functions φ_ν belong to the family $\mathcal{F}_I(\delta)$. ∎

Corollary *Let θ be an element of S_I^∞. If the minimal polynomial of θ has a reciprocal factor of degree two, then θ is a limit point of a sequence of elements of S_I^∞.*

Proof. Suppose $P = P_1 \cdot P_2$ where P_1 is a reciprocal polynomial of degree two; then by Lemma 11.8 there exists a polynomial A_1 such that the rational fraction A_1/Q_1 belongs to the family $\mathcal{F}_I(\delta)$, where (I_1, I_2) is the partition associated with the product $P_1 \cdot P_2$. Then the rational function A_1P_2/Q satisfies the conditions of the theorem. ∎

Remark

The fact that the function A/Q satisfies the condition (A_p) for an absolute value $p \in I^-$ does not imply that the same is true for φ_ν. For example, consider the rational function

$$\frac{A(x)}{Q(x)} = \frac{q - (2q+1)x + qx^2}{q - 2(q+1)x + qx^2},$$

which satisfies condition (A_p) for every prime divisor p of q. By constructing the correponding sequence (φ_ν) we obtain

$$Q_\nu(X) = q - 2(q+1)X + qX^2 + X^{\nu+2}(q - (2q+1)X + qX^2).$$

This polynomial has $(\nu+2)$ zeros on $C_p(0,1)$, so the highest common divisor of P_ν and Q_ν is at most of degree four. Hence φ_ν possesses at least one pole on $C_p(0,1)$, and cannot satisfy condition (A_p)

Definition 11.7.3 *We denote by S_I^{**} the set of elements $\theta \in S_I^\infty$ to which can be associated a function A/Q of the family $\mathcal{F}_I(\delta)$ satisfying the following conditions:*

$$|A(0)|_p \geq |Q(0)|_p |\theta|_p, \ \text{for} \ \ p \in I^-, \ \ \text{and} \ \ |A(0)|_\infty \geq |Q(0)|_\infty \ \ \text{if} \ \ \infty \in I.$$

Theorem 11.7.4. *The set S_I^{**} is closed in A_I, and an element $\theta \in S_I^{**}$ belongs to the derived set if and only if one can associate to θ a function $A/Q \in \mathcal{F}_I(\delta)$, where A is distinct from P and Q and such that the above inequalities hold.*

The above results are simpler when the set I contains only one element. The case $I = \{\infty\}$ corresponds to Pisot and Salem numbers (Chapters 5–8); we now deal with the case $I = \{p\}$, where p is a non-archimedean absolute value of $\mathbf{Q}$.

An element $\alpha \in U_{\{p\}}$ is associated to an irreducible polynomial P:

$$P(X) = \varepsilon(p^t X^s + q_{s-1}X^{s-1} + \cdots + q_0) \quad \text{where} \ \ |q_0|_\infty \leq p^t.$$

Since the polynomial P is irreducible, the set $D_{\{p\}}$ is empty. Hence there exists a partition of $U_{\{p\}}$ into two sets $S_{\{p\}}^\infty$ and $T_{\{p\}}$; $S_{\{p\}}^\infty$ is called **the set of Chabauty numbers** in $\mathbf{Q}_p$, and $T_{\{p\}}$ is called **the set of Salem p-adic numbers**.

Let α be an element of $S_{\{p\}}^p$. Besides the above conditions, the coefficients of the polynomial P satisfy $|q_j|_p < 1$ for $j = 0, 1, \ldots, s-2$. Hence, if $\alpha \notin S_{\{p\}}^\infty$, P is a reciprocal polynomial of degree two: $P(X) = p^t X^2 + aX + p^t$, where $|a|_p = 1$ and $|a|_\infty < 2p^t$.

Hence every element of $S_{\{p\}}^p$ whose minimal polynomial is of degree greater than two belongs to $S_{\{p\}}^\infty$. Then we can show the following result.

Theorem 11.7.5. *The set $S_{\{p\}}^\infty$ is closed in $\mathbf{Q}_p$.*

Proof. We give here a proof that uses Lemma 11.6. Beyond its historic interest, the proof has the advantage of showing where the difficulties one meets with the sets S_I^∞ when I contains more than one element are rooted. It is also possible to prove this theorem with the help of the family $\mathcal{F}_I(\delta)$.

Let θ be an element of $S^{\infty}_{\{p\}}$. By (11.6.3) there exists an element $\lambda \in \mathbf{Z}_p$ such that the following inequalities hold:

$$|\lambda\theta|_p \geq 1, \qquad \sum_{n=0}^{\infty} \varepsilon_{\infty}^2(\lambda\theta^n) \leq p^{2t}.$$

Consider a sequence (θ_ν) of elements of $S^{\infty}_{\{p\}}$ converging to a number θ. One can assume that $|\theta_\nu|_p = |\theta|_p = p^t$ for every $\nu \in \mathbf{N}$.

We can associate to this sequence a sequence (λ_ν) of elements of $\mathbf{Z}_p$ satisfying $|\lambda|_p \geq p^{-t}$. From this sequence we can extract a convergent subsequence whose limit λ satisfies the inequalities

$$p^{-t} \leq |\lambda|_p < 1, \quad \text{and} \quad \sum_{n=0}^{\infty} \varepsilon_{\infty}^2(\lambda\theta^n) \leq p^{2t}.$$

Hence the series $\sum\limits_{n=0}^{\infty} E(\lambda\theta^n)X^n$ is associated to a function of bounded characteristicin $D_\infty(0,1) \times D_p(0,1)$. It is then rational and we can deduce that θ belongs to $S^{\infty}_{\{p\}}$ and λ to $Q^{\{p\}}[\theta]$. ∎

Theorem 11.7.6. *An element $\theta \in S^{\infty}_{\{p\}}$ belongs to the derived set if and only if there exists a polynomial $A \in \mathbf{Z}[X]$, distinct from P and Q and such that the rational function A/Q belongs to $\mathcal{F}_{\{p\}}(\delta)$.*

One can also obtain a characterization of the second derived set.

A proof analogous to that of Lemma 11.7 shows that a function belonging to a family $\mathcal{F}_{\{p\}}(\delta)$ exists for every element $\theta \in S^{\{\infty,p\}}_{\{p\}}$. Hence we deduce

Theorem 11.7.7. *The set $S^{\{\infty,p\}}_{\{p\}}$ is closed in $\mathbf{Q}_p$.*

11.8 Limit points of the sets T_I

As in the real case, we are unable to determine all the limit elements of T_I. Here the problem seems even harder, since our knowledge of the derived set of S^{∞}_I is fragmentary. We can however prove the following theorem, which is a generalization of Theorem 6.4.1.

Theorem 11.8. *Every element of S_I^∞ is a limit of at least one sequence of elements of T_I.*

Proof. Let $\theta \in S_I^\infty$. We assume, to begin with, that the associated polynomials P and Q are relatively prime. Set

$$R_n(X) = X^n P(X) + Q(X), \quad n \in \mathbf{N};$$

R_n is a reciprocal polynomial.

The proof is in two steps:

1) *For n large enough R_n has as zero an element $\tau_n \in T_I$.* By considering the methods used for T in Chapter 6 together with the Newton polygon of R_n we have

- For $p \in I$ the zeros of R_n except $\tau_{n,p}$ and $1/\tau_{n,p}$ belong to $C_p(0,1)$.
- For $p \notin I$ the zeros of R_n belong to $C_p(0,1)$.

Moreover the polynomial R_n is not necessarily irreducible, but the irreducible polynomial which has $\tau_{n,\infty}$ as zero cannot always be of degree two for all n. It follows that τ_n belongs to T_I for an infinitely many n.

2) *The sequence (τ_n) converges to θ.* We have the two equalities

$$P(\tau_n) = -\tau_n^{s-n} P(1/\tau_n) \tag{1}$$
$$P(\tau_n) = q(\tau_n - \theta)(\tau_n - \theta^{(2)}) \cdots (\tau_n - \theta^{(s)}); \tag{2}$$

we proceed then as for the set T (Theorem 6.4.1).

Now suppose that P and Q are not relatively prime. Two cases can arise: either P is itself a reciprocal polynomial of degree two, or it has such a factor. In both cases I contains the archimedean absolute value; then by Theorem 11.7.3, θ can be regarded as a limit of a sequence of elements of S_I^∞, which has at most a finite number of terms whose minimal polynomial is reciprocal and of degree two, or which has such a polynomial as a factor. The proof ends as in the real case. ∎

Notes

The results of this chapter form a part of the history of generalizations of Pisot and Salem numbers. We recall the principal episodes. After Salem's theorem on the closure of the set S became known, it seemed natural to define closed sets of k-tuples of algebraic integers. The results and difficulties are explained in Chapter 9.

The first non-real Pisot set was introduced by Chabauty in 1950 [2], and is called the **set of p-adic Pisot numbers** or **set of Chabauty numbers**. Its properties are close to those of the set S, and are discussed in §11.7 as a particular case of set U_I.

In 1962 Pisot defined the sets S_q discussed in Chapter 9.

Also in 1962 Bateman and Duquette introduced Pisot elements in a field of formal power series, which will be dealt with in Chapter 12.

Since the properties concerning the distribution modulo 1 of sequences $(\lambda\alpha^n)$ cannot be generalized to the sets S_q, in 1964 Bertrandias [1] defined Pisot elements in a ring of adeles of $\mathbf{Q}$, with characterizations analogous to those of classical Pisot numbers. The study of these sets, together with that of the corresponding Salem elements (Decomps-Guilloux [3]), is the subject of the main part of this chapter.

Some results can be generalized to adeles of an algebraic number field ([5], [6], [7], [8], [9]), but these will not be dealt with in this book.

Finally, in 1968 Rauzy tried to give a general definition of Pisot elements that allowed him to establish a relation between the classical Pisot numbers and the Pisot elements of $\mathbf{R}\{1/x\}$. We generalized this definition slightly, which will be expounded in the conclusion of Chapter 12.

The set U_I constitutes a generalization of $S \cup T$. It was studied and caracterized by Decomps-Guilloux [3] (Theorems 11.1 and 11.2.1) and this characterization was recently improved by Grandet-Hugot [7] (Theorem 11.2.1).

The sets $S_I^{p'}$ were introduced by Bertrandias [1], who proved Theorem 11.2.5 and characterized the sets by the distribution properties of the sequence $(\varepsilon(\lambda\theta^n))$ analogous to those of Pisot numbers. Other characterizations state sufficient conditions for a real number to be a Pisot number, and can be found in [7] (Theorems 11.5.2 and 11.5.3).

The closed subsets of S_I^{∞} were first studied in the ring of adeles of an algebraic field ([5], [6], [7], [8], [9]). We have adapted the results to the simpler case of the adeles of $\mathbf{Q}$.

The set B_I was introduced in 1967 by Bertrandias (unpublished), and it is the first known closed set among the discussed sets. Its study is interesting because it shows the difficulties met within the construction of closed sets. These difficulties are similar to those encountered in the sets of k-tuples of algebraic integers.

References

[1] F. Bertrandias-Besson, Ensembles remarquables d'adèles algébriques, *Bull. Soc. Math. France, Mémoire* 4, (1962).

[2] C. Chabauty, Sur la répartition modulo 1 de certaines suites p-adiques, *C.R.A.S.* 231, (1950), 465-466.

[3] A. Decomps-Guilloux, Généralisation des nombres de Salem aux adèles, *Acta Arith.* 16, (1970), 265-314.

[4] M. Grandet-Hugot, Etude de certaines suites $(\lambda\alpha^n)$ dans les adèles, *Ann. Ec. Norm. Sup. 3è série*, 83, (1966), 171-185.

[5] M. Grandet-Hugot, P.V. eléments dans un corps de nombres algébriques, *Acta Arith.* 20, (1972), 203-214.

[6] M. Grandet-Hugot, Fonctions à caractéristique bornée et P.V éléments. *Acta Arith.* 34, (1979), 349-360.

[7] M. Grandet-Hugot, Nouvelles caractérisations des nombres de Pisot dans un anneau d'adèles, *Acta Arith.* 52, (1989), 229-239.

[8] H.G. Senge, Closed sets of algebraic numbers, *Duke Math. J.* 34, (1967), 307-325.

[9] C.J. Smyth, Closed sets of algebraic numbers in complete fields, *Mathematika*, 17, (1970), 199-206.

CHAPTER 12

PISOT ELEMENTS IN A FIELD OF FORMAL POWER SERIES

Suppose k is an arbitrary commutative field; in this chapter we define and study sets of algebraic elements over $k[x]$ analogous to the sets U and S.

Here the situation is quite different from that of the previous chapters: the sets are dense, the characterizations simpler, no Salem elements. Moreover the results can be easily extended to adeles of $k(x)$.

12.0 Generalities and notation

Let k be a commutative field. We set $\mathcal{Z} = k[x]$ and $\mathcal{F} = k(x)$. The following theorem shows that there exists a set of absolute values on $\mathcal{F}$.

Theorem 12.0.1.

1. If k is a finite field with q elements, every absolute value on $\mathcal{F}$ is ultrametric, trivial on k and of one of the following forms:

(i) For $a \in \mathcal{Z}$, $|a| = q^{\deg a}$ is the ∞-adic absolute value.

(ii) Let v be a prime polynomial in $\mathcal{Z}$. If a is relatively prime to v we set $|a|_v = 1$, and if $a = v^h b$ with b relatively prime to v we set $|a|_v = q^{-h}$. These are the v-adic absolute values.

2. If k is an infinite field we may define on $\mathcal{F}$ the same absolute values as in case 1, ultrametric and trivial on k. Here however there exist other absolute values.

In all these cases we denote by $\mathcal{V}$ the set of absolute values.

Theorem 12.0.2. *The set $\mathcal{V}$ satisfies the product formula, that is, for all $x \in \mathcal{F}, x \neq 0$, we have $\prod_{v \in \mathcal{V}} |x|_v = 1$.*

For every $v \in \mathcal{V}$, we denote by $\mathcal{F}_v$ the completion of $\mathcal{F}$ for the absolute value v, by $\mathcal{Z}_v$ the valuation ring, by Γ_v the value group of $\mathcal{F}_v$ and by $\mathcal{C}_v$ the completion of algebraic closure of $\mathcal{F}_v$.

For $a \in \mathcal{F}_v$ and $r \in \Gamma_v$ we set

$$D_v(a,r) = \{x \in \mathcal{F}_v; \quad |x - a|_v < r\}$$
$$\hat{D}_v(a,r) = \{x \in \mathcal{F}_v; \quad |x - a|_v \leq r\}.$$

The fields $\mathcal{F}_v$ are locally compact if and only if$k$ is a finite field.

We will now study $\mathcal{F}_\infty$. It can easily be shown that $\mathcal{F}_\infty$ is the field $k\{t^{-1}\}$ of the formal Laurent series $x = \sum_{n=-\infty}^{h} a_n t^n$, $a_n \in \mathcal{F}$; h is called the degree of x.

The Artin decomposition of x is defined by the following result.

Theorem 12.0.3. *Every element $x \in \mathcal{F}_\infty$ can be written in a unique way: $x = E(x) + \varepsilon(x)$ with $E(x) \in \mathcal{Z}$ and $|\varepsilon(x)| < 1$.*

We now define uniform distribution modulo $\mathcal{Z}$ in $\mathcal{F}_\infty$.

Definition 12.0.1. *A sequence (x_n) of elements of $\mathcal{F}_\infty$ is said to be* **uniformly distributed modulo** $\mathcal{Z}$ *if, for every $h \in \mathbf{N}$ and every $\beta \in \mathcal{F}_\infty$, we have*

$$\lim_{N \to +\infty} \frac{1}{N} A(N; h, \beta) = q^{-h}$$

where $A(N; h, \beta) = \text{card}(\{n \in \mathbf{N}; n \leq N \text{ and } |x_n - \beta| < q^{-h}\})$.

If k is a finite field, we can prove **Koksma's theorem**. At present we only state a corollary:

Theorem 12.0.4. *Let $\alpha \in \mathcal{F}_\infty \setminus \mathcal{Z}_\infty$; the sequence $(x\alpha^n)$ is uniformly distributed modulo $\mathcal{Z}$ for almost all $x \in \mathcal{F}_\infty$* (in the sense of a Haar measure).

All the extensions of $\mathcal{F}$ considered here are separable and contained in $\mathcal{F}_\infty$. They are simple extensions and have an integer basis.

In particular, if k is a perfect field,[†] every algebraic extension of $\mathcal{F}$ contained in $\mathcal{F}_\infty$ is separable.

12.1 Definition of the sets $\mathcal{U}$ and $\mathcal{S}$

Definition 12.1.1. *We denote by $\mathcal{U}$ the set of elements $\alpha \in \mathcal{F}_\infty$ satisfying $|\alpha| > 1$, which are algebraic integers over $\mathcal{Z}$ and whose remaining conjugates belong to the disk $\hat{D}_\infty(0,1)$.*

The set $\mathcal{U}$ contains $\mathcal{Z}$ and is hence non-empty. Moreover the definition implies that every element of $\mathcal{U}$ is separable on $\mathcal{F}$. Here the generalization of the notion of Pisot numbers is easy, but we are unable generalize the notion of Salem numbers.

Definition 12.1.2. *We denote by $\mathcal{S}$ the set of elements $\theta \in \mathcal{U}$, whose remaining conjugates belong to the disk $D_\infty(0,1)$. This set is called the* **Pisot set** *of $\mathcal{F}_\infty$.*

By considering the Newton polygon of the minimal polynomial of an element $\alpha \in \mathcal{U}$ (resp. of an element $\theta \in \mathcal{S}$) we obtain the following characterization.

Theorem 12.1.1. *An element α (resp. θ) in $\mathcal{F}_\infty$ belongs to the set $\mathcal{U}$ (resp. $\mathcal{S}$) if and only ifits minimal polynomial can be written as $P(X) = X^s + q_1X^{s-1} + \cdots + q_s$, $q_j \in \mathcal{Z}$ for $j = 1, \ldots, s$, with $|q_1| = |\alpha| > 1$ (resp. $|q_1| = |\theta| > 1$) $|q_j| \leq |\alpha|$ (resp. $|q_j| < |\theta|$) for $j = 2, \ldots, s$.*

This implies that the minimal polynomial of a $\mathcal{U}$-element is irreducible and an element $\alpha \in \mathcal{U}$ is necessarily separable over $\mathcal{F}$.

We obtain the following theorem by applying Minkowski's theorem to a field of formal power series.

Theorem 12.1.2. *Every finite separable extension of $\mathcal{F}$ that is included in $\mathcal{F}_\infty$ can be generated by a Pisot element.*

Corollary 1. *If k is a perfect field, every finite extension of $\mathcal{F}$ contained in $\mathcal{F}_\infty$ can be generated by a Pisot element.*

[†] A field k is perfect if it is of characteristic zero or if it is of characteristic p with $k^p = k$.

Corollary 2. *Suppose k is an arbitrary commutative field. Then $\mathcal{F}_\infty$ contains Pisot elements of all degrees in $\mathcal{F}$.*

12.2 Characterizations of the sets $\mathcal{U}$ and $\mathcal{S}$

The following characterizations recall on the one hand those of the sets U and S, and on the other those of the sets U_I and S_I^J. In particular they justify calling $\mathcal{S}$ a Pisot set.

Theorem 12.2.1. *An element $\alpha \in \mathcal{F}_\infty$, satisfying $|\alpha| > 1$, belongs to $\mathcal{U}$ if and only ifthere exists a non-zero element $\lambda \in \mathcal{F}_\infty$ such that, for n large enough, $|\varepsilon(\lambda\alpha^n)| < |\alpha|^{-2}$. Then λ belongs to $\mathcal{F}(\alpha)$.*

The proof uses the following lemma.

Lemma 12.2. *Let (ξ_n) be a sequence of elements of $\mathcal{F}_\infty$ and (u_n) a sequence of elements of $\mathcal{Z}$. We suppose (ξ_n) satisfies a recurrence relation*

$$\xi_{n+r} + a_{r-1}\xi_{n+r-1} + \cdots + a_0\xi_n = 0 \qquad (n \geq N);$$

where a_j, $(j = 0, \ldots, r-1)$ are fixed elements of $\mathcal{F}_\infty$ satisfying $\max|a_j| > 1$ and, for n large enough,

$$|u_n - \xi_n| < \frac{1}{\max|a_j|^2};$$

then the series $\sum_{n\in\mathbf{N}} u_n X^n$ is rational.

Proof. We set $A = \max|a_j|$. From the assumptions it follows that there exists an integer N such that $|u_n - \xi_n| < A^{-2}$ for $n \geq 2$. We then define ε_n by the equalities

$$\begin{aligned}
\varepsilon_n &= u_n + a_{r-1}u_{n-1} + \cdots + a_0u_{n-r} \\
&= u_n - \xi_n + a_{r-1}(u_{n-1} - \xi_{n-1}) + \cdots + a_0(u_{n-r} - \xi_{n-r}) \\
\eta_n &= \varepsilon_n + a_{r-1}\varepsilon_{n-1} + \cdots + a_0\varepsilon_{n-r}.
\end{aligned}$$

Hence we get the following inequalities

$$|\varepsilon_n| < A^{-1} \text{ for } n \geq N; \quad |\eta_n| < 1 \text{ for } n > N + 2r.$$

Let D_n be the Kronecker determinant of order $n+1$ associated with the series $\sum_{n\in\mathbf{N}} u_n X^n$: $D_n = \det(u_{i+j})$. For $i \geq N+r$ and $j \geq N+r$ we manipulate the rows and columns. In the end we obtain $D_n = \det(\delta_{i,j})$ where

$$\begin{aligned} \delta_{i,j} &= u_{i+j} && \text{if } i < N+r,\ j < N+r \\ \delta_{i,j} &= \eta_{i+j} && \text{if } i \geq N+r,\ j \geq N+r \\ \delta_{i,j} &= \varepsilon_{i+j} && \text{in the other cases.} \end{aligned}$$

Then if $M = \max\limits_{0\leq i\leq N+2r-2} |u_i|$, we have $|D_n| \leq M^{N+r} A^{-n+N+r-1}$. Hence for n large enough we have $|D_n| < 1$ and hence $D_n = 0$ because $D_n \in \mathcal{Z}$. The series $\sum_{n\in\mathbf{N}} u_n X^n$ is rational. ∎

Proof of the theorem. The proof of the necessary condition is close to that done for the sets U and U_I (Theorems 5.2.4 and 11.2.3) and we will not reproduce it here.

In order to prove that the condition is sufficient, we consider a pair (λ, α) of elements of $\mathcal{F}_\infty$ for which we have $|\alpha| > 1, \lambda \neq 0$, $|\varepsilon(\lambda\alpha^n)| < |\alpha|^{-2}$ for $n \geq n_0$.

Then the sequences (u_n) and $(\lambda\alpha^n)$ satisfy the assumptions of the lemma, hence the series $\sum_{n\in\mathbf{N}} u_n X^n$ is rational, and Fatou's lemma allows us to complete the proof. ∎

The characterization of $\mathcal{S}$ follows immediatly from the previous result.

Theorem 12.2.2. *An element $\theta \in \mathcal{F}_\infty$, satisfying $|\theta| > 1$, belongs to $\mathcal{S}$ if and only ifthere exists an element $\lambda \neq 0$ in $\mathcal{F}_\infty$ such that $\lim\limits_{n\to+\infty} \varepsilon(\lambda\theta^n) = 0$; moreover λ belongs to $\mathcal{F}(\theta)$.*

This condition can be replaced by the following: *the sequence $(\lambda\alpha^n)$ has a finite number of limit points modulo $\mathcal{Z}$.*

The results of this section show in particular that, if k is a finite field, the sets $\mathcal{S}$ and $\mathcal{U}$ are included in the exceptional set of Koksma's theorem on uniform distribution modulo $\mathcal{Z}$.

12.3 Limit points of the sets $\mathcal{U}$ and $\mathcal{S}$

The following theorem shows the main difference between the sets S and $\mathcal{S}$.

Theorem 12.3.1. *The sets $\mathcal{S}$ and $\mathcal{U}$ are dense in $\mathcal{F}_\infty/\mathcal{Z}_\infty$.*

Proof. Let α be an element of $\mathcal{F}_\infty$ satisfying $|\alpha| > 1$ and not belonging to $\mathcal{S}$. We are going to construct a sequence of elements of $\mathcal{S}$ converging to α. For this, consider the sequence (P_n) with coefficients in $\mathcal{Z}$ defined by

$$P_1(X) = X - E(\alpha) \quad P_n(X) = XP_{n-1}(X) - E(\alpha)P_{n-1}(\alpha).$$

Every polynomial P_n satisfies the conditions of Theorem 12.1.1. P_n is irreducible on $\mathcal{F}$ and has as zero an element θ_n of $\mathcal{S}$ satisfying $|\theta_n| = |\alpha|$. Moreover the sequence (θ_n) is not stationary because we have $P_n(\theta_{n-1}) = -E(\alpha)P_{n-1}(\alpha) \neq 0$. The equality $P_n(\alpha) = \varepsilon_{n-1}(\alpha)$ implies $|\theta_n - \alpha| < |\alpha|^{-n+1}$, hence $\alpha = \lim\limits_{n\to+\infty} \theta_n$. Suppose α does not belong to $\mathcal{U}$; then there exists a sequence of elements of $\mathcal{U}$ converging to α. This sequence can be constructed explicitly by replacing Artin's decomposition by a decomposition of the form $\alpha = E'(\alpha) + \varepsilon'(\alpha)$, with $E'(\alpha) \in \mathcal{Z}$ and $|\varepsilon'(\alpha)| \leq 1$. ■

Remark

Actually we have proved that the sets $\mathcal{S}$ and $\mathcal{U}$ are dense on the circle $C(0, |\alpha|)$. Though $\mathcal{S}$ is dense, it possesses closed subsets, and is the union of closed subsets.

Definition 12.3. *Let r be a positive real number. We denote by $\mathcal{S}_r$ the subset of $\mathcal{S}$ of elements θ whose remaining conjugates in $\mathcal{C}_\infty$ have absolute value at most equal to q^{-r}.*

Then $\mathcal{S} = \bigcup\limits_{r>0} \mathcal{S}_r$.

Lemma 12.3. *A element $\theta \in \mathcal{F}_\infty$ satisfying $|\theta| > 1$ belongs to $\mathcal{S}_r$ if and only ifthere exists a non-zero element $\lambda \in \mathcal{F}_\infty$ such that $|\varepsilon(\lambda\theta^n)| \leq cq^{-nr}$ for $n \geq n_0$.*

Proof. The assumptions imply θ belongs to $\mathcal{S}$ and λ to $\mathcal{F}(\theta)$. Hence the series $\sum\limits_{n\in\mathbf{N}} u_nX^n$ is rational and the poles of the associated function, other than $1/\theta$, have an absolute value greater than q^r. Hence θ belongs to $\mathcal{S}_r$. ■

Theorem 12.3.2. *The sets $\mathcal{S}_r$ are closed in $\mathcal{F}_\infty$.*

Proof. Let (θ_ν) be a sequence of elements of $\mathcal{S}_r$ converging to a limit θ; then

$$u_{n,\nu} = E(\theta_\nu^n) + \sum_{j=2}^{s_\nu} \theta_\nu^{(j)^n},$$

where s_ν is the degree of θ_ν. Hence we have

$$|\varepsilon(\theta_\nu^n)| = \sum_{j=2}^{s_\nu} |\theta_\nu^{(j)}|^n \leq q^{-m}.$$

Moreover, for every integer n, there exists an index $\nu_0(n)$ such that $\nu > \nu_0 \Rightarrow |\theta_\nu^n - \theta^n| < q^{-m}$, hence

$$|E(\theta^n) - E(\theta_\nu^n)| \leq \max\{|\theta^n - \theta_\nu^n|, |\varepsilon(\theta^n) - \varepsilon(\theta_\nu^n)|\} < 1.$$

This implies $E(\theta^n) = E(\theta_\nu^n)$ for $\nu > \nu_0(n)$, hence $|\varepsilon(\theta_\nu^n)| \leq q^{-m}$, and according to the lemma, θ belongs to $\mathcal{S}_r$. ■

12.4 Relation between the sets S and $\mathcal{S}$

We have seen that the classical Pisot numbers and their generalizations in an adele ring of $\mathbf{Q}$ or a field of formal series present many analogies. It seems natural to think of giving them a common definition.

We can also seek to establish a relation between real Pisot numbers and Pisot elements in the field $\mathbf{Q}\{x^{-1}\}$, which represent algebraic functions meromorphic at infinity.

Here we present recent advances in two directions. We only state the results, referring the reader to the bibliography for the demonstrations.

Notation

Let A be a Dedekind ring and K its quotient field. We suppose that there exists over K a family V of absolute values satisfying the product formula and including the set I of absolute values associated to the essential valuations of A. We furnish K with a pseudo-absolute value by setting $\|x\| = \sup_{v \in I} |x|_v$ for $x \in K$; then if x belongs to A and is not zero, $\|x\| \geq 1$.

For every $v \in V$, we denote by K_v the completion of K and by C_v the completion of the algebraic closure of K. Then the completion of K for the pseudo-valuation is the ring $\mathcal{A} = \prod_{v \in V} K_v$.

Examples:

1. If $A = \mathbf{Z}$ then $I = \{\infty\}$ and $\mathcal{A} = \mathbf{Q}$.
2. If $A = k(x)$, where k is an arbitrary commutative field, then $\mathcal{A} = \mathcal{F}$.
3. If $A = Q^I$, where I is the set of ultrametric absolute values of $\mathbf{Q}$, then $\mathcal{A}$ is the ring A_I of I-adeles of $\mathbf{Q}$.

Definition 12.4.1. *Let A be a Dedekind ring and let γ be a real positive number. We denote by $\mathcal{S}(A,\gamma)$ the set of elements $\theta \in \mathcal{A}$ that are algebraic integers over A satisfying $\|\theta\| > 1$ and whose remaining conjugates in $\prod\limits_{v \in I} C_v$ have absolute value not greater than $\|\theta\|^{-\gamma}$.*

Definition 12.4.2. *The set $\mathcal{S}(A) = \bigcup\limits_{\gamma>0} \mathcal{S}(A,\gamma)$ is called the* **Pisot set** *of A.*

Remark

The sets $\mathcal{S}(\mathbf{Z})$ and $\mathcal{S}(k[x])$ are Pisot sets respectively in $\mathbf{R}$ and in $\mathcal{F}_\infty$; but the set $\mathcal{S}(Q^I)$ is distinct from S_I^∞ because $\mathcal{S}(Q^I) = \bigcup\limits_{J \subset I} S_I^J$.

Theorem 12.4.1. *An element $\theta \in \mathcal{A}$ satisfying $\|\theta\| > 1$ belongs to the set $\mathcal{S}(A,\gamma)$ if and only if $\inf\limits_{a \in A} \|\theta^n - a\| < c\|\theta\|^{-n\gamma}$, where c is a non-zero constant.*

Corollary *An element $\theta \in \mathcal{A}$ satisfying $\|\theta\| > 1$ is a Pisot element if and only ifthere exist two constants c and ρ, $0 < \rho < 1$, such that $\inf\limits_{a \in A} \|\theta^n - a\| < c\rho^n$.*

In the real case we find the first characterization of Pisot numbers by Thue [6].

Theorem 12.4.2. *For every $\gamma > 0$ the set $\mathcal{S}(A,\gamma)$ is discrete in $\mathcal{A}$.*

The proof uses the following lemma.

Lemma 12.4. *The degree of an element $\theta \in \mathcal{S}(A,\gamma)$ over K does not exceed $1 + 1/\gamma$.*

Now we suppose $A = \mathbf{Z}$. Let $\mathcal{Z}'$ be the set of linear combinations with integer coefficients of the binomial polynomials; it is known that $\mathcal{Z}'$ is the set of polynomials $P \in \mathbf{Z}[X]$ satisfying $P(\mathbf{Z}) \subset \mathbf{Z}$. If we furnish $\mathcal{Z}'$ with the absolute value ∞, then $\mathcal{Z}'$ possesses the properties of a ring A and we can define the sets $\mathcal{S}(\mathcal{Z}',\gamma)$ and prove the following result.

Theorem 12.4.3. *If γ is an irrational number, then the following assertions are equivalent:*

(i) f belongs to $\mathbf{C}\{x^{-1}\}$ and $f(n)$ to $\mathcal{S}(\mathbf{Z},\gamma)$ for every integer n large enough.

(ii) f belongs to the set $\mathcal{S}(\mathcal{Z}',\gamma)$.

If γ is rational, we can apply this theorem to every irrational number arbitrarily close to γ, which allows us to state the following result.

Corollary *The following assertions are equivalent:*

(i) f belongs to $\mathbf{C}\{x^{-1}\}$ and $f(n)$ is a Pisot number for every integer n large enough.

(ii) f belongs to the set $\mathcal{S}(\mathcal{Z}',\gamma)$.

The conclusion is that if θ is a Pisot element of $\mathbf{C}\{x^{-1}\}$, then the function $z \mapsto \theta(z)$ is an algebraic function meromorphic at infinity. Moreover, if the minimal polynomial of θ has all its coefficients in $\mathcal{Z}$, then for n large enough $\theta(n)$ is a Pisot number.

Conversely, let θ be a function meromorphic at infinity such that $\theta(n)$ is a Pisot number for n large enough; then θ is a Pisot element of $\mathbf{C}\{x^{-1}\}$.

Notes

In 1962 Bateman and Duquette [1] introduced and characterized Pisot elements in a field of formal power series (Theorems 12.1.1 and 12.1.2). The study of the sets $\mathcal{S}$ and $\mathcal{U}$ was resumed in 1967 by Grandet-Hugot ([3], [4]). In particular she showed that $\mathcal{S}$ and $\mathcal{U}$ are dense (Theorem 12.3.1) and extended these results to the *adeles of* $\mathcal{F}$. At the same time Rauzy defined the sets $\mathcal{S}_r$ and proved they were closed (Theorem 12.3.2, unpublished). Finally, in 1968, he established a relationship between Pisot sets and the set $\mathcal{S}$ (§12.4) [5], which suggests that there exists a common definition for all Pisot elements.

We should also mention that as regards classical Pisot numbers, the Pisot elements of a field of formal power series intervene in the study of periodic expansions through the Jacobi-Perron algorithm (cf. Dubois [2]).

References

[1] P. Bateman and A. Duquette, The analogue of Pisot-Vijayaraghavan numbers in fields of power series, *Ill. J. Math.*, 6, (1962), 594-406.

[2] E. Dubois, Algorithme de Jacobi-Perron dans un corps de séries formelles, *Thèse 3è cycle, Caen,* (1970).

[3] M. Grandet-Hugot, Sur une propriété des nombres de Pisot dans un corps de séries formelles, *C.R.A.S.*, 266, ser. A, (1967), A39-A41.

[4] M. Grandet-Hugot, Eléments algébriques remarquables dans un corps de séries formelles, *Acta Arith.* 14, (1968), 177-184.

[5] G. Rauzy, Algébricité des fonctions méromorphes prenant certaines valeurs algébriques, *Bull. Soc. Math. France*, 96, (1968), 197-208.

[6] A. Thue, Über eine Eigenschaft die keine tranzendente Grosse haben kann, *Skrifter Vidensk Krislina*, 2, (1912), 1-15

CHAPTER 13

PISOT SEQUENCES, BOYD SEQUENCES AND LINEAR RECURRENCE

We first prove two theorems that are very useful for studying the convergence properties of certain rational sequences used in the succeeding sections.

13.0 Convergence theorems

Let $\mathbf{b}_n = (a_n, \ldots, a_{n+k-1})$, a k-tuple of integers corresponding to a sequence of $\mathbf{R}^k$-diffeomorphisms $(F_n)_{n\in\mathbf{N}}$. If Φ_n denotes the reciprocal diffeomorphism of F_n, we shall prove that under certain conditions the sequence $\Phi_n(\mathbf{b}_n)$ converges to an element $\boldsymbol{\beta}$ of $\mathbf{R}^k$, and also find a bound for $\boldsymbol{\varepsilon}_n = F_n(\boldsymbol{\beta}) - \mathbf{b}_n$.

The space $\mathbf{R}^k$ is considered with the sup norm.

If $\mathbf{x} = (x_1, \ldots, x_k) \in \mathbf{R}^k$ and $\mathbf{c} = (c_1, \ldots, c_n)$ denotes a k-tuple of real positive numbers, $B(\mathbf{x}, \mathbf{c})$ (resp. $\overline{B}(\mathbf{x}, \mathbf{c})$) is the open (resp. closed) paving with center $\mathbf{x}$ and side length $2\mathbf{c}$.

Theorem 13.0.1. *Let $A = (A_j)_{1\le j\le k}$ be an open paving of $\mathbf{R}^k$ and $(F_n)_{n\in\mathbf{N}}$ a sequence of diffeomorphisms from A to $\mathbf{R}^k$. Assume that the components $F_{n,j}$ of F_n, $1 \le j \le k$, satisfy*

$$F_{n,j} = F_{n-1,j+1}, \quad 1 \le j \le k-1, \quad n \ge 1. \tag{1}$$

If $W = (W_j)_{1\le j\le k}$ is such that $\overline{W} \subset A$, we denote $V_n = F_n(W)$.

Assume that the reciprocal diffeomorphisms Φ_n of F_n satisfy

$$|\, D_k \Phi_{n,j}(F_n(\boldsymbol{\alpha}))\,| \le \psi_{n,j}, \tag{2}$$

$$\sum_{m=1}^{\infty} \psi_{m,j} < +\infty,$$

for every $\boldsymbol{\alpha} \in A$ and j, $1 \le j \le k$.

Denote, for $l \geq 0$, $\Psi_{l,j} = \sum_{m=l+1}^{\infty} \psi_{m,j}$ *and* $\Psi_l = (\Psi_{l,j})_{1 \leq j \leq k} \in (\mathbf{R}^+)^k$.

Assume finally that there exists a $\mathbf{b}_0 \in V_0$, $\mathbf{b}_0 = (a_0, \ldots, a_{k-1})$, $a_i \in \mathbf{Z}$, $0 \leq i \leq k-1$, *such that*

$$B_0 = B(\Phi_0(\mathbf{b}_0), \frac{1}{2}\Psi_0) \subset W.$$

Then,

i) for $n \geq 1$, *every* $\mathbf{b}_n \in F_n(A)$, $\mathbf{b}_n = (a_n, \ldots, a_{n+k-1})$, $a_i \in \mathbf{Z}$, $n \leq i \leq n+k-1$, *such that*

$$\left\| \mathbf{b}_n - F_n \circ \Phi_{n-1}(\mathbf{b}_{n-1}) \right\| \leq \frac{1}{2}, \tag{3}$$

belongs to V_n;

ii) for every j, $1 \leq j \leq k$, *the sequence* $(\Phi_{n,j}(\mathbf{b}_n))$ *tends to* $\beta_j \in A_j$ *and the* β_j *satisfy*

$$|\, \beta_j - \Phi_{0,j}(\mathbf{b}_0) \,| \leq \frac{1}{2} \Psi_{0,j}.$$

Proof. Assume by induction that there exist $\mathbf{b}_m \in V_m$, $0 \leq m \leq n-1$, such that

$$B_{m-1} = B(\Phi_{m-1}(\mathbf{b}_{m-1}), \frac{1}{2}\Psi_{m-1}) \subset B_{m-2} \cdots \subset B_0 \subset W.$$

Since $\mathbf{b}_{n-1} \in V_{n-1} = F_{n-1}(W)$, $\Phi_{n-1}(\mathbf{b}_{n-1}) \in W \subset A$.

Denoting $\widetilde{\mathbf{b}}_n = F_n \circ \Phi_{n-1}(\mathbf{b}_{n-1})$, we have

$$\Phi_n(\widetilde{\mathbf{b}}_n) = \Phi_{n-1}(\mathbf{b}_{n-1}). \tag{4}$$

From (1) and (4) we deduce

$$\widetilde{b}_{n,j} = F_{n,j}(\Phi_n(\widetilde{\mathbf{b}}_n)) = F_{n,j}(\Phi_{n-1}(\mathbf{b}_{n-1})) = F_{n-1,j+1}(\Phi_{n-1}(\mathbf{b}_{n-1})) = b_{n-1,j+1},$$

for $1 \leq j \leq k-1$.

Thus $\widetilde{b}_{n,j} \in \mathbf{Z}$, $1 \leq j \leq k-1$, since $b_{n-1,j+1} \in \mathbf{Z}$, $0 \leq j \leq k-1$.

We now define

$$\mathbf{b}_n = (\widetilde{b}_{n,1}, \ldots, \widetilde{b}_{n,k-1}, a_{n+k-1}) = (a_{n-1}, \ldots, a_{n-1+k-1}, a_{n+k-1})$$

and choose a_{n+k-1} from (3) by

$$\left\|\mathbf{b}_n - \widetilde{\mathbf{b}}_n\right\| = \left\|\mathbf{b}_n - F_n \circ \Phi_{n-1}(\mathbf{b}_{n-1})\right\| = |\, a_{n+k-1} - \widetilde{b}_{n,k}\,| \leq \frac{1}{2}\,.$$

Since $\mathbf{b}_n$ and $\widetilde{\mathbf{b}}_n$ belong to $F_n(A)$ and have the same components except for the last one, we deduce from the mean-value theorem that, for $1 \leq j \leq k$,

$$|\,\Phi_{n,j}(\mathbf{b}_n) - \Phi_{n,j}(\widetilde{\mathbf{b}}_n)\,| \leq \left\|\mathbf{b}_n - \widetilde{\mathbf{b}}_n\right\| \sup_{z \in I} |\, D_k \Phi_{n,j}(z)\,|$$

where $I = [\mathbf{b}_n, \widetilde{\mathbf{b}}_n]$, and from (2) and (3)

$$|\,\Phi_{n,j}(\mathbf{b}_n) - \Phi_{n,j}(\widetilde{\mathbf{b}}_n)\,| \leq \frac{1}{2}\,\psi_{n,j}. \tag{5}$$

If $\mathbf{x} = (x_1, \ldots, x_k) \in B_n = B(\Phi_n(\mathbf{b}_n), \frac{1}{2}\Psi_n)$, then we get for $1 \leq j \leq k$:

$$|\, x_j - \Phi_{n,j}(\mathbf{b}_n)\,| < \frac{1}{2}\,\Psi_{n,j} = \frac{1}{2} \sum_{m=n+1}^{\infty} \psi_{m,j}, \tag{6}$$

and from (5) and (6)

$$|\, x_j - \Phi_{n,j}(\widetilde{\mathbf{b}}_n)\,| < \frac{1}{2}\,\psi_{n,j} + \frac{1}{2}\,\Psi_{n,j} = \frac{1}{2}\,\Psi_{n-1,j};$$

thus, $$B_n \subset B(\Phi_n(\widetilde{\mathbf{b}}_n), \frac{1}{2}\,\Psi_{n-1}) = B(\Phi_{n-1}(\mathbf{b}_{n-1}), \frac{1}{2}\,\Psi_{n-1}) = B_{n-1}$$

and $$B_n \subset B_{n-1} \subset \cdots \subset B_0 \subset W.$$

From $B_n \subset W$, we deduce that $\mathbf{b}_n \in V_n$, that is, i).

Putting (5) in the form

$$|\,\Phi_{n,j}(\mathbf{b}_n) - \Phi_{n-1,j}(\mathbf{b}_{n-1})\,| \leq \frac{1}{2}\,\psi_{n,j} \tag{7}$$

and since the series $\sum_{n \geq 0} \psi_{n,j}$ converges, the sequence $(\Phi_{n,j}(\mathbf{b}_n))_n$ is a Cauchy sequence in W_j and thus tends to $\beta_j \in A_j$.

From (7) we get

$$|\,\Phi_{n,j}(\mathbf{b}_n) - \Phi_{0,j}(\mathbf{b}_0)\,| \leq \frac{1}{2} \sum_{m=1}^{n} \psi_{m,j}$$

and for n going to infinity

$$|\beta_j - \Phi_{0,j}(\mathbf{b}_0)| \leq \frac{1}{2}\Psi_{0,j}, \qquad 1 \leq j \leq k.$$

This completes the proof of ii). ■

Theorem 13.0.2. *We keep the same assumptions as in Theorem 13.0.1. Let* $\beta = \lim_{n \longrightarrow +\infty} \Phi_n(\mathbf{b}_n)$, $\varepsilon_n = F_n(\beta) - \mathbf{b}_n$ *and* $G_{n,l} = F_n \circ \Phi_{n+l}$, $l \geq 1$. *If there exists* j_0, $1 \leq j_0 \leq k$ *such that*

$$|D_k G_{n,l,j_0}(a_{n+l}, \ldots, a_{n+l+k-2}, a_{n+l+k-1} + u)| \leq s_l \tag{8}$$

with $\sum_{l \geq 1} s_l < +\infty$ *and* $|u| < 1/2$, *then the sequence* $(\varepsilon_n)_{n \in \mathbf{N}}$ *is bounded.*

Proof. Let

$$\beta_n = \Phi_n(\mathbf{b}_n), \; n \geq 0; \tag{9}$$

then $\varepsilon_n = F_n(\beta) - F_n(\beta_n)$.

But, by assumption, we have

$$\mathbf{b}_n = \widetilde{\mathbf{b}}_n + x_n \qquad \text{with } \left\|\omega_n\right\| \leq 1/2 \quad \text{by (3)}.$$

Therefore

$$\mathbf{b}_n = F_n \circ \Phi_{n-1}(\mathbf{b}_{n-1}) + \omega_n = F_n(\beta_{n-1}) + \omega_n, \qquad n \geq 1;$$

that is,

$$\beta_{n-1} = \Phi_n(\mathbf{b}_n - \omega_n). \tag{10}$$

Setting now

$$\begin{aligned}\varepsilon_n = F_n(\beta_{n+1}) - F_n(\beta_n) + F_n(\beta_{n+2}) - F_n(\beta_{n+1}) + \cdots \\ + F_n(\beta_{n+l}) - F_n(\beta_{n+l-1}) + F_n(\beta) - F_n(\beta_{n+l}),\end{aligned}$$

and using the fact that β_n tends to β, we deduce

$$\varepsilon_n = \sum_{l \geq 1} F_n(\beta_{n+l}) - F_n(\beta_{n+l-1}).$$

Now from (9) and (10) we have

$$\varepsilon_n = \sum_{l \geq 1} [F_n \circ \Phi_{n+l}(\mathbf{b}_{n+l}) - F_n \circ \Phi_{n+l}(\mathbf{b}_{n+l} - \omega_{n+l})],$$

i.e.,
$$\varepsilon_n = \sum_{l\geq 1}[G_{n,l}(\mathbf{b}_{n+l}) - G_{n,l}(\mathbf{b}_{n+l} - \boldsymbol{\omega}_{n+l})].$$

Since only the last component of $\boldsymbol{\omega}_{n+l}$ is non-zero, then by the mean value theorem we have, from (8),

$$|\,\varepsilon_{n,j_0}\,| \leq \frac{1}{2}\sum_{l\geq 1} s_l, \quad n \geq 1. \tag{11}$$

But from (1) and relations $b_{n,j} = b_{n-1,j+1}$, $1 \leq j \leq k-1$, $n \geq 1$, and since $\boldsymbol{\varepsilon}_n = F_n(\boldsymbol{\beta}) - \mathbf{b}_n$, we get

$$\varepsilon_{n,j} = \varepsilon_{n-1,j+1}, \quad 1 \leq j \leq k-1, \quad n \geq 1.$$

Therefore,
$$\left\|\boldsymbol{\varepsilon}_n\right\| = \sup_{j_0-1\leq l\leq k-j_0} |\,\varepsilon_{n-l,j_0}\,|.$$

Using (11), we deduce that the sequence $(\boldsymbol{\varepsilon}_n)$ is a bounded sequence. ■

13.1 Pisot sequences

In Chapter 5, we showed how certain algebraic properties of α can be deduced from the distributional properties modulo 1 of the sequence $(\lambda\alpha^n)_{n\in\mathbf{N}}$, $\lambda > 0$ and $\alpha > 1$. In particular, by Theorem 5.6.1, the set of pairs $((\lambda,\alpha),\lambda) > 0$, $\alpha > 1$, satisfying

$$\left\|\lambda\alpha^n\right\| \leq \frac{1}{2(\alpha+1)^2}$$

for n large, is countable. The proof uses the property

$$|u_{n+2} - \frac{u_{n+1}^2}{u_n}| < \frac{1}{2} \quad \text{for } n \geq n_0 \tag{1}$$

of the sequence $u_n = \lambda\alpha^n + \left\|\lambda\alpha^n\right\|$. We remark that

$$\lim_{n\longrightarrow+\infty} \frac{u_{n+1}}{u_n} = \alpha \quad \text{and} \quad \lim_{n\longrightarrow+\infty} \frac{u_n^{n+1}}{u_{n+1}^n} = \lambda. \tag{2}$$

Conversely, we are interested in sequences of positive integers satisfying relations of type (1): When do the sequences $\left(\frac{u_{n+1}}{u_n}\right)$ and $\left(\frac{u_n^{n+1}}{u_{n+1}^n}\right)$ converge, and when they do, is $\lim_{n\longrightarrow+\infty} \frac{u_{n+1}}{u_n}$ an algebraic number?

Definition 13.1. *A Pisot sequence $E(a_0, a_1)$ is a sequence of rational integers derived from two positive integers a_0 and a_1 by the relation*

$$-\frac{1}{2} < a_{n+1} - \frac{a_n^2}{a_{n-1}} \leq \frac{1}{2}, \qquad n \geq 1.$$

The theorems of §13.0 lead us to the following theorem.

Theorem 13.1.1. *Let $E(a_0, a_1)$ be a Pisot sequence, with a_0 and a_1 rational positive integers satisfying $a_1 > a_0 + \frac{3}{2}\sqrt{\frac{3}{2}a_0}$. Then $\lim_{n \longrightarrow +\infty} \frac{a_{n+1}}{a_n} = \gamma > 1$ and*

$$\left|\frac{a_1}{a_0} - \gamma\right| < \frac{9}{8(a_1 - a_0)}, \quad \textit{i.e.}\ \left|\frac{a_1}{a_0} - \gamma\right| < \frac{1}{2}\sqrt{\frac{3}{2}a_0}. \tag{3}$$

Moreover $\lim_{n \longrightarrow +\infty} \frac{a_n^{n+1}}{a_{n+1}^n} = \mu$ *and*

$$|a_0 - \mu| < \min\left(\frac{9}{8((a_1/a_0) - 1)^2}, \frac{a_0}{3}\right). \tag{4}$$

Furthermore, the set of γ generated by all a_0 and a_1 satisfying the given conditions is everywhere dense on the real line, and, setting $\mu\gamma^n = a_n + \varepsilon_n$, we have for n sufficiently large

$$|\varepsilon_n| \leq c' \quad \textit{with} \quad c' > \frac{1}{2(\gamma - 1)^2}. \tag{5}$$

Proof. Define $A \subset \mathbf{R}^2$ by $A = \{(\lambda, \alpha);\ \lambda > \nu > 0,\ \alpha > 1 + \tau,\ \tau > 0\}$ and the maps $(F_n)_{n \in \mathbf{N}}$ from A to $\mathbf{R}^2$ by

$$F_n : \quad A \longrightarrow \mathbf{R}^2$$

$$\boldsymbol{\alpha} = (\lambda, \alpha) \mapsto (\lambda\alpha^n, \lambda\alpha^{n+1}).$$

Since the Jacobian determinant of F_n equals $\lambda\alpha^{2n}$, the F_n are diffeomorphisms whose reciprocal diffeomorphisms Φ_n are given by

$$\Phi_n : \quad F_n(A) \longrightarrow A$$

$$(u, v) \mapsto (u^{n+1}/v^n, v/u).$$

Moreover we have the relations

$$D_2\Phi_{n,1}(u_n, u_{n+1}) = -n(\frac{u_n}{u_{n+1}})^{n+1}, \qquad D_2\Phi_{n,2}(u_n, u_{n+1}) = \frac{1}{u_n};$$

thus

$$D_2\Phi_{n,1}(F_n(\boldsymbol{\alpha})) = -\frac{n}{\alpha^{n+1}}, \qquad D_2\Phi_{n,2}(F_n(\boldsymbol{\alpha})) = \frac{1}{\lambda\alpha^n};$$

and hence,

$$|\, D_2\Phi_{n,1}(F_n(\boldsymbol{\alpha}))\,| \leq \frac{n}{(1+\tau)^{n+1}}, \qquad |D_2\Phi_{n,2}(F_n(\boldsymbol{\alpha}))| = \frac{1}{\nu(1+\tau)^n}.$$

Therefore

$$\Psi_0 = (\Psi_{0,1}, \Psi_{0,2}) = \Big(\sum_{n\geq 1} \frac{n}{(1+\tau)^{n+1}}, \sum_{n\geq 1} \frac{1}{\nu(1+\tau)^n}\Big) = \Big(\frac{1}{\tau^2}, \frac{1}{\nu\tau}\Big).$$

If we choose rational integers a_0 and a_1 such that $a_0 \geq \nu + \dfrac{1}{2\tau^2}$ and $\dfrac{a_1}{a_0} \geq 1 + \tau + \dfrac{1}{2\nu\tau}$, then $\mathbf{b}_0 = (a_0, a_1) \in F_0(W)$ and $B_0 \subset W$, where $W = \{(x,y) \in \mathbf{R}^2;\ x > \nu + \dfrac{1}{2\tau^2},\ \ y > 1 + \tau + \dfrac{1}{2\nu\tau}\}$ satisfies $W \subset \overline{W} \subset A$.

We suppose now, for induction purposes, the elements $\mathbf{b}_j$, $0 \leq j \leq n-1$, defined as in Theorem 13.0.1.i).

Since $\mathbf{b}_{n-1} \in F_{n-1}(A)$, then $a_{n-1} = \lambda\alpha^{n-1}$ and $a_n = \lambda\alpha^n$, $\lambda > \nu$, $\alpha > 1+\tau$.

But $\mathbf{b}_{n-1} \in V_{n-1} = F_{n-1}(W)$, so $\dfrac{a_n}{a_{n-1}} \geq 1 + \tau + \dfrac{1}{2\nu\tau}$; thus we get

$$\frac{a_n^2}{a_{n-1}} > \nu(1+\tau)^{n+1} + \nu\,\frac{(1+\tau)^n}{2\nu\tau} > \nu(1+\tau)^{n+1} + \frac{1}{2\tau} + \frac{n}{2}. \tag{6}$$

Since $(a_n)_{n\in\mathbf{N}}$ is a Pisot sequence, the integer a_{n+1} satisfies

$$-\frac{1}{2} < a_{n+1} - \frac{a_n^2}{a_{n-1}} \leq \frac{1}{2},$$

and we deduce from (6)

$$a_{n+1} \geq \nu(1+\tau)^{n+1} + \frac{1}{2\tau} > \nu(1+\tau)^{n+1},$$

that is $\mathbf{b}_n \in F_n(A)$; hence $\mathbf{b}_n \in V_n = F_n(W)$ by Theorem 13.0.1.i).

Now, from Theorem 13.0.1.ii), it follows that

$$\lim_{n\longrightarrow+\infty} \Phi_{n,1}(\mathbf{b}_n) = \lim_{n\longrightarrow+\infty} \frac{a_n^{n+1}}{a_{n+1}^n} = \mu$$

$$\lim_{n\longrightarrow+\infty} \Phi_{n,2}(\mathbf{b}_n) = \lim_{n\longrightarrow+\infty} \frac{a_{n+1}}{a_n} = \gamma,$$

with $|\,\mu - \Phi_{0,1}(\mathbf{b}_0)\,| \leq \dfrac{1}{2}\Psi_{0,1}$, that is,

$$|\,\mu - a_0\,| \leq \frac{1}{2\tau^2} \tag{7}$$

and $|\,\gamma - \Phi_{0,2}(\mathbf{b}_0)\,| \leq \dfrac{1}{2}\Psi_{0,2}$, that is,

$$|\gamma - \frac{a_1}{a_0}\,| \leq \frac{1}{2\nu\tau}\,. \tag{8}$$

We now have to determine W, i.e., real numbers $\nu > 0$ and $\tau > 0$ such that

$$a_0 = \nu + \frac{1}{2\tau^2}$$

$$\frac{a_1}{a_0} = 1 + \tau + \frac{1}{2\nu\tau}\,.$$

But the map from $(\mathbf{R}_+)^2$ to $(\mathbf{R}_+)^2$ defined by

$$x = \nu + \frac{1}{2\tau^2}$$

$$y = \tau + \frac{1}{2\nu\tau}$$

can be inversed, since its Jacobian determinant $(1 - \dfrac{1}{\tau^2\nu})(1 + \dfrac{1}{2\nu\tau})$ is non-zero if $\tau^2\nu > 1$, or equivalently if $xy^2 > 27/8$, since $x/y = \nu/\tau$ and $y^2x = (8M^3 + 12M^2 + 6M + 1)/(8M^2)$, where $M = \tau^2\nu$.

Thus, if $a_0(\dfrac{a_1}{a_0} - 1)^2 > \dfrac{27}{8}$ i.e., $a_1 > a_0 + \dfrac{3}{2}\sqrt{\dfrac{3}{2}a_0}$, it is possible to find $\nu > 0$ and $\tau > 0$ depending on a_0 and a_1 in the manner described above.

These relations can be written

$$a_0 = \frac{2\tau^2\nu + 1}{2\tau^2\nu}\,\nu \qquad \text{and} \qquad \frac{a_1}{a_0} - 1 = \frac{2\tau^2\nu + 1}{2\tau^2\nu}\,\tau.$$

Since $\tau^2\nu > 1$, it follows that

$$a_0 < \frac{3}{2}\nu, \quad \frac{a_1}{a_0} - 1 < \frac{3}{2}\tau \text{ and then } \frac{1}{2\nu\tau} < \frac{9}{8(a_1 - a_0)},$$

that is $\frac{1}{2\nu\tau} < \frac{1}{2}\sqrt{\frac{3}{2a_0}}$, since $a_1 - a_0 \geq \frac{3}{2}\sqrt{\frac{3}{2}a_0}$, which, together with (8), implies (3).

From the identity $a_0 = \frac{2\tau^2\nu + 1}{2\tau^2}$ and since $\tau^2\nu > 1$, we get

$$\frac{1}{2\tau^2} < \frac{a_0}{3}. \tag{9}$$

Finally, by squaring both sides of the positive inequality $\frac{a_1}{a_0} - 1 < \frac{3}{2}\tau$, we get

$$\frac{1}{2\tau^2} < \frac{9}{8(\frac{a_1}{a_0} - 1)^2}. \tag{10}$$

From (9) and (10) we deduce $\frac{1}{2\tau^2} < \min\left(\frac{9}{8(\frac{a_1}{a_0} - 1)^2}, \frac{a_0}{3}\right)$, and from (7) we deduce (4).

The inequality $\frac{a_{n+1}}{a_n} \geq 1 + \tau + \frac{1}{2\nu\tau}$ implies $\lim_{n \longrightarrow +\infty} \frac{a_{n+1}}{a_n} = \gamma > 1$, and (8) implies that the γ generated by all a_0 and a_1 satisfying the given conditions are everywhere dense on the real line.

A simple calculation of $G_{n,l}(a_{n+l}, a_{n+l+1}) = F_n \circ \Phi_{n+l}(a_{n+l}, a_{n+l+1})$ gives

$$G_{n,l}(a_{n+l}, a_{n+l+1}) = \left(\frac{a_{n+l}^{l+1}}{a_{n+l+1}^{l}}, \frac{a_{n+l}^{l}}{a_{n+l+1}^{l-1}}\right),$$

which implies

$$|\, D_2 G_{n,l,1}(a_{n+l}, a_{n+l+1} + u)\,| = l\left|\frac{a_{n+l}}{a_{n+l+1} + u}\right|^{l+1}.$$

Since $\lim_{n \longrightarrow +\infty} \frac{a_{n+l+1}}{a_{n+l}} = \gamma$, we obtain

$$l\left|\frac{a_{n+l}}{a_{n+l+1} + u}\right|^{l+1} \leq s_l = \frac{l}{(\gamma - \eta)^{l+1}},$$

for $\eta > 0$ fixed and arbitrarily small and for l sufficiently large.

And now (5) follows from Theorem 13.0.2. ■

Remark 13.1. If $a_{n+1} - a_n \geq 2\sqrt{a_n}$, i.e., if $a_n\left|\frac{a_{n+1}}{a_n} - 1\right|^2 \geq 4$, we can deduce from the preceding proof the inequality $\frac{1}{2\nu\tau} < \frac{0.76}{a_{n+1} - a_n}$, which implies $|\gamma - \frac{a_{n+1}}{a_n}| \leq \frac{0.76}{a_{n+1} - a_n}$.

The next theorem can be proved directly.

Theorem 13.1.2. *Let $E(a_0, a_1)$ be a Pisot sequence corresponding to integers a_0 and a_1 such that $a_1 \geq a_0 > 0$. Then* $\lim_{n \longrightarrow +\infty} \frac{a_{n+1}}{a_n} = \gamma \geq 1$.

Proof. If $l_n = a_{n+1} - a_n$, we shall prove that $\frac{l_{n+1}^2}{a_{n+1}} > h + \eta$ if $\frac{l_n^2}{a_n} = h + \eta$, where $h \in \mathbf{N}^*$ and $-(1/2) \leq \eta < (1/2)$. Since $(a_n)_{n \in \mathbf{N}}$ is a Pisot sequence, the assumption implies $a_{n+2} = 2l_n + a_n + h$, i.e., $a_{n+2} - a_{n+1} = l_{n+1} = l_n + h$. Hence

$$\frac{l_{n+1}^2}{a_{n+1}} = \frac{(l_n + h)^2}{a_n + l_n}.$$

But $\frac{l_n^2}{h + \eta} < \frac{(l_n + h)^2}{h + \eta} - l_n$ i.e., $l_n(h - \eta) + h^2 > 0$ for every h, $h \geq 1$; and since $a_n = \frac{l_n^2}{h + \eta}$ we have $a_n + l_n \leq \frac{(l_n + h)^2}{h + \eta}$, which is the desired result.

Therefore, if $a_1 \geq a_0 + \sqrt{\frac{a_0}{2}}$, i.e., $\frac{l_0^2}{a_0} \geq \frac{1}{2}$ we can write $\frac{l_0^2}{a_0} = h + \eta$, with $h \geq 1$. Hence $l_{n+1} \geq l_n + 1$.

Summing twice, we get

$$a_n \geq a_0 + nl_0 + n(n-1)/2. \tag{11}$$

It follows from the definition of a Pisot sequence that

$$\left|\frac{a_{n+1}}{a_n} - \frac{a_n}{a_{n-1}}\right| \leq \frac{1}{2a_n}, \tag{12}$$

and from (11), we deduce $\lim_{n \longrightarrow +\infty} \frac{a_{n+1}}{a_n} = \gamma$.

The inequality $\gamma \geq 1$ follows from $l_{n+1} \geq l_n + 1$.

If $a_1 < a_0 + \sqrt{\dfrac{a_0}{2}}$, i.e., $\dfrac{l_0^2}{a_0} < \dfrac{1}{2}$, we get $l_1 = l_0$ by a similar argument.

Assuming now, by induction, $l_{n+1} = l_n$, the relations $\dfrac{l_{n+1}^2}{a_{n+1}} = \dfrac{l_n^2}{a_{n+1}} \leq \dfrac{l_n^2}{a_n} = \dfrac{l_0^2}{a_0}$ imply $l_{n+2} = l_{n+1}$.

Thus the Pisot sequence $E(a_0, a_1)$ is either constant (if $a_0 = a_1$), or in an arithmetic progression (if $a_1 > a_0$).

In both cases, we have $\lim\limits_{n \longrightarrow +\infty} \dfrac{a_{n+1}}{a_n} = 1$. ■

The following useful corollary is a direct consequence of Theorems 13.1.1 and 13.1.2.

Corollary 13.1. *Let $E(a_0, a_1)$ be a Pisot sequence and denote $\gamma_n = \dfrac{a_{n+1}}{a_n}$.*

Then

i) we have $\lim\limits_{n \longrightarrow +\infty} \dfrac{a_{n+1}}{a_n} \gamma > 1$ if and only if $a_1 \geq a_0 + \sqrt{\dfrac{a_0}{2}}$;

ii) if $a_1 \geq a_0 + 2\sqrt{a_0}$, then $a_{n+1} \geq a_n + 2\sqrt{a_n}$ for $n > 0$, $\lim\limits_{n \longrightarrow +\infty} \gamma_n = \gamma > 0$, and $\lim\limits_{n \longrightarrow +\infty} a_n \gamma^{-n} = \mu$; also

$$|\gamma - \gamma_n| < \frac{1}{a_{n+1} - a_n} \tag{13}$$

$$|\mu\gamma^n - a_n| \leq \frac{1}{(\gamma - 1)(\phi_n - 1)} \tag{14}$$

where $\phi_n = \inf\{\gamma_m, \ m \geq n\}$.

Proof. The inequality $a_1 < a_0 + \sqrt{\dfrac{a_0}{2}}$ implies $\gamma = 1$ by the proof of Theorem 13.1.2.

Assuming then $a_1 \geq a_0 + \sqrt{\dfrac{a_0}{2}}$, we must show that $\gamma > 1$.

According to the proof of Theorem 13.1.2, and with the same notation, we have that $l_{n+1} \geq l_n + 1$, for every $n \geq 0$. Furthermore, there exists n_0 such that

$$l_{n_0+1} \geq l_{n_0} + 2.$$

For the inequalities

$$l_{n+1}^2 - \frac{3}{2}\,a_{n+1} \geq (l_n+1)^2 - \frac{3}{2}\,(l_n+a_n) > l_n^2 - \frac{3}{2}\,a_n$$

imply that the sequence $(l_n^2 - \frac{3}{2}\,a_n)$ is strictly increasing in $\mathbf{Z}[\frac{1}{2}]$.

Thus there exists $n_0 \geq 0$ such that $l_{n+1}^2 - \frac{3}{2}\,a_{n+1} > 0$, for every $n \geq n_0$; that is, $\frac{l_n^2}{a_n} > \frac{3}{2} = 2 - \frac{1}{2}$, which implies, as in the proof of Theorem 13.1.2, $l_{n+1} \geq l_n + 2$ for $n \geq n_0$.

Summing, we deduce that, for $n \geq n_0$,

$$l_{n+1} \geq l_{n_0} + 2(n+1-n_0),$$

and again by summation,

$$a_{n+2} - a_{n_0+1} \geq (n+1-n_0)(a_{n_0+1} - a_{n_0}) + (n+1-n_0)(n+2-n_0);$$

finally

$$a_{n+1} - a_{n_0+1} \geq (n-n_0)^2 \qquad \text{for } n > n_0.$$

From (12), it follows that

$$|\gamma - \frac{a_{n_0+2}}{a_{n_0+1}}| \leq \frac{1}{2a_{n_0+2}} + \frac{1}{2a_{n_0+3}} + \cdots = \frac{1}{2}\sum_{j=2}^{\infty} \frac{1}{a_{n_0+j}}.$$

Thus

$$\gamma \geq \frac{a_{n_0+2}}{a_{n_0+1}} - \frac{1}{2}\sum_{n=1}^{\infty} \frac{1}{a_{n_0+1}+n^2} \geq \frac{a_{n_0+2}}{a_{n_0+1}} - \frac{\pi}{4\sqrt{a_{n_0+1}}} > 1,$$

since $\frac{a_{n_0+2}}{a_{n_0+1}} > 1 + \sqrt{\frac{3}{2a_{n_0+1}}}$.

ii) It follows directly from the proof of Theorem 13.1.2 that $a_1 \geq a_0 + 2\sqrt{a_0}$ implies $a_{n+1} \geq a_n + 2\sqrt{a_n}$ for $n > 0$.

The relations $\lim_{n \longrightarrow +\infty} \gamma_n = \gamma > 1$ and $\lim_{n \longrightarrow +\infty} a_n \gamma^{-n} = \mu$ are deduced from Theorem 13.1.1, since $a_1 \geq a_0 + 2\sqrt{a_0} > a_0 + \frac{3}{2}\sqrt{\frac{3}{2}\,a_0}$.

Inequality (13) follows from Remark 13.1.

Finally, writing (13) as

$$|a_m\gamma^{-m} - a_{m+1}\gamma^{-(m+1)}| \leq \frac{a_m}{\gamma^{m+1}(a_{m+1}-a_m)} \leq \frac{1}{\gamma^{m+1}(\phi_m - 1)} \quad \text{for } m \geq n,$$

(14) follows immediately. ∎

13.2 Linear recurrence and Pisot sequences

We recall that a sequence $(u_n)_{n\in\mathbf{N}}$ is recurrent if there exist $s+1$ rational integers $q_0, q_1, \ldots, q_s$ such that

$$q_0u_n + q_1u_{n-1} + \cdots + q_su_{n-s} = 0 \quad \text{for } n \geq n_0 \geq s.$$

Given a Pisot sequence $E(a_0, a_1)$, $0 \leq a_0 \leq a_1$, there corresponds to $E(a_0, a_1)$ a limit point $\gamma(a_0, a_1)$ defined by $\lim_{n\longrightarrow+\infty} \frac{a_{n+1}}{a_n} = \gamma(a_0, a_1) \geq 1$. We shall denote by E the set of all limit points $\gamma(a_0, a_1)$.

In this section we shall first prove the inclusion $U = S \cup T \subset E$, which shows that Pisot sequences associated to elements of U are recurrent.

Conversely, Flor [14] proved that every limit point γ, $\gamma > 1$, $\gamma \in E$, of a recurrent Pisot sequence is either a Salem or a Pisot number.

However, not all Pisot sequences are recurrent, and we shall prove that $E(14, 23)$ is not recurrent.

Furthermore, we shall see that limit points corresponding to non-recurrent Pisot sequences are everywhere dense in $\left[\frac{1+\sqrt{5}}{2}, +\infty\right[$.

Theorem 13.2.1. *Let γ be a Pisot or a Salem number. Then there exists a recurrent Pisot sequence $E(a_0, a_1)$ such that $\gamma = \gamma(a_0, a_1)$.*

Proof. First let γ be a *Pisot number*.

By Theorems 5.4.1 and 5.6.1 there exists $\lambda > 0$ such that $u_n = E(\lambda\gamma^n)$ is a Pisot sequence for $n \geq n_0$.

Now the relation

$$\lambda\gamma^n = E(\lambda\gamma^n) + \varepsilon(\lambda\gamma^n) = u_n + \varepsilon_n$$

implies $\gamma = \gamma(u_{n_0}, u_{n_0+1})$.

Since $\dfrac{1}{1-\lambda\gamma z} = \sum_{n\geq 0} \lambda\gamma^n z^n$ is meromorphic in the complex plane, and since by Theorem 5.4.1 the series $\sum_{n\geq 0} \varepsilon_n z^n$ converges in $D(0,R)$, $R>1$, therefore if a series $\sum_{n\geq 0} u_n z^n$ with rational integer coefficients is meromorphic in $D(0,R)$, $R>1$, then by a theorem of Borel's, $\sum u_n z^n$ is a rational function A/Q, where A and Q belong to $\mathbf{Z}[z]$, are relatively prime and $Q(0)=1$.

Therefore a sequence $(u_n)_{n\geq n_0}$ corresponding to a Pisot number γ is recurrent.

Now let γ be a *Salem number*.

From Theorems 5.5.1 and 5.6.1, there exists $\lambda > 0$ such that $u_n = E(\lambda\theta^n)$ is a Pisot sequence for $n \geq n_0$. By the proof of Theorem 5.5.2 (sufficient part), the series $\sum_{n\geq 0} u_n z^n$ is a rational function; hence the Pisot sequence $(u_n)_{n\geq n_0}$ is recurrent.

Let $E(a_0,a_1)$ be a Pisot sequence with limit point $\gamma(a_0,a_1)>1$.

If $E(a_0,a_1)$ satisfies a linear recurrence, then $f(z) = \sum_{n\geq 0} a_n z^n$ is a rational function $f = A/R$, A and R relatively prime, with integer coefficients and $R(0)=1$ by Fatou's theorem.

Thus we have

$$a_n = g_1 a_{n-1} + \cdots + g_s a_{n-s}\ , \quad n \geq p \tag{1}$$

and $R(z) = 1 - g_1 z - \cdots - g_s z^s, \quad g_i \in \mathbf{Z}$.

Flor [14] proved that if R^* is the reciprocal polynomial of R, then R^* has a root equal to $\gamma(a_0,a_1) > 1$ and all the other roots of R^* lie in $\overline{D(0,1)}$. Therefore $\gamma(a_0,a_1)$ is either a Pisot or a Salem number. ∎

Definition 13.2. *A recurrent Pisot sequence $E(a_0,a_1)$ is called S-recurrent (resp. T-recurrent) if the limit point $\gamma(a_0,a_1)$ is a Pisot (resp. Salem) number.*

It is generally easier to prove S-recurrence, so we shall give a criterion for testing T-recurrence of $E(a_0,a_1)$, provided $\gamma(a_0,a_1) > \dfrac{\sqrt{5}+1}{2}$.

Suppose $E(a_0,a_1)$ is T-recurrent and let

$$f(z) = \sum_{n=0} a_n z^n = B(z) + \frac{E(z)}{D(z)} \tag{2}$$

where B, D, E belong to $\mathbf{Z}[z]$ and $\deg(E) < \deg(D) = s$, $s \leq p$.

Since $E(a_0, a_1)$ is T-recurrent, then the reciprocal polynomial D^* of D has the roots $\gamma = \gamma(a_0, a_1)$, γ^{-1}, $\alpha_3, \ldots, \alpha_s$ with $\gamma > 1$ and $|\alpha_i| = 1$, $i \geq 3$.

Since all the roots of D are simple, we can write, by (2),

$$a_n = \lambda\gamma^n + \mu\gamma^{-n} + \sum_{k=3}^{s} \beta_k \alpha_k^n + b_n, \quad n \geq 0,$$

$$b_n = 0 \text{ for } n \geq p - s.$$

If $b_n = 0$ for all $n \geq 0$, the recurrence is said to be *pure*.

We shall use the following notation:

$$a_n = \lambda\gamma^n + \varepsilon_n, \quad n \geq 0, \tag{3}$$

$$\delta_n = \sum_{k=3}^{s} \beta_k \alpha_k^n, \quad n \in \mathbf{Z}, \tag{4}$$

$$c_n = \lambda\gamma^n + \mu\gamma^{-n} + \delta_n, \quad n \in \mathbf{Z}. \tag{5}$$

Notice that the c_n that satisfy (1) are rational integers.

Theorem 13.2.2. (Criterion for T-recurrence) *Suppose that the Pisot sequence $E(a_0, a_1)$ is T-recurrent.*

Then, for all $n \in \mathbf{Z}$, c_n satisfies

$$|c_{n-1} - \Delta(c_n, c_{n+1}, c_{n+2})| \leq \frac{1+\gamma}{2\gamma^2} ,$$

where $\gamma = \gamma(a_0, a_1)$ and $\Delta(x, y, z) = \dfrac{x\gamma(2+\gamma^2) - y(1+2\gamma^2) + z\gamma}{\gamma^2}$.

Moreover, if $\gamma > \dfrac{1+\sqrt{5}}{2}$ and $a_n > \varepsilon + \dfrac{\varepsilon + 4\varepsilon^2}{\gamma^2 - \gamma - 1}$, where $\varepsilon = \sup_{n\geq 0} \left|\varepsilon_n\right|$, then $a_n = c_n$.

Proof. From (3) and the definition of a Pisot sequence, we deduce

$$|\lambda\gamma^{n-1}(\gamma^2\varepsilon_{n-1} - 2\gamma\varepsilon_n + \varepsilon_{n+1}) + \varepsilon_{n+1}\varepsilon_{n-1} - \varepsilon_n^2| \leq \frac{\lambda\gamma^{n-1} + \varepsilon_{n-1}}{2}. \tag{6}$$

Now, if $\varepsilon = \sup_{n\geq 0} | \varepsilon_n |$, we have

$$| \gamma^2 \varepsilon_{n-1} - 2\gamma\varepsilon_n + \varepsilon_{n+1} | \leq \frac{1}{2} + \frac{\varepsilon + 4\varepsilon^2}{2\lambda\gamma^{n-1}} \cdot \tag{7}$$

Since $\varepsilon_n = \mu\gamma^{-n} + \delta_n + b_n$, it follows that

$$\gamma^2\varepsilon_{n-1} - 2\gamma\varepsilon_n + \varepsilon_{n+1} = \tag{8}$$
$$\gamma^2\delta_{n-1} - 2\gamma\delta_n + \delta_{n+1} + \gamma^2 b_{n-1} - 2\gamma b_n + b_{n+1} + \mu\gamma^{-(n-1)}(\gamma - \gamma^{-1})^2.$$

Since $b_n = 0$ for $n > p$, we obtain from (7) and (8):

$$\limsup_{n\longrightarrow+\infty} | \gamma^2\delta_{n-1} - 2\gamma\delta_n + \delta_{n+1} | =$$

$$\limsup_{n\longrightarrow+\infty} | \gamma^2\varepsilon_{n-1} - 2\gamma\varepsilon_n + \varepsilon_{n+1} | \leq \frac{1}{2} \cdot$$

Moreover, by (5), δ_n is a linear combination of α_j^n, where $\alpha_j = \exp(2\pi i\omega_j)$. Then by Dirichlet's theorem, there exist infinitely many integers q such that

$$\|q\omega_j\| < q^{-1/s}, \quad 3 \leq j \leq s,$$

where $\|x\|$ is the distance from the real number x to the nearest integer. Therefore, if q is large enough,

$$| \alpha_j^n - \alpha_j^{n+q} | = | \alpha_j^n(1 - \alpha_j^q) | = | 1 - \alpha_j^q | = 2 \sin \frac{\|q\omega_j\|}{2} \leq q^{-1/s}.$$

So $| \delta_n - \delta_{n+q} |$ can be made arbitrarily small.

And now, from

$$| \gamma^2\delta_{n+q-1} - 2\gamma\delta_{n+q} + \delta_{n+q+1} | \leq \frac{1}{2},$$

we deduce

$$| \gamma^2\delta_{n-1} - 2\gamma\delta_n + \delta_{n+1} | \leq \frac{1}{2}, \quad n \in \mathbf{Z}. \tag{9}$$

Writing $D = \gamma^2(c_{n-1} - \Delta(c_n, c_{n+1}, c_{n+2}))$ in the form

$$D = (\gamma^2 c_{n-1} - 2\gamma c_n + c_{n+1}) - \gamma(\gamma^2 c_n - 2\gamma c_{n+1} + c_{n+2}),$$

we get, by (5),

$$D = (\gamma^2\delta_{n-1} - 2\gamma\delta_n + \delta_{n+1}) - \gamma(\gamma^2\delta_n - 2\gamma\delta_{n+1} + \delta_{n+2}),$$

and it follows from (9) that

$$|D| \leq \frac{1}{2}(\gamma+1).$$

Finally, we have

$$|c_{n-1} - \Delta(c_n, c_{n+1}, c_{n+2})| \leq \frac{1+\gamma}{2\gamma^2}.$$

Now calculating with a_n instead of c_n, we derive from (7):

$$|a_{n-1} - \Delta(a_n, a_{n+1}, a_{n+2})| =$$

$$\frac{1}{\gamma^2}\left[\gamma^2\varepsilon_{n-1} - 2\gamma\varepsilon_n + \varepsilon_{n+1} - \gamma(\gamma^2\varepsilon_n - 2\gamma\varepsilon_{n+1} + \varepsilon_{n+2})\right]$$

$$\leq \frac{1+\gamma}{2\gamma^2} + \frac{\varepsilon+4\varepsilon^2}{\lambda\gamma^{n+1}}.$$

Thus, if $\dfrac{1+\gamma}{\gamma^2} + \dfrac{\varepsilon+4\varepsilon^2}{\lambda\gamma^{n+1}} < 1$, the relations $a_m = c_m$ for $m \geq n$ imply $a_{n-1} = c_{n-1}$.

But if $\gamma^2 - \gamma - 1 > 0$, i.e., $\gamma > \dfrac{1+\sqrt{5}}{2}$, the last inequality can be written $\lambda\gamma^{n-1} > \dfrac{\varepsilon+4\varepsilon^2}{\gamma^2-\gamma-1}$, and this is true if $a_{n-1} > \varepsilon + \dfrac{\varepsilon+4\varepsilon^2}{\gamma^2-\gamma-1}$. ∎

Remark 13.2. The previous criterion is not effective if $\gamma \leq \dfrac{1+\sqrt{5}}{2}$, since then the integers c_n are not known.

Proposition 13.2.1. *The Pisot sequence $E(14,23)$ is not T-recurrent.*

Proof. Suppose $E(14,23)$ T-recurrent and let $\gamma = \gamma(14,23)$.

A calculation by computer gives $\gamma = 1.652757892644\ldots$.

We deduce from Corollary 13.1

$$\phi_n \geq \gamma - \frac{1}{a_1 - a_0} \geq \gamma - \frac{1}{9},$$

and since $\gamma > 1.65$, we have

$$\varepsilon \leq 1/[(\gamma-1)(\gamma - \frac{10}{9})].$$

It follows that

$$\varepsilon \leq 2.8548771, \quad \varepsilon + 4\varepsilon^2 \leq 35.45617 \quad \text{and} \quad \gamma^2 - \gamma - 1 > 0.0725.$$

Since the first terms of $E(14,23)$ are

$$a_0 = 14 \quad a_1 = 23 \quad a_2 = 38 \quad a_3 = 63 \quad a_4 = 104 \quad a_5 = 172 \quad a_6 = 284$$

$$a_7 = 469 \quad a_8 = 775 \quad a_9 = 1281 \quad a_{10} = 2117,$$

we deduce from Theorem 13.2.2 that $a_n = c_n$ if $a_n > 491.9$, that is, if $n \geq 8$. Now, since $\gamma > 1.65$, we get from Theorem 13.2.2

$$|\, c_{n-1} - \Delta(c_n, c_{n+1}, c_{n+2}) \,| \leq \frac{1+\gamma}{2\gamma^2} \leq 0.4866\ldots \tag{10}$$

Thus we can compute c_7 from $c_8 = a_8$, $c_9 = a_9$ and $c_{10} = a_{10}$.

And we get

$$c_7 = a_7 = 469 \quad c_6 = a_6 = 284 \quad c_5 = a_5 = 172 \quad c_4 = a_4 = 104$$

$$c_3 = a_3 = 63 \quad c_2 = a_2 = 38 \quad c_1 = a_1 = 23 \quad c_0 = a_0 = 14 \quad c_{-1} = 9 \quad c_{-2} = 7$$

$$c_{-3} = 7 c_{-4} = 9 \quad c_{-5} = 13 \quad c_{-6} = 20 \quad c_{-7} = 32 \quad c_{-8} = 52 \quad c_{-9} = 85$$

$$c_{-10} = 140 \quad c_{-11} = 231.$$

So we have $\Delta(c_{-9}, c_{-10}, c_{-11}) = 381.49718\ldots.$

Since the integer c_{-12} satisfies (10) for $n = -11$, we obtain a contradiction. This completes the proof. ∎

Proposition 13.2.2. *The Pisot sequence $E(14,23)$ is not recurrent.*

Proof. Since $E(14,23)$ is not T-recurrent by Proposition 13.2.1, we only have to prove that $\gamma = \gamma(14,23)$ is not a Pisot number.

Suppose the contrary, i.e., $\gamma \in S$.

Since the smallest quadratic Pisot number is greater than 2 and $\gamma < 1.66$, we deduce that γ is not quadratic. Thus there corresponds to γ a rational function P/Q, where $P = \varepsilon Q^*$, $\varepsilon = \pm 1$ is such that $\varepsilon Q^*(0) = P(0) > 0$, $Q(0) = 1$, $1/\gamma$ being the only pole of P/Q lying in $\overline{D(0,1)}$. Now let the expansion of P/Q in the neighborhood of the origin be

$$\frac{P(z)}{Q(z)} = \sum_{n \geq 0} b_n z^n, \quad b_i \in \mathbf{Z}.$$

Since $f(z) = (1-\gamma z)P(z)/Q(z)$ is analytic in $\overline{D(0,1)}$ and satisfies $|f(z)| \leq |1-\gamma z|$ if $|z| = 1$, we deduce from Parseval's formula

$$\frac{1}{2\pi}\int_0^{2\pi} |f(e^{i\theta})|^2 \, d\theta = b_0^2 + (b_1 - b_0\gamma)^2 + \cdots + (b_n - b_{n-1}\gamma)^2 + \cdots$$
$$\leq 1 + \gamma^2.$$

Therefore, if γ were a Pisot number, there would exist an integer sequence $(b_n)_{n\geq 0}$ satisfying

$$b_0^2 + (b_1 - b_0\gamma)^2 + \cdots + (b_n - b_{n-1}\gamma)^2 + \cdots \leq 1 + \gamma^2. \tag{11}$$

If $b_n = 0$ for $n \geq n_0$, (11) would imply $b_0 = 1$, $b_n = 0$ for $n \geq 1$ and $P = Q$, which contradicts the fact that P/Q has the pole $1/\gamma$ in $D(0,1)$. Thus $(b_n)_{n\geq 0}$ is an infinite sequence of integers.

Now we claim that $(b_n)_{n\geq 0}$ is an increasing sequence.

Otherwise, there would exist a smallest integer k, $k > 0$, such that $b_k < b_{k-1}$. Since $1 \leq b_0 \leq b_{k-1}$, (12) would imply

$$(b_k - b_{k-1}\gamma)^2 < (b_k - b_{k-1}\gamma)^2 + \sum_{n=k}^{\infty}(b_{n+1} - b_n\gamma)^2 \leq \gamma^2, \tag{12}$$

and
$$b_{k-1}\gamma \leq \gamma + b_k; \qquad \text{so } b_k > 0.$$

But
$$b_{k-1}\gamma - b_k = (b_{k-1} - b_k)\gamma + b_k(\gamma - 1) > (b_{k-1} - b_k)\gamma \geq \gamma,$$

which contradicts (12).

Therefore, if γ were a Pisot number, there would exist an infinite increasing sequence of positive integers satisfying (11). Furthermore, if $b_0, \ldots, b_j$ are known, then there are only finitely many possible b_{j+1}. So the integers b_n satisfying (12) can be arranged in a tree. The tree is finite, since it can be calculated in a finite time by a computer. Now we have a contradiction with the assumption. Therefore $\gamma(14, 23)$ is not a Pisot number; this completes the proof. ∎

Theorem 13.2.3. *The limit numbers $\gamma(a_0, a_1)$ corresponding to non-recurrent Pisot sequences are dense in the interval $[\frac{\sqrt{5}+1}{2}, +\infty[$.*

Proof. Let a_0 be a positive integer. If ρ belongs to $[\frac{\sqrt{5}+1}{2}, +\infty[$ and is neither a Pisot number nor an algebraic number of degree less than 4, we write

$$a_0\rho = a_1 + \xi, \quad a_0\rho^2 = b + \eta, \quad \frac{a_0(1+\rho^2)}{\rho} = c + \zeta,$$

where a, b, c are integers and ξ, η, ζ real numbers in $I = [-\frac{1}{2}, \frac{1}{2}[$.

Since ρ belongs to $[\frac{\sqrt{5}+1}{2}, +\infty[$, there exists α, $0 < \alpha < \frac{1}{2}$ such that $\frac{1+\rho}{2\rho^2} < \frac{1}{2} - \alpha$. Since ρ, ρ^2 and $\frac{1+\rho^2}{\rho}$ are irrational, then by the Kronecker-Weyl theorem, the 3-tuples (ξ, η, ζ) are uniformly distributed in I^3 if a_0 tends to infinity.

Therefore there exist infinitely many a_0 such that

$$\left|\eta - 2\rho\xi\right| < \alpha/2 \text{ and } \frac{1}{2} - \frac{\alpha}{2} < \left|\zeta + \frac{\xi(1+\rho^2)}{\rho^2} - \frac{\eta}{\rho}\right| < \frac{1}{2}. \tag{13}$$

Restricting ourselves now to such a_0, we have

$$\frac{a_1^2}{a_0} = \frac{(a_0\rho - \xi)^2}{a_0} = b + \eta - 2\rho\xi + \frac{\xi^2}{a_0}.$$

Then by (13), $\left|\eta - 2\rho\xi + \frac{\xi^2}{a_0}\right| < \frac{1}{2}$, for a_0 large enough.

Thus $a_2 = b$, where a_2 is the third term of $E(a_0, a_1)$.

Therefore we have

$$\begin{aligned}\frac{a_0(a_0+a_2)}{a_1} &= \frac{a_0(a_0 + a_0\rho^2 - \eta)}{a_0\rho - \xi} = \frac{a_0^2(1+\rho^2) - \eta a_0}{a_0\rho}[1 + \frac{\xi}{a_0\rho} + (\frac{\xi}{a_0\rho})^2 + \cdots] \\ &= c + \zeta + \frac{\xi(1+\rho^2)}{\rho^2} - \frac{\eta}{\rho} + O(\frac{1}{a_0}),\end{aligned}$$

and for a_0 large enough, (13) implies

$$\left\|\frac{a_0(a_0+a_2)}{a_1}\right\| > \frac{1}{2} - \frac{\alpha}{2}. \tag{14}$$

But by Theorem 13.2.2, if a_0 is large and $E(a_0, a_1)$ is a T-recurrent Pisot sequence, then $E(a_0, a_1)$ is pure recurrent; so we have $a_0 = c_0$, $a_1 = c_1$, $a_2 = c_2$ and

$$\begin{aligned}A &= a_2a_0 - a_1^2 - (a_{-1}c_{-1} - a_0^2)\\ &= \lambda(\gamma^2\delta_0 - 2\gamma\delta_1 + \delta_2) - \lambda\gamma^{-1}(\gamma^2\delta_{-1} - 2\gamma\delta_0 + \delta_1) + \mu(\gamma^{-2}\delta_0 + \delta_2 - 2\gamma^{-1}\delta_1)\\ &\quad - \mu(\gamma^{-1}\delta_{-1} + \gamma\delta_1 - 2\delta_0) + \delta_2\delta_0 - \delta_1^2 - (\delta_1\delta_{-1} - \delta_0^2).\end{aligned}$$

Now, since $\varepsilon_n = \mu\gamma^{-n} + \delta_n$, we have

$$\limsup_{m\longrightarrow+\infty} |\,\delta_m\,| = \limsup_{m\longrightarrow+\infty} |\,\varepsilon_m\,|,$$

and from Corollary 13.1, it follows that $|\,\delta_n\,| \leq \dfrac{1}{(\gamma-1)^2}$.

We deduce from (9) and the last inequality that

$$|\,A\,| \leq \lambda\,\frac{1-\gamma^{-1}}{2} + |\mu|\,\frac{\gamma^{-2}\,(\gamma+1)^3}{(\gamma-1)^2} + \frac{4}{(\gamma-1)^4}$$

and by (7) and (8), if $n = 1$, then

$$\left|\mu(\gamma-\gamma^{-1})^2\right| \leq 1 + \frac{\varepsilon + 4\varepsilon^2}{2\lambda}\;,$$

where $\varepsilon = \sup_n |\varepsilon_n|$.

Thus we get

$$|\,A\,| \leq \frac{\lambda(1+\gamma^{-1})}{2} + \frac{\gamma+1}{\gamma-1}\,\frac{1}{(\gamma-1)^3}\,(1 + \frac{\varepsilon+4\varepsilon^2}{2\lambda}) + \frac{4}{(\gamma-1)^4}\,.$$

Since $A = a_0(a_0 + a_2) - a_1(a_1 + c_{-1})$ and $\lambda = \dfrac{a_1 - \varepsilon_1}{\gamma}$, we deduce that

$$\left|\,\frac{a_0(a_0+a_2)}{a_1} - (a_1 + c_{-1})\right| \leq$$

$$\frac{\gamma+1}{2\gamma^2} + \frac{1}{a_1}\Big[\,\frac{\varepsilon(\gamma+1)}{2\gamma^2} + \frac{\gamma+1}{(\gamma-1)^4}\,(1 + \frac{\varepsilon+4\varepsilon^2}{2\lambda}) + \frac{4}{(\gamma-1)^4}\,\Big].$$

Therefore if a_0 is large enough, and since c_{-1} is integer, we have

$$\left\|\,\frac{a_0(a_0+a_2)}{a_1}\,\right\| \leq \frac{\gamma+1}{2\gamma^2} + \frac{\alpha}{4}\;. \tag{15}$$

Now if $\rho = \lim_{m\longrightarrow+\infty} \dfrac{a_{1,m}}{a_{0,m}}$, we obtain that $\lim_{m\longrightarrow+\infty} \gamma(a_{0,m}, a_{1,m}) = \rho$ from Corollary 13.1. Hence, for m large enough we deduce from (15) that

$$\left\| \frac{a_{0,m}(a_{0,m}+a_{2,m})}{a_{1,m}} \right\| \le \frac{\rho+1}{2\rho^2} + \frac{\alpha}{2} < \frac{1}{2} - \frac{\alpha}{2},$$

which contradicts (14).

Therefore the various Pisot sequences $E(a_{0,m}, a_{1,m})$ that satisfy

$\lim_{n \longrightarrow +\infty} \frac{a_{1,m}}{a_{0,m}} = \rho$ are not T-recurrent for m sufficiently large.

Moreover, since the set S of Pisot numbers is closed, if ρ is not a Pisot number, we deduce that the $\gamma(a_{0,m}, a_{1,m})$ are not Pisot numbers for $a_{0,m}$ large enough. Therefore the corresponding Pisot sequences $E(a_{0,m}, a_{1,m})$ are not S-recurrent. This completes the proof. ■

13.3 Boyd sequences

In the last section, we saw how important Pisot sequences are for the study of the set T of Salem numbers. For, if to each element of $E \cap]a, b[$, $1 \le a < b$, there correspond only non-recurrent Pisot sequences, then it follows that $T \cap]a, b[= \emptyset$ and therefore Inf $T > 1$. (Otherwise, the existence of $\tau \in T$, $1 < \tau < (b/a)$ implies $\tau^m \in T \cap]a, b[$ for an integer m such that $(\log a / \log \tau) < m (\log b / \log \tau)$.)

This is why the study of $E \cap]a, b[$ for $b < (1+\sqrt{5})/2$ is so useful, since the elements of $S \cap]a, b[$ are known and of finite number. Unfortunately, the known T-recurrence criterion is not effective in this case.

It is for this reason that Boyd introduced Boyd sequences and especially the so-called "geometric" sequences. In this section we prove first the existence of certain limits corresponding to "geometric" Boyd sequences. Then if F is the set of these limits, the sets E and F have similar properties; thus $U = S \cup T \subset F$, and if $\gamma \in F$, $\gamma > 1$, corresponds to a recurrent Boyd sequence, then γ is either a Pisot or a Salem number.

More precisely, we establish that to any Salem number or quadratic Pisot number there corresponds a "geometric" pure recurrent Boyd sequence. Finally, we give a criterion for pure T-recurrence.

Definition 13.3.1. *A Boyd sequence $F(a_0, a_1, a_2)$ is a sequence of integers a_n satisfying*

$$-\frac{1}{2} < a_{n+1} + a_n - \frac{a_{n+1}}{a_n}(a_{n+1} + a_{n-1}) \le \frac{1}{2}, \quad n \ge 1. \tag{1}$$

Thus a_{n+2} is uniquely determined unless $a_n = 0$. If $a_n = 0$ for $n \ge 2$, let $a_{n+2} = -a_{n-2}$.

Definition 13.3.2. *A Boyd sequence $F(a_0, a_1, a_2)$ is said to be "geometric" if $a_n > 0$ from a certain rank on and* $\liminf_{n \longrightarrow +\infty} \left(\frac{a_{n+1}}{a_n} \right) > 1$.

Lemma 13.3. *Let $(a_n)_{n \geq 0}$ be a "geometric" Boyd sequence. Then we have*

$$\lim_{n \longrightarrow +\infty} \frac{a_{n+1}}{a_n} = \gamma > 1 \quad \text{and} \quad \lim_{n \longrightarrow +\infty} \frac{a_n^{n+1}}{a_{n+1}^n} = \lambda > 0.$$

Moreover, the sequence $\varepsilon_n = \lambda \gamma^n - a_n$ is bounded.

Proof. Let $\gamma_n = \dfrac{a_{n+1}}{a_n}$ and $\lambda_n = \dfrac{a_n^{n+1}}{a_{n+1}^n}$. We then have the relations

$$\lambda_n = a_n \gamma_n^{-n} = a_{n+1} \gamma_n^{-(n+1)}.$$

We first show that $b_n = a_{n+1} - \dfrac{a_n^2}{a_{n-1}}$ is a bounded sequence.

Since $(a_n)_{n \geq 0}$ is a "geometric" Boyd sequence, there exists a real number α, $\alpha > 1$, and a positive integer m such that $\gamma_n \geq \alpha$ for $n \geq m$. From (1) we deduce

$$|\, b_{n+1} - \gamma_{n-1}^{-1} b_n \,| \leq (1/2) \quad \text{and} \quad |\, b_{n+1} \,| \leq \alpha^{-1} \Big| b_n \Big| + (1/2), \quad n > m.$$

And from

$$|\, b_{n+1} \,| < \alpha^{m-n} \,|\, b_{m+1} \,| + \frac{\alpha}{2(\alpha - 1)} \leq |\, b_{m+1} \,| + \frac{\alpha}{2(\alpha - 1)} = C, \quad n > m,$$

it follows that (b_n) is a bounded sequence.

The relation $b_n = a_{n+1}(\gamma_{n+1} - \gamma_n)$ now implies

$$|\, \gamma_{n+1} - \gamma_n \,| \leq \frac{C}{a_{n+1}} \leq C \, \frac{\alpha^{m-n-1}}{a_m}, \quad n > m.$$

Therefore (γ_n) is a Cauchy sequence and converges to a γ, $\gamma > 1$ such that

$$|\, \gamma - \gamma_n \,| \leq \frac{C}{a_n} \frac{1}{\alpha - 1}, \quad n > m. \tag{2}$$

This implies that $\lim_{n \longrightarrow +\infty} \dfrac{a_{n+1}}{a_n} = \gamma$. From

$$|\, \lambda_{n+1} - \lambda_n \,| = a_{n+1} \left| \frac{1}{\gamma_{n+1}^{n+1}} - \frac{1}{\gamma_n^{n+1}} \right| \leq$$

$$\frac{n+1}{\alpha^{n+2}}\, a_{n+1}\big|\gamma_{n+1}-\gamma_n\big| = (n+1)\,\frac{\big|b_{n+1}\big|}{\alpha^{n+2}} \leq \frac{C(n+1)}{\alpha^{n+2}}$$

we see that (λ_n) is a Cauchy sequence converging to a $\lambda \geq 0$.

Let us prove now that $\lim_{n\longrightarrow+\infty} (a_n\gamma^{-n}) = \lambda$.

It follows from (2) that

$$|\, a_n\gamma^{-n} - a_{n+1}\gamma^{-(n+1)}\,| \leq \frac{C}{\gamma^{n+1}(\alpha-1)}\,; \tag{3}$$

therefore

$$\lim_{n\longrightarrow+\infty} a_n\gamma^{-n} = l.$$

And now the inequality

$$|\, a_n\gamma^{-n} - a_n\gamma_n^{-n}\,| \leq n\,\frac{a_n}{\alpha^{n+1}}\,|\,\gamma-\gamma_n\,| \leq \frac{Cn}{\alpha^{n+1}(\alpha-1)}$$

implies $l = \lambda$.

Finally, we deduce from (3): $|\,\varepsilon_n\,| \leq C/(\alpha-1)(\gamma-1)$; therefore (ε_n) is a bounded sequence, which implies $\lambda > 0$. ■

Theorem 13.3.1. *1) If $a_1 > 0$ and $a_2 > 2a_1$, the Boyd sequence $F(0, a_1, a_2)$ is "geometric" and the set of γ, where $\gamma = \lim_{n\longrightarrow+\infty} \frac{a_{n+1}}{a_n}$, is dense in $[1, +\infty[$.*

2) For $a_1 \geq a_0$ and $a_2 - 2a_1 + a_0 > \sqrt{2a_1}$, the Boyd sequence $F(0, a_1, a_2)$ is "geometric." Moreover, for $n \geq 0$ we have

$$d_n = a_{n+1} - 2a_n + a_{n-1} > \sqrt{2a_n}$$

and

$$\left|\gamma+\gamma^{-1} - \frac{a_{n+1}+a_{n-1}}{a_n}\right| < \frac{1}{d_n + \sqrt{d_n^2 - 2a_n}}\,.$$

Proof. We show first by induction

$$a_{n+1} - 2a_n + a_{n-1} \geq n \tag{4}$$

$$\frac{a_{n+1}}{a_n} \geq \frac{n+1}{n}\,. \tag{5}$$

If $n = 1$, relations (4) and (5) are satisfied.

Suppose then that (4) and (5) hold for all integers less than or equal to n.

We deduce from (1)

$$a_{n+2} + a_n > \frac{a_{n+1}}{a_n}(a_{n+1} + a_{n-1}) - \frac{1}{2};$$

and from (4) and (5)

$$a_{n+2} + a_n > 2a_{n+1} + n + 1 - (1/2);$$

and since $a_{n+2} + a_n$ is an integer,

$$a_{n+2} + a_n \geq 2a_{n+1} + n + 1.$$

Thus (4) holds for $n + 1$ and so does (5).

Now we get by summation

$$a_{n+1} - a_n + a_0 - a_1 = a_{n+1} - a_n - a_1 \geq n(n+1)/2.$$

Summing again we deduce

$$a_{n+1} \geq (n+1)a_1 + \frac{1.2 + 2.3 + \cdots + n(n+1)}{2} \quad ;$$

hence we have

$$a_{n+1} \geq (n+1)a_1 + C_{n+2}^3 \tag{6}$$

and

$$a_{n+1} > C_{n+2}^3 = O(n^3). \tag{7}$$

Now, from (1) it follows that

$$\left| \frac{a_{n+2} + a_n}{a_{n+1}} - \frac{a_{n+1} + a_{n-1}}{a_n} \right| \leq \frac{1}{2a_{n+1}}; \tag{8}$$

thus we get from (7) that $(a_{n+2} + a_n)/a_{n+1}$ is a Cauchy sequence converging to ρ.

From (7) and (8) we obtain

$$\left| \rho - \frac{a_{n+1} + a_{n-1}}{a_n} \right| \leq \sum_{m \geq n} \frac{3}{m(m+1)(m+2)} \leq \tag{9}$$

$$\frac{3}{n} \sum_{m \geq n} \frac{1}{(m+1)(m+2)} \leq \frac{3}{n(n+1)};$$

and from (4) we deduce $\rho \geq 2$.

If $\rho = 2$, it follows from (4) and (9) that

$$a_n \geq \frac{n^2(n+1)}{3} > 2C_{n+1}^3$$

and from (8) that

$$\left|2 - \frac{a_{n+1} + a_{n-1}}{a_n}\right| \leq \frac{3}{2n(n+1)};$$

hence $a_n > 2^2 C_{n+1}^3$. Now it is easy to see that, by induction, $a_n > 2^k C_{n+1}^3$ for every k, which, however, is impossible.

Therefore, since $\lim\limits_{n\to+\infty} \dfrac{a_{n+1} + a_{n-1}}{a_n} > 2$ and $\dfrac{a_{n-1}}{a_n} < 1$, the Boyd sequence (a_n) is "geometric."

Now let k be an integer. For $1 \leq n \leq k$ take $a_n > na_1$, and for $n > k$ take $a_n > C_{n+1}^3$. We deduce then from (8)

$$\begin{aligned}\left|\rho - \frac{a_2}{a_1}\right| &< \sum_{n=2}^{k}(2na_1)^{-1} + \sum_{n=k+1}^{\infty} \frac{3}{(n-1)n(n+1)}\\ &< \frac{\log k}{2a_1} + \frac{3}{k(k+1)}.\end{aligned}$$

Choosing $k = [a_1^{1/2}]$, it follows that

$$\left|\gamma + \gamma^{-1} - \frac{a_2}{a_1}\right| < \frac{\log a_1}{4a_1} + \frac{3}{(a_1 - a_1^{1/2})}.$$

Therefore the set of limits γ is dense in $[1, +\infty[$.

2) By assumption,

$$d_1 = a_2 - 2a_1 + a_0 \geq \frac{a_1}{2c} + c \quad \text{where } c = a_1/\left(d_1 + \sqrt{d_1^2 - 2a_1}\right).$$

Assume now by induction that for $i \leq n$

$$d_i = a_{i+1} - 2a_i + a_{i-1} \geq \frac{a_i}{2c} + c \quad \text{and} \quad a_i \geq a_{i-1}, \quad i \geq 1.$$

From (1) we deduce

$$-\frac{1}{2} < a_{n+2} - 2a_{n+1} + a_n - \frac{a_{n+1}}{a_n}(a_{n+1} - 2a_n + a_{n-1}) \leq \frac{1}{2};$$

hence

$$d_{n+1} \geq \frac{a_{n+1}}{a_n} d_n - \frac{1}{2} \geq \frac{a_{n+1}}{a_n} \left(\frac{a_n}{2c} + c\right) - \frac{1}{2} \ ;$$

and finally

$$d_{n+1} \geq \frac{a_{n+1}}{2c} + \frac{a_{n+1}}{a_n} c - \frac{1}{2} .eqno(10)$$

Now, from $d_n \geq (a_n/(2c)) + c$ and $a_n \geq a_{n-1}$, it follows that

$$a_{n+1} \geq a_n\left(\frac{1}{2c} + 1\right) + c. \tag{11}$$

Then we deduce from (10)

$$d_{n+1} \geq \frac{a_{n+1}}{2c} + c \geq \sqrt{2a_{n+1}}$$

and from (11)

$$a_{n+1} \geq a_n \qquad \text{and} \qquad \lim_{n \longrightarrow +\infty} inf \frac{a_{n+1}}{a_n} > 1.$$

Therefore it follows from Lemma 13.3 that

$$\lim_{n \longrightarrow +\infty} \frac{a_{n+1}}{a_n} = \gamma > 1.$$

Finally, we deduce from (8) and $a_k \geq (1 + \frac{1}{2c_n})^{k-n} a_n$, where $k \geq n$ and $c_n = a_n/(d_n + \sqrt{d_n^2 - 2a_n})$

$$\left|\gamma + \gamma^{-1} - \frac{a_{n+1} + a_{n-1}}{a_n}\right| \leq 1/\left(d_n + \sqrt{d_n^2 - 2a_n}\right).$$

■

Proposition 13.3.

i) Let $(a_n)_{n>0}$ be a "geometric" Boyd sequence. If λ and γ are the limits defined in Lemma 13.3, we set $a_n = \lambda\gamma^n + \varepsilon_n$. Then the inequality

$$\limsup_{n \longrightarrow +\infty} |\, \varepsilon_n \gamma^2 - 2(\varepsilon_{n+1} + \varepsilon_{n-1})\gamma + \varepsilon_{n+2} + 2\varepsilon_n - (\varepsilon_{n-1}/\gamma) \,| \leq (1/2) \tag{12}$$

holds.

ii) Conversely, if $a_n = \lambda\gamma^n + \varepsilon_n$ where, $\lambda > 0$, $\gamma > 1$, and ε_n is bounded, and if we have

$$\limsup_{n \longrightarrow +\infty} |\, \varepsilon_n \gamma^2 - 2(\varepsilon_{n+1} + \varepsilon_{n-1})\gamma + \varepsilon_{n+2} + 2\varepsilon_n - (\varepsilon_{n+1}/\gamma) \,| < (1/2), \tag{13}$$

then $(a_n)_{n\geq 0}$ *is a Boyd sequence for* n *greater than some fixed* n_0.

iii) The inequality

$$\limsup_{n\longrightarrow+\infty} |\,\varepsilon_n\,| < \frac{\gamma}{2(\gamma+1)^3} \tag{14}$$

implies (13).

Proof. Let $D_n(a) = a_n(a_{n+2}+a_n) - a_{n+1}(a_{n+1}+a_{n-1})$. We deduce from (1) that $|\,D_n(a)\,| \leq (a_n/2)$. Since

$$D_n(a) = \lambda\gamma^n(\varepsilon_n\gamma^2 - (2\varepsilon_{n+1}+\varepsilon_{n-1})\gamma + \varepsilon_{n+2} + 2\varepsilon_n - (\varepsilon_{n+1}/\gamma)) + D_n(\varepsilon),$$

i) and ii) follow immediately.

The proof of iii) is arrived at through an easy calculation. ■

Theorem 13.3.2.

i) With the notation of Lemma 13.3, let $(a_n)_{n\geq 0}$ *be a "geometric" Boyd sequence. If* (a_n) *is a recurrent sequence, then* γ *is either a Pisot or a Salem number. Moreover the poles of the rational function* $f = \sum_{n\geq 0} a_n z^n$ *are simple.*

ii) Conversely, if γ *denotes a Pisot or a Salem number, there exists a Boyd sequence* $(a_n)_{n\geq 0}$ *such that* $\gamma = \lim_{n\longrightarrow+\infty} \frac{a_{n+1}}{a_n}$.

iii) If γ *is a Salem number or a quadratic Pisot number, there exists a Boyd sequence* $F(0, a_1, a_2)$ *such that* $\lim_{n\longrightarrow+\infty} \frac{a_{n+1}}{a_n} = \gamma$. *Moreover* (a_n) *is a pure recurrent sequence.*

Proof.

i) Since (a_n) is a recurrent sequence, denote $\sum_{n\geq 0} a_n z^n = f(z) = \frac{A(z)}{Q(z)}$ where A and Q belong to $\mathbf{Z}[z]$ and satisfy $Q(0) = 1$.

Now, from $a_n = \lambda\gamma^n + \varepsilon_n$ we get

$$\sum_{n\geq 0} a_n z^n = f(z) = \frac{\lambda}{1-\gamma z} + \sum_{n\geq 0} \varepsilon_n z^n.$$

Hence the roots of $P = Q^*$ are algebraic integers and $1/\gamma$ is a simple pole of f and the only one in $D(0,1)$. Since (ε_n) is a bounded sequence, the other poles of f are on or outside the unit circle, and the ones on the unit circle are simple.

Therefore γ is either a Pisot or a Salem number and f has only simple poles.

ii) Conversely, if $\gamma \in U = S \cup T$, there exists $\lambda \in \mathbf{Q}(\gamma)$ such that

$\left\|\lambda\gamma^n\right\| < \dfrac{\gamma}{2(\gamma+1)^3}$, $n \geq 0$, by Theorems 5.4.1, 5.5.1 and 5.6.1.

Therefore we deduce from Proposition 13.3 that (a_n) is a Boyd sequence for n large enough.

iii) If γ is a Salem or a quadratic Pisot number, γ is a root of a monic reciprocal polynomial T of degree $2m$ with integer coefficients.

Let us then write $T(z) = z^m R(\zeta)$, where $\zeta = z + z^{-1}$. Define now the monic polynomial B of degree $m-1$ and $A(z) = z^{m-1}B(z)$ such that $z\dfrac{A(z)}{T(z)}$ is the generating function of a Boyd sequence $F(0, a_1, a_2)$, i.e., such that we have in the neighborhood of the origin

$$z\frac{A(z)}{T(z)} = a_1 z + a_2 z^2 + \cdots + a_n z^n + \cdots. \tag{15}$$

If we denote by γ, γ^{-1}, α_2, $\overline{\alpha}_2$, $\ldots$, α_m, $\overline{\alpha}_m$ the zeros of T and by $\rho = \gamma + \gamma^{-1}, \rho_2 = \alpha_2 + \overline{\alpha}_2, \ldots, \rho_m = \alpha_m + \overline{\alpha}_m$ the zeros of R, then we deduce from (15)

$$a_n = \lambda\gamma^n + \mu\gamma^{-n} + \sum_{k=2}^{m}(\beta_k\alpha_k^n + \overline{\beta}_k\overline{\alpha}_k^n), \tag{16}$$

where $\lambda = (-\gamma A(\gamma^{-1})/T'(\gamma^{-1})) = (\gamma - \gamma^{-1})^{-1}B(\rho)/R'(\rho) = -\mu$ and $\beta_k = (\alpha_k - \overline{\alpha}_k)^{-1}B(\rho_k)/R'(\rho_k) = -\overline{\beta}_k$.

Therefore we get

$$a_n = \lambda\gamma^n - \lambda\gamma^{-n} + \delta_n \quad \text{where} \quad \delta_n = \sum_{k=2}^{m}\beta_k(\alpha_k^n - \overline{\alpha}_k^n),$$

and (a_n) will be a Boyd sequence if (17) is satisfied:

$$\begin{aligned} &\big|\delta_n\gamma^2 - (2\delta_{n+1} + \delta_{n-1})\gamma + 2\delta_n + \delta_{n+2} - (\delta_{n+1}/\gamma) \\ &- \gamma^{-2n}\big(\delta_n\gamma^{-2} - (2\delta_{n+1} + \delta_{n-1})\gamma^{-1} + 2\delta_n + \delta_{n+2} - \delta_{n+1}\gamma\big) \\ &+ \lambda^{-1}\gamma^{-n}\big(\delta_n(\delta_n + \delta_{n+2}) - \delta_{n+1}(\delta_{n+1} + \delta_{n-1})\big)\big| \\ &\leq (1 - \gamma^{-2n} + \delta_n(\lambda\gamma^n)^{-1})/2. \end{aligned} \tag{17}$$

For (17) to be satisfied, it suffices that β_i , $2 \leq i \leq m$, be small.

Let $R_\rho(\zeta) = \dfrac{R(\zeta)}{\zeta - \rho} = \zeta^{m-1} + c_1\zeta^{m-2} + \cdots + c_{m-1}$, $c_i \in \mathbf{R}$, $1 \leq i \leq m-1$. Then by Dirichlet's theorem, there exist arbitrarily large integers a_1, n_1, ..., n_{m-1} such that

$$|\, a_1 c_k - n_k \,| < a_1^{-1/(m-1)}, \quad 1 \leq k \leq m-1.$$

Let $B(\zeta) = a_1\zeta^{m-1} + n_1\zeta^{m-2} + \cdots + n_{m-1}$.

Since $R_\rho(\rho_k) = 0$, we have

$$|B(\rho_k)/R'(\rho_k)| = \left| \frac{B(\rho_k) - a_1 R_\rho(\rho_k)}{R'(\rho_k)} \right| < \frac{2^m a_1^{-1/(m-1)}}{R'(\rho_k)}.$$

So if a_1 is chosen large enough, then $\beta_2, \ldots, \beta_m$ will satisfy (17).

Finally since $\deg(zA(z)) = 2m-1$ and $\deg(T(z)) = 2m$, the sequence (a_n) is a pure recurrent sequence.

This completes the proof. ■

We end this section with a criterion for T-recurrence.

Theorem 13.3.3. **(Criterion for T-recurrence)** *Let $(a_n)_{n\geq 0}$ be a T-recurrent Boyd sequence $F(a_0, a_1, a_2)$. Then, for $n \in \mathbf{Z}$, the following inequalities hold:*

$$|\, c_{n-1} - \Delta(c_n, c_{n+1}, c_{n+2}) \,| \leq 1/(2\gamma),$$

where $\gamma = \gamma(a_0, a_1, a_2)$ and $\Delta(x, y, z) = [x\gamma(2+\gamma^2) - y(1+2\gamma^2) + z\gamma]/\gamma^2$.

Moreover, if $a_n > \varepsilon + \dfrac{\varepsilon + 8\varepsilon^2}{\gamma^2}$ and $\varepsilon = \sup_{n\geq 0} |\,\varepsilon_n\,|$, we have $a_n = c_n$.

Proof. The proof is similar to that of Theorem 13.2.2.

The last theorem unfortunately is not effective, since ε is finite but not computable. Nevertheless Boyd has proved that the Boyd sequence $F(0, 30, 61)$ is not a recurrent sequence [10].

Notes

In 1938 Ch. Pisot published his thesis [16] in which he studied parametrized sequences with analytic methods. These included sequences now called Pisot sequences. Furthermore Pisot proved that the sequences $E(2, a_1)$ and $E(3, a_1)$ are recurrent and that the associated limits $\gamma(2, a_1)$ and $\gamma(3, a_1)$ are Pisot numbers.

That the set S of Pisot numbers is closed in the real line was proved by Salem in 1944 [17], so the set E of limits associated to Pisot sequences strictly includes the set S.

Flor [13] proved in 1960 that every number of E associated to a recurrent Pisot sequence is either a Salem or a Pisot number and that all the roots of the minimal recurrence polynomial are simple.

The problem then arose: Are all Pisot sequences recurrent?

Galyean [15] noted in 1971 that the Pisot sequence $E(4, 13)$ satisfies no recurrence relation of degree less than 100.

In 1977 Boyd [4] used a criterion for T-recurrence to show that $E(14, 23)$ and $E(31, 51)$ are not recurrent. More recently, in 1985 [8] he proved that the rationals p/q such that $p/q > q/2$ (3/2, for instance) do not belong to E.

Corollary 13.1 has been improved by Boyd in [9], where he slightly improves the criterion for T-recurrence of Pisot sequences.

Pisot and Boyd sequences have been generalized by Cantor [11]–[12] and more recently by Bertin [2]–[3].

Many problems still remain open:

- Are almost all Pisot sequences and "geometric" Boyd sequences non-recurrent?
- If γ is a Pisot or a Salem number such that $\gamma = \gamma(a_0, a_1) \in E$ (resp. $\gamma = \gamma(a_0, a_1, a_2) \in F$), is the Pisot sequence $E(a_0, a_1)$ (resp. Boyd sequence $F(a_0, a_1, a_2)$) recurrent?
- Are the algebraic numbers of E and F Pisot or Salem numbers?

References

[1] M.J. Bertin, Nouvelles applications d'un théorème de Pisot, *Groupe d'étude en théorie analytique des nombres,* Publications de l'I.H.P. 1 ère et 2 ième année.

[2] M.J. Bertin, Généralisation des suites de Pisot et de Boyd, *Acta Arithm.,* 57, (1991), 211-223.

[3] M.J. Bertin, Généralisation des suites de Pisot et de Boyd II, *Acta Arithm.,* 59.3, (1991), 11-15.

[4] D.W. BOYD, Pisot sequences which satisfy no linear recurrence, *Acta Arithm.*, 32, (1977), 89-98.

[5] D.W. BOYD, Some integer sequences related to Pisot sequences, *Acta Arithm.*, 34, (1979), 295-305.

[6] D.W. BOYD, Pisot sequences, Pisot numbers and Salem numbers, *Soc. Math. de France. Astérisque,* 61, (1979), 35-42.

[7] D.W. BOYD, On linear recurrence relations satisfied by Pisot sequences, *Acta Arithm.*, 47, (1986), 13-27.

[8] D.W. BOYD, Which rationals are ratios of Pisot sequences? *Canad. Math. Bull.*, 28, (1985), n°3, 343-349.

[9] D.W. BOYD, Pisot sequences which satisfy no linear recurrence II, *Acta Arithm.*, 48, (1987), n°2, 191-195.

[10] D.W. BOYD, Non-recurrence of F(0, 30, 61), *Private letter, April 23, 1987.*

[11] D.G. CANTOR, On families of Pisot E-sequences, *Ann. Sc. Ec. Norm. Sup.*, (4), 9, (1976), 283-308.

[12] D.G. CANTOR, Investigations of T-numbers and E-sequences in Computers in Number Theory, Ed. by A.O.L. Atkins and B.J. Birch *Academic Press N.Y.*, (1971).

[13] M.J.DE LEON, Pisot sequences, *J. Reine Angw. Math.*, 249, (1971), 20-30.

[14] P.FLOR, Uber eine Klasse von Folgen naturlicher Zahlen, *Math. Ann.*, 140, (1960), 299-307.

[15] P.GALYEAN, On linear recurrence relations for E-sequences, *Thesis, University of California,* Los Angeles (1971).

[16] CH.PISOT, La répartition modulo 1 et les nombres algébriques, *Scuola. Norm. Sup. Pisa,* 7, (1938), 205-248.

[17] R.SALEM, *Algebraic Numbers and Fourier Analysis,* Heath Mathematical Monographs, Boston, (1963).

CHAPTER 14

GENERALIZATIONS OF PISOT AND BOYD SEQUENCES

The purpose of this chapter is to extend some of the results obtained in the previous chapters to sequences of rationals and to sequences of polynomials.

In the proofs we will give only what is fairly different from the real case; the reader can easily reconstitute the rest.

The notation was introduced in for the rationals in Chapter 10 and for polynomialsin Chapter 12. However, in A_I we sometimes modify the fundamental domain as follows: let ω be an interval of the form $]a, a+1]$ or $[a, a+1[$, the associated fundamental domain is $F_I(\omega) = \omega \times \prod_{p \in I^-} \mathbf{Z}_p$, and Artin's decomposition is written $x = E_\omega(x) + \varepsilon_\omega(x)$ where $E_\omega(x) \in Q^I$, $\varepsilon_\omega(x) \in F_I(\omega)$; moreover we set $\mu(\omega) = \sup\{|x|/x \in \omega\}$.

14.1 Convergence theorems in A_I

Since the mean value theorem is not valid in general in a valued field, we will impose on the functions Lipschitz conditions rather than differentiability conditions.

Notation

We denote by k a natural integer and, for $\mathbf{x} = (x^{(1)}, \ldots, x^{(k)}) \in A_I^k$, we set $\|\mathbf{x}\|_p = \max_j |x^{(j)}|_p$, $p \in I$; then Artin's decomposition in A_I^k can be written $\mathbf{x} = \mathbf{E}_\omega(\mathbf{x}) + \boldsymbol{\varepsilon}_\omega(\mathbf{x})$, with $\mathbf{E}_\omega(\mathbf{x}) = (E_\omega(x^{(j)}))_{j=1\ldots,k}$ and $(\varepsilon_\omega(x^{(j)}))_{j=1,\ldots,k}$.

For $p \in I$ and $\mathbf{r}_p = (r_p^{(1)}; \ldots, r_p^{(k)}) \in \Gamma_p^k$ we set $\mathbf{B}_p(\mathbf{x}_p, \mathbf{r}_p) = \prod_{j=1}^{k} D_p(x^{(j)}, r_p^{(j)})$, and if $\mathbf{r} = (\mathbf{r}_p)_{p \in I}$, $\mathbf{B}_I(\mathbf{x}, \mathbf{r}) = \prod_{p \in I} \mathbf{B}_p(\mathbf{x}_p, \mathbf{r}_p)$. The fundamental theorem can then be stated as follows:

Theorem 14.1.1. *We consider a sequence* $(\mathbf{F}_n)$ *of functions defined on an open set* $\Delta \subset A_I^k$ *with values in* A_I^k *and satisfying:* $F_n^{(j)} = F_{n-1}^{(j+1)}$, *i.e., every function* $\mathbf{F}_n$ *can be written* $\mathbf{F}_n = (f_n, f_{n+1}, \dots, f_{n+k-1})$; $f_n : \Delta \mapsto A_I^k$. *We suppose the following conditions satisfied:*

(i) *For every* $n \in \mathbf{N}$, $\mathbf{F}_n$ *is a homeomorphism of* Δ *on an open set* $\mathbf{D}_n \subset A_I^k$; *we denote by* Φ_n *the inverse homeomorphism.*

(ii) *There exists a non empty subset* J *of* $\{1, \dots k\}$ *such that*

a) *For every* $j \in J$ *and* $p \in I$ *the coordinate functions* $\Phi_{n,p}^{(j)}$ *are Lipschitz of constant* $\varphi_{n,p}^{(j)}$ *with respect to the last variable and* $\lim_{n \to +\infty} \varphi_{n,p}^{(j)} = 0$ *for* $p \in I^-$ *and* $\sum_{n=0}^{\infty} \varphi_{n,\infty}^{(j)} < +\infty$ *if* $\infty \in I$.

We set then $\Psi_{h-1,p}^{(j)} = \sup_{m \geq h} \varphi_{m,h}^{(j)}$ *for* $p \in I^-$, $\Psi_{h-1,\infty}^{(j)} = \sum_{m=h}^{\infty} \varphi_{m,\infty}^{(j)}$ *if* $\infty \in I$, *and* $\Psi_h = (\Psi_{h,p}^{(j)})_{p \in I,\ j=1,\dots,k}$.

b) *For* $j \notin J$, *we have* $\Delta^{(j)} = A_I$. *We set* $\varphi_{n,p}^{(j)} = +\infty$ *for* $p \in J$. *Then, begining with an element* $\mathbf{b}_0 = (a_0, a_1, \dots, a_{k-1}) \in \mathbf{D}_0 \cap (Q^I)^k$ *such that* $\mathbf{B}_I(\Phi_0(\mathbf{b}_0), \Psi_0) \subset \Delta$, *we can define a sequence* $(\mathbf{b}_n)$ *of elements of* $(Q^I)^k$ *by the relation*

$$\mathbf{b}_n = \mathbf{E}_\omega(\mathbf{F}_n \circ \Phi_{n-1}(\mathbf{b}_{n-1})), \tag{1}$$

and, for $j \in J$ *the sequence* $(\Phi_n^{(j)}(\mathbf{b}_n))$ *converges to an element* $\beta^{(j)} \in \Delta^{(j)}$. *Moreover we have the following inequalities:*

$$\begin{cases} |\beta^{(j)} - \Phi_0^{(j)}(\mathbf{b}_0)|_p \leq \Psi_{0,p}^{(j)} & \text{for } \ p \in I^- \\ |\beta^{(j)} - \Phi_0^{(j)}(\mathbf{b}_0)|_\infty \leq \mu(\omega)\Psi_{0,\infty}^{(j)} & \text{if } \ \infty \in I\ . \end{cases} \tag{2}$$

The reader can easily reconstruct the proof of this theorem from the proof of Theorem 13.0.1.

In order to obtain directly the recurrence relation satisfied by the sequence (a_n), we set $\mathbf{G}_m^n = \mathbf{F}_m \circ \Phi_n$ and we denote by g_n the last component of $\mathbf{G}_{n+1}^n$. Then the relation (1) is equivalent to

$$a_{n+k} = E_\omega(g_n(a_n, a_{n+1}, \dots, a_{n+k-1})). \tag{3}$$

In the rest of this section, we suppose $J = \{1, 2, \dots, k\}$. Then one can prove the following results.

Theorem 14.1.2. *Let $\gamma \in \Delta$ and $\mathbf{W}$ be a neighborhood of γ whose closure is contained in Δ. Then there exists an index $n_0 \geq 0$ and an element $\mathbf{b}_{n_0} \in \mathbf{F}_{n_0}(\mathbf{W}) \cap (Q^I)^k$ such that the following inclusion is satisfied:*

$$\mathbf{B}_I(\Phi_{n_0}(\mathbf{b}_{n_0}); \Psi_{n_0}) \subset \mathbf{W}.$$

This, together with inequalities (2), allows us to show that the set of elements β that are limits of the sequences $(\Phi_n(\mathbf{b}_n))$ is dense in Δ.

Theorem 14.1.3. *With the same hypothesis and notations as above, we set $\varepsilon_n = \mathbf{F}_n(\beta) - \mathbf{b}_n$. If there exists an index j for which every function $G_{n,p}^{n+h,j}$ is Lipschitz of constant $s_{h,p}$ (not dependent on n) with respect to the last variable, with $s_{h,p} = O(1)$ for $p \in I^-$ and $\sum_{h=1}^{\infty} s_{h,\infty} < +\infty$ if $\infty \in I$, then $(\|\varepsilon_n\|)$ is bounded.*

Remark

If $J \neq \{1, \ldots, k\}$, this result is still true for some components of ε_n.

14.2 Pisot sequences in A_I

The purpose of the next two sections is the application of Theorem 14.1.1 to certain particular cases.

We take $k = 2$ and denote by $(\mu_p, \nu_p)_{p \in I}$ pairs of positive real numbers. Then Δ is defined as the set of $(x, y) \in A_I$ that satisfy the inequalities

$$|x|_p > 1 + \mu_p,\ |y|_p > \nu_p \text{ for } p \in I^-, \qquad x_\infty > 1 + \mu_\infty \text{ if } \infty \in I.$$

For $n \in \mathbf{N}$ we set $f_n(x, y) = x^n y$. Hence $\mathbf{F}_n(x, y) = (x^n y, x^{n+1} y)$.

We can easily verify that the sequence $(\mathbf{F}_n)$ satisfies the hypothesis of Theorem 14.1.1 with $J = \{1, 2\}$, and that for every $(u, v) \in \mathbf{D}_n$ we have

$$|vu^{-1}|_p > 1 + \mu_p,\ |u^{n+1}v^{-n}|_p > \nu_p \text{ for } p \in I, \quad (u_\infty > 0,\ v_\infty > 0 \text{ if } \infty \in I).$$

Hence

$$\Phi_n(u, v) = (vu^{-1}, u^{n+1}v^{-n}), \quad G_n^{n+h}(u, v) = (u^{n+h}v^{-h}, u^h v^{-(h+1)}),$$
$$g_n(u, v) = v^2 u^{-1}.$$

Then we obtain the following equations:

$$\Psi_{0,p}^{(1)} = \frac{1}{\nu_p(1+\mu_p)}, \qquad \Psi_{0,p}^{(2)} = \frac{1}{(1+\mu_p)^2} \qquad \text{for } p \in I^-$$

$$\Psi_{0,\infty}^{(1)} = \frac{1}{\nu_\infty(1+\mu_\infty)}, \qquad \Psi_{0,\infty}^{(2)} = \frac{1}{\mu_\infty^2} \qquad \text{if } \infty \in I.$$

Our definition of Pisot sequences depends on the interval ω. This choice has no incidence on the general properties of the sequences, and only allows us to obtain, more or less easily, certain inequalities. In practice we will only take the intervals $]-1/2, 1/2]$ and $[0, 1[$, the first because it is traditionally used in the real case, and the second because it leads to much simpler results.

Definition 14.2.1. *We call a* **Pisot** ω**-sequence** *in* Q^I *a sequence* (a_n) *of elements of* Q^I *defined by:*

- *a given pair* (a_0, a_1)*;*
- *the recurrence relation:* $a_{n+2} = E_\omega(a_n^2/a_{n+1})$, $n \geq 1$.

Such a sequence is written $\mathcal{E}_\omega(a_0, a_1)$. For $\omega =]-1/2, 1/2]$, we will simply say **Pisot sequence** and we denote it $\mathcal{E}(a_0, a_1)$, and for $\omega = [0, 1[$ we will say **Pisot 0-sequence** and denote it $\mathcal{E}_0(a_0, a_1)$.

Definition 14.2.2. *We say that a Pisot* ω*-sequence is geometric if*

$$\liminf \left| \frac{a_{n+1}}{a_n} \right|_p > 1 \quad \textit{for } p \in I.$$

Lemma 14.2. *Let* (a_n) *be a geometric Pisot* ω*-sequence. Then the sequences* (a_{n+1}/a_n) *and* (a_n^{n+1}/a_{n+1}^n) *converge and their limits* α *and* λ *satisfy*

$$|a_n - \lambda\alpha^n|_p \leq \frac{1}{|\alpha|_p^2} \quad \textit{for } p \in I^-, \quad |a_n - \lambda\alpha^n|_\infty \leq \frac{\mu(\omega)}{\alpha_\infty - 1} \quad \textit{if } \infty \in I.$$

Proof. The assumptions imply the existence, for $p \in I$, of a number $k_p > 1$ such that $\left| \frac{a_{n+1}}{a_n} \right| \geq k_p$, for n large enough. One deduces the convergence of the series $\sum \frac{1}{|a_n|_p}$ and the existence of the two limits satisfying the inequalities. The calculations are of the same nature as those of §13.3. ∎

For $\omega =]-1/2, 1/2]$, the method used in the demonstration of Theorem 13.1.1 gives the following statement.

Theorem 14.2.1. *Let (a_0, a_1) be a pair of elements of Q^I satisfying*

$$|a_1|_p > |a_0|_p > 1 \quad \textit{for} \quad p \in I, \qquad a_0 > 1,\ a_1 > a_0 + \frac{3}{2}\sqrt{\frac{3a_0}{2}} \quad \textit{if} \quad \infty \in I.$$

Then the sequence $\mathcal{E}(a_0, a_1)$ is geometric. Moreover we have, if $\infty \in I$,

$$\left|\alpha - \frac{a_0}{a_1}\right| < \frac{1}{2}\sqrt{\frac{3}{2a_0}}$$

$$|a_n - \lambda\alpha^n|_p \leq 1 \quad \textit{for} \quad p \in I^- \quad \textit{and} \quad |a_n - \lambda\alpha^n|_\infty = O(1).$$

We remark that if I does not contain the archimedean absolute value, the condition $|a_1|_p > |a_0|_p > 1$ for $p \in I$ implies by itself the convergence of the sequence $\mathcal{E}(a_0, a_1)$ for every interval ω.

Theorem 14.2.2. *Suppose I contains the archimedean absolute value. Let (a_0, a_1) be a pair of elements of Q^I satisfying $|a_1|_p > |a_0|_p > 1$ for $p \in I$ and $a_1 > a_0 + \sqrt{2a_0}$. Then the sequence $\mathcal{E}(a_0, a_1)$ is geometric.*

Proof. Consider the relation

$$a_{n+1} \geq ra_n + s. \qquad (R_n)$$

We wish to determine the constants r and s ($r > 1$ and $s > 0$) for which (R_0) implies (R_n) for every $n \in \mathbf{N}$.

Now (R_{n-1}) implies $a_{n+1} \geq ra_n + rs - 1/2$, hence (R_n) if $s \geq 1/2(r-1)$. If $s = 1/2(r-1)$, then (R_0) becomes $a_1 \geq ra_0 + 1/2(r-1)$, and the expression on the right has a minimum at $r = 1 + \sqrt{2a_0}$. The result follows. ∎

If $\omega = [0, 1[$ direct study is considerably simplified, as is shown by the following statement.

Theorem 14.2.3. *Let (a_0, a_1) be a pair of elements of Q^I satisfying $|a_1|_p > |a_0|_p > 1$ for $p \in I$. Then the sequence $\mathcal{E}_0(a_0, a_1)$ is geometric and*

$$|a_n - \lambda\alpha^n|_p \leq \frac{1}{|\alpha|_p^2} \quad \textit{for} \quad p \in I^-, \qquad 0 \leq a_n - \lambda_\infty\alpha_\infty^n \leq \frac{a_1^2}{(a_1 - a_0)^2} \quad \textit{if} \quad \infty \in I.$$

As in the real case there exists a relation between Pisot ω-sequences and the sets U_I:

Theorem 14.2.4. *Every geometric and recurrent Pisot ω-sequence can be associated to a pair (α, λ) where $\alpha \in U_I$ and $\lambda \in \mathbf{Q}_I[\alpha]$. Conversely, every $\alpha \in U_I$ can be associated to a geometric and recurrent Pisot ω-sequence.*

We will not pursue this subject further here.

In conclusion, we show how to define a Pisot ω-sequence starting with two rational numbers.

Definition 14.2.3. *Let (a_0, a_1) be a pair of rational numbers satisfying $a_1 > a_0 > 1$. We set $I = \{p \in \mathbf{P} / |a_1|_p > |a_0|_p > 1\}$, and call Pisot ω-sequence $\mathcal{E}_\omega(a_0, a_1)$ the sequence (a_n) defined by $a_{n+1} = E_\omega(\frac{a_n^2}{a_{n-1}})$, where the Artin decomposition is effectuated in A_I.*

This sequence can be studied in the same way as the previous ones.

14.3 Boyd sequences in A_I

These sequences are obtained by an application of the results of § 14.1 to the particular case: $k = 3$, $f_n : (x, y, z) \mapsto yx^n + zx^{-n}$. We denote (μ_p, ν_p) a pair of strictly positive real numbers and define the open set $\Delta \subset A_I^3$ by the inequalities

$$|x|_p > 1 + \mu_p,\ |y|_p > \nu_p \quad \text{for} \quad p \in I; \quad x_\infty > 1 + \mu_\infty,\ y_\infty > \nu_\infty \quad \text{if} \quad \infty \in I;$$

then $J = \{1, 2\}$.

As in the real case, we determine the function Φ_n by solving the system

$$\begin{cases} u = & yx^n & + & zx^{-n} \\ v = & yx^{n+1} & + & zx^{-(n+1)} \\ w = & yx^{n+2} & + & zx^{-(n+2)} \end{cases} \tag{1}$$

whose Jacobian is $J = -(x-1)^2 x^{-3} v$. In order to solve the system (1) we must determine the roots of equation

$$vx^2 - (u + w)x + v = 0. \tag{2}$$

In A_I the discriminant of (2) is $\delta^2 = (u + w)^2 - 4v^2$.

For $p \in I^-$ equation (2) has two roots in $\mathbf{Q}_p$ if and only ifwe have

$$|u + w|_p > |v|_p \tag{3}$$

and this condition implies that δ_p^2 is a square in $\mathbf{Q}_p$. We denote δ_p its square root, which satisfies $|u+v+\delta_p|_p > 2|v|_p$.

If I contains the archimedean absolute value, the definition of Δ implies $u_\infty > 0$, $v_\infty > 0$, $w_\infty > 0$ for n large enough. Then the condition for (2) to have two real roots is written

$$u_\infty + w_\infty - 2v_\infty > 0. \tag{4}$$

These considerations can be summarized in the following way:

If the following conditions are satisfied:

$$|u+w|_p > |v|_p \ \ \textit{for} \ \ p \in I^-, \ u_\infty + w_\infty - 2v_\infty > 0 \ \ \textit{if} \ \ \infty \in I, \tag{5}$$

then the sequence $(\mathbf{F}_n)$ *satifies the assumptions of Theorem* 14.1.1 *with* $g_n(u,v,w) = wu^{-1}(u+w) - v$.

The components of the functions Φ_n and the constants $\varphi_{n,p}^{(j)}$ can be computed as in §13.3, but we will not do this, a direct computation being simpler.

Definition 14.3.1. *A sequence* (a_n) *of elements of* Q^I *is called a* **Boyd** ω-**sequence** *in* Q^I *if it is defined by*

- *a given triplet* (a_0, a_1, a_2)
- *the recurrence relation*

$$a_{n+2} = E_\omega\left(\frac{a_{n+1}}{a_n}(a_{n+1} + a_{n-1}) - a_n\right) \ \ \textit{if} \ \ a_n \neq 0$$
$$a_{n+2} = a_{n-2} \ \ \textit{if} \ \ a_n = 0.$$

Such a sequencce is denoted $\mathcal{F}_\omega(a_0, a_1, a_2)$. *For* $\omega = [0,1[$, *the sequence is called a Boyd 0-sequence, and denoted* $\mathcal{F}_0(a_0, a_1, a_2)$.

Definition 14.3.2. *A Boyd* ω*-sequence* (a_n) *is called* **geometric** *if* $\liminf \left|\frac{a_{n+1}}{a_n}\right|_p > 1$ *for* $p \in I$.

The following result is obtained in the same way as in the real case.

Theorem 14.3.1. *Let (a_n) be a geometric Boyd ω-sequence. Then the sequences (a_{n+1}/a_n) and (a_n^{n+1}/a_{n+1}^n) converge and their limits α and λ satisfy the inequalities*

$$|a_n - \lambda\alpha^n|_p \leq |\alpha|^{-2} \text{ for } p \in I^-, \qquad |a_n - \lambda\alpha^n|_\infty \leq \frac{\mu(\omega)}{\alpha_\infty - 1} \text{ if } \infty \in I.$$

We remark that if I does not contain the archimedean absolute value then every Boyd ω-sequence is a Pisot ω-sequence.

We easily obtain a condition for a Boyd 0-sequence to be geometric:

Theorem 14.3.2. *Let (a_0, a_1, a_2) be a triplet of elements of Q^I satisfying*

$$|a_2|_p > |a_1|_p > |a_0|_p \geq 1 \text{ for } p \in I^-$$
$$a_2 > a_1 > a_0 \text{ and } a_2 - 2a_1 + a_0 > 0 \text{ if } \infty \in I.$$

Then the sequence $\mathcal{F}_0(a_0, a_1, a_2)$ is geometric.

Proof. The assumptions imply that the triplet (a_0, a_1, a_2) satisfies Condition (5) for $p \in I^-$. We will prove by induction that the sequence $(|a_n|_p)$ is increasing and that

$$\left|\frac{a_{n+2}}{a_{n+1}}\right|_p = \left|\frac{a_{n+1}}{a_n}\right|_p = \ldots = \left|\frac{a_1}{a_0}\right|_p > 1.$$

If I contains the archimedean absolute value, we set $d_n = a_{n+1} - 2a_n + a_{n-1}$. Then by hypothesis $d_1 > 0$, it can be shown that the sequences (a_n) and (d_n) are increasing, and we deduce that $\mathcal{F}_0(a_0, a_1, a_2)$ is geometric. ∎

As for the relationship between geometric recurrent Boyd sequences and the sets U_I, we are in a situation similar to that found in the previous chapter.

14.4 Pisot and Boyd sequences in a field of formal power series

The extension of the notion of Pisot or Boyd sequence to a field of formal power series seems natural, but the results are different and simpler than in the preceeding cases.

Definition 14.4.1. *We call a* **Pisot sequence** *in $\mathcal{Z}$ every sequence (a_n) of elements of $\mathcal{Z}$ defined by:*

- *a given pair (a_0, a_1),*
- *the recurrence relation*

$$a_{n+2} = E(\frac{a_{n+1}^2}{a_n}) \text{ for } n \geq 0.$$

Such a sequence will be denoted $\mathcal{E}(a_0, a_1)$.

Definition 14.4.2. *We call* **Boyd sequence** *in $\mathcal{Z}$ every sequence (a_n) of elements of $\mathcal{Z}$ defined by:*

- *a given triplet (a_0, a_1, a_2),*
- *the recurrence relation*

$$a_{n+2} = E\left(\frac{a_{n+1}}{a_n}(a_{n+1} + a_{n-1}) - a_n\right) \quad \text{if} \quad a_n \neq 0$$
$$a_{n+2} = -a_{n-2} \quad \text{if} \quad a_n = 0.$$

Such a sequence will be denoted $\mathcal{F}(a_0, a_1, a_2)$

Here we cannot distinguish between these two types of sequences because a simple calculation shows that

(i) *If the triplet (a_0, a_1, a_2) satisfies $|a_2| > |a_1| > |a_0] \geq 1$ then $\mathcal{F}(a_0, a_1, a_2)$ is identical to a Pisot sequence from some rank on.*

(ii) *If the pair (a_0, a_1) satisfies $|a_1| > |a_0] \geq 1$, then the sequences $\mathcal{E}(a_0, a_1)$ and $\mathcal{F}(a_0, a_1, a_2)$ are identical.*

This remark explains why we only study Pisot sequences. We easily prove the following result.

Theorem 14.4.1. *Let (a_0, a_1) be a pair of elements of $\mathcal{Z}$ satisfying $|a_1| > |a_0| \geq 1$. Consider the sequence $\mathcal{E}(a_0, a_1)$; the sequences (a_{n+1}/a_n) and (a_n^{n+1}/a_{n+1}^n) converge in $\mathcal{F}_\infty$ and their limits α and λ satisfy the inequalities $|\alpha| > 1$ and $|a_n - \lambda\alpha^n| < |\alpha|^{-2}$.*

From this assertion and Theorem 12.2.1 we deduce:

Theorem 14.4.2. *Every Pisot sequence in $\mathcal{Z}$ is recurrent and the associated pair (α, λ) is such that $\alpha \in \mathcal{U}$ and $\lambda \in \mathcal{F}(\alpha)$. Conversely every $\alpha \in \mathcal{U}$ can be associated to a Pisot sequence.*

The *adeles* of $\mathcal{F}$ can be studied in the same way as Pisot elements.

Notes

Pisot sequences in Q^I were introduced by Decomps-Guilloux [2], who proved Theorem 14.2.2. The rest of this chapter is so far unpublished.

The assertions of §14.4 have not been published. However we should mention that Pisot sequences in $\mathcal{Z}$ were studied by Rauzy around 1968.

Cantor [1] has proposed another definition of Pisot sequences in $\mathcal{Z}$, but his subject is not within the scope of this book, and we refer the reader to Cantor's paper.

References

[1] D.G. Cantor, On families of Pisot E-sequences, *Ann. Scient. Ec. Norm. Sup. 4è série* 9, (1976), 233-309.

[2] A. Decomps-Guillox, Généralisation des nombres de Salem aux adèles, *Acta Arith.* 16, (1970), 265-314.

CHAPTER 15

THE SALEM–ZYGMUND THEOREM

15.1 Introduction

The Salem–Zygmund theorem, about sets of uniqueness in the theory of trigonometric series, is certainly the result that has given Pisot numbers most of their renown, at least among analysts.

This result, which was stated in 1943 by Salem but only received a complete proof twelve years later, gives a necessary and sufficient condition for sets to have a certain property related to trigonometric series in terms of the real parameter defining the set belonging or not belonging to the class S of Pisot numbers. The striking fact here was the appearance of an arithmetic condition where a measure condition would have been expected.

Besides various generalizations of this theorem, various questions of Fourier analysis connected to spectral synthesis or to almost periodic functions also led to results involving Pisot numbers. This subject has been exposed in the classical books of Salem [21], Kahane and Salem [7] and thoroughly studied in Meyer's treatise [18], which is almost entirely devoted to it.

That is why we shall restrict ourselves, in this chapter, to proving the Salem–Zygmund theorem itself. The proof clearly illustrates, without too many technical tricks, the way Pisot numbers occur in Fourier analysis. In particular the use of adeles in the proof of the necessary condition gives at the same time a straightforward extension of the theorem to p-adic fields (results of Bertrandias [2], see also [26]) and, we hope, some idea of the methods used by Meyer for other problems.

The interested reader is referred to [18] and [17] for results on spectral synthesis in relation to Pisot numbers, and to [26] for the analog in the p-adic case.

For the sake of completeness we give in §15.5 a result of Senge and Strauss published after Meyer's book, and which answers a question left open by Salem [21, p. 62].

15.2 Sets of uniqueness

15.2.1 Uniqueness and convergence of trigonometric series

For a detailed historical account see [1]; see also [7] and [18] for further references, and, for a modern point of view, [8].

The word "uniqueness" has its origin in the following problem: Let $f(t)$ be the sum of a trigonometric series: $f(t) = \sum_{n\in\mathbf{Z}} e^{\mathrm{int}}$ for all $t \in \mathbf{R}$. Are the coefficients c_n uniquely determined by f? The answer, which goes back to G. Cantor ([4]), is "Yes":

Theorem 15.2.1 *If the trigonometric series $\sum_{n\in\mathbf{Z}} c_n e^{\mathrm{int}}$ converges to zero for all $t \in [0, 2\pi]$ then the sequence $(c_n)_{n\in\mathbf{Z}}$ is identically zero.*

Now a very natural question to ask is to what extent the coefficients c_n of the series $\sum_{n\in\mathbf{Z}} c_n\ e^{\mathrm{int}}$ are determined by its sum on only a subset of $[0, 2\pi]$.

More precisely, we define

Definition 15.2.1 *A subset E of $[0, 2\pi]$ is called set of uniqueness if every trigonometric series that converges to zero on the complement of E is identically zero. Otherwise E is called a set of multiplicity.*

For example, it is known (see for instance [21, p. 44]) that

- *Every denumerable set is a set of uniqueness.* (Cantor–Young),
- *Every set of positive Lebesgue measure is a set of multiplicity,*

but

- *There exist closed sets of multiplicity of measure zero* ([13]).

Nevertheless a general classification of subsets of $[0, 2\pi]$ in terms of sets of uniqueness and sets of multiplicity is not known, and, in fact, cannot be expected to be easily formulated, as follows from recent results in the subject ([8]).

15.2.2 Compact sets of uniqueness in **R** *and distributions*

If we restrict ourselves to compact sets, and use the notion of distribution (in the sense of L. Schwartz), we obtain the following simple criterion:

Proposition 15.2.1 *For a compact subset E of $[0, 2\pi]$ the two following properties are equivalent:*

1. *E is a set of uniqueness.*
2. *Every distribution supported by E that has a Fourier transform (defined as a function on $\mathbf{R}$) vanishing at infinity is identically zero.*

For the proof see [18, p. 81].

In the following we shall consider only compact sets. We shall drop the original point of view of representing functions by trigonometric series and take the second property above as the definition of uniqueness.

Definition 15.2.2 *A compact set of $\mathbf{R}$ is called a set of uniqueness if it supports no non-trivial distribution whose Fourier transform tends to zero at infinity. If it supports such a distribution it is called a set of multiplicity (and a set of strict multiplicity if the distribution is a measure).*

Remark: From this definition it follows clearly that the class of compact sets of uniqueness of $\mathbf{R}$ is invariant by translation and dilation.

15.2.3 Sets of uniqueness in $\mathbf{Q}_p$

Definition 15.2.2 can be immediately extended to every locally compact abelian group by using the notion of tempered distributions and Fourier transforms on these groups. (See [3]).

We recall that in $\mathbf{Q}_p$ and in finite extentions of $\mathbf{Q}_p$ things are simple: the space that plays the role of Schwartz's space $\mathcal{S}$ is just the space of compactly supported and locally constant functions (see [6] for an elementary account).

It will be noticed that all the distributions concerned by the previous Definition 15.2.2 have bounded Fourier transforms (because they are continuous and tend to zero at infinity). Hence they can be defined, by Fourier transforms, as continuous linear forms on the space of Lebesgue integrable functions on the dual group. In this context these distribution are called pseudo-measures (see [18, p. 87] for this point of view).

15.3 Symmetric perfect sets

15.3.1 Sets E_θ and the Salem–Zygmund theorem

We shall now consider a family of compact sets indexed by a real parameter, very similar to Cantor's triadic set.

Definition 15.3.1 *Let θ be a real number, $\theta > 2$, and let E_θ be the set of all sums $\sum_{k\geq 0}\varepsilon_k\theta^{-k}$ where each coefficient ε_k can take the value 0 or 1.*

For $\theta = 3$, E_θ is exactly Cantor's set constructed on the interval $[0, \theta/(\theta-1)]$; for other values of θ, E_θ can be constructed in the same way by means of a sequence of "dissections" of some interval of $\mathbf{R}$. (See [7]).

We can now state the Salem–Zygmund theorem:

Theorem 15.3.1 *The set E_θ is a set of uniqueness if and only if θ is a Pisot number ($\theta > 2$).*

First we give the proof of the necessary condition. We begin by constructing a measure μ_θ supported by E_θ which has a Fourier transform vanishing at infinity when $\theta \notin S$; hence, in this case, E_θ will be a set of strict multiplicity.

For each integer $N \geq 1$, let $\mu_{\theta,N}$ be the uniformly distributed probability measure on the set of finite sums

$$E_{\theta,N} = \{\sum_{k=0}^{N}\varepsilon_k\theta^{-k}; \qquad \varepsilon_k = 0 \text{ or } 1\}.$$

This measure, which is the convolution product of the uniform probabilities on the sets $\{0, \theta^{-k}\}$, has the Fourier transform

$$\begin{aligned}\hat{\mu}_{\theta,N}(\xi) &= \int_{\mathbf{R}} e^{-2i\pi\xi x}\, d\mu_{\theta,N}(x)\\ &= \prod_{k=0}^{N}\frac{1}{2}(1 + e^{-2i\pi\xi\theta^{-k}}).\end{aligned}$$

Proposition 15.3.1 *The sequence $\mu_{\theta,N}$ converges weakly to a limit μ_θ. The measure μ_θ (called the L-measure of E_θ) is supported by E_θ and has the Fourier* *transform*

$$\hat{\mu}_\theta(\xi) = \prod_{k=0}^{\infty} \frac{1}{2}(1 + e^{-2i\pi\xi\theta^{-k}})$$
$$= \exp(-i\pi \sum_{k=0}^{\infty} \xi\theta^{-k}) \prod_{k=0}^{\infty} \cos(\pi\xi\theta^{-k}).$$

This proposition is an immediate consequence, by the Lévy-Cramer theorem, of the locally uniform convergence of the infinite product.

15.3.2 The necessary condition

Proposition 15.3.2 *The function $\hat{\mu}_\theta(\xi)$ tends to zero at infinity if and only if θ is not a Pisot number.*

Proof: This is a consequence of Pisot's theorem on the distribution modulo 1 of the sequence $\lambda\theta^k$ (Theorem 5.4.2). We have only to consider the behavior at infinity of the function

$$\Gamma(\xi) = \prod_{k=0}^{\infty} \cos(\pi\xi\theta^{-k}).$$

Let us set, for $n \geq 1$

$$P_n(u) = \prod_{k=0}^{n} \cos(\pi u\theta^k).$$

First, when $\theta \notin S$, let us write, for $\theta^n \leq \xi < \theta^{n+1}$,

$$\xi = \theta^n u$$

with $1 \leq u < \theta$.

We have

$$\Gamma(\xi) = \cos(\pi u\theta^n) \cdots \cos(\pi u\theta)\,\Gamma(u)$$
$$= P_n(u)\,\Gamma(u).$$

Hence

$$|\Gamma(\xi)| \leq |P_n(u)|.$$

Now it is sufficient to prove that $\lim_{n\to\infty} P_n(u) = 0$ uniformly for $u \in [1, \theta]$ to deduce that $\lim_{\xi\to\infty} \Gamma(\xi) = 0$.

But if θ is not a Pisot number we have, for all $u \neq 0$,

$$\sum_{k=1}^{\infty} \sin^2(\pi u \theta^k) = +\infty$$

and so the infinite product $\prod_{k=1}^{\infty} \cos(\pi u \theta^k)$ is divergent. Consequently if $\theta \notin S$, $|P_n(u)|$ tends to zero on $[0,1]$ and, as it is a decreasing sequence of functions, the convergence is uniform by Dini's theorem.

Conversely, if $\theta \in S$, we have $\sum_{k=1}^{\infty} \sin^2(\pi \theta^k) < \infty$, and so

$$P_n(1) \to A \neq 0.$$

As

$$\Gamma(\theta^n) = P_n(1)\,\Gamma(1)$$

it is clear that $\Gamma(\xi)$ does not tend to 0 as ξ tends to infinity.

15.3.3 Generalization to other fields

As we have seen, the proof of Proposition 15.3.2 was entirely based on the relation between the convergence of the series $\sum \sin^2 \lambda\theta^n$ and θ belonging or not to the class S of Pisot numbers. This can be generalized to Pisot numbers in other fields: the field $\mathbf{C}$ of complex numbers, the field $\mathbf{Q}_p$ of p-adic numbers, finite extensions of $\mathbf{Q}_p$, the ring of adeles, and so on.

We have, for example, the following statements:

Proposition 15.3.3 *Let θ be a complex number, $|\theta| > 1$, θ not real and let E_θ be the set $E_\theta = \{\sum_{k\geq 1} \varepsilon_k \theta^{-k}\ ,\ \varepsilon_k = 0 \text{ or } 1\}$; the Fourier transform $\hat{\mu}_\theta(\xi)$ of the L-measure of the set E_θ tends to 0 at infinity in $\mathbf{C}$ if and only if θ is a "complex Pisot number" (i.e., an algebraic integer whose conjugates over $\mathbf{Q}$ not conjugate over $\mathbf{R}$ have absolute values less than 1).*

Proposition 15.3.4 *Let θ be a p-adic number, $|\theta|_p > 1$, and let E_θ be the set $E_\theta = \{\sum_{k\geq 1} \varepsilon_k \theta^{-k}\ ,\ \varepsilon_k = 0 \text{ or } 1\}$; the Fourier transform $\hat{\mu}_\theta(\xi)$ of the L-measure of the set E_θ tends to 0 at infinity in $\mathbf{Q}_p$ if and only if θ belongs to S_p^0 (Pisot–Chabauty numbers).*

For finite extensions of $\mathbf{Q}_p$ the statement is similar to the statement in $\mathbf{C}$: we need merely take an appropriate definition of Pisot numbers.

Definition 15.3.2 *If E is a finite extension of $\mathbf{Q}_p$ an element θ of E will be called a "Pisot number of E" if*

- *θ is algebraic over $\mathbf{Q}$, and of degree over $\mathbf{Q}_p$ equal to its degree over $\mathbf{Q}$;*
- *$|\theta|_p > 1$ but all its conjugates over $\mathbf{Q}$ not conjugate over $\mathbf{Q}_p$ have absolute values not greater than 1;*
- *for all p'-adic absolute values ($p' \neq p$) on $\mathbf{Q}(\theta)$, θ has an absolute value not greater than 1;*
- *for all archimedian absolute values on $\mathbf{Q}(\theta)$, θ has an absolute value less than 1.*

With these Pisot numbers a statement exactly similar to Proposition 15.3.2 can be given for the sets E_θ in finite extensions of $\mathbf{Q}_p$ with these Pisot numbers.([26]).

For a statement in adeles see [2].

15.4 The sufficient condition for the Salem–Zygmund theorem

15.4.1 The generalized Piatecki-Shapiro criterion

The following statement is a generalization of the classical criterions of Rajchman and of Piatecki-Shapiro ([20], [19], see [7, p. 58-60]):

Proposition 15.4.1 [Y. Meyer] *Let E be a compact subset of $\mathbf{R}$. Suppose that there exist: a locally compact abelian group G, a homomorphism h of $\mathbf{R}$ into G whose image is everywhere dense in G, a compact subset K of G, $K \neq G$ and a sequence $(t_j)_{j\geq 1}$ of reals converging to infinity and such that $h(t_jE) \subset K$ for all $j \geq 1$. Then E is a set of uniqueness.*

The idea of the proof is the following: arguing by contradiction, let us suppose that there exists a non-zero distribution T with support in E and whose Fourier transform vanishes at infinity. By multiplying T, if necessary, by an exponential, we can suppose that $\widehat{T}(0) = 1$.

Now consider the homomorphisms of $\mathbf{R}$ into the group G, defined by $h_j(x) = h(t_jx)$. These homomorphisms can be thought of as progressively faster "windings" of $\mathbf{R}$ on G. For all j, the image S_j of the distribution T by h_j is a distribution on G with support in K. From this it follows that the sequence S_j cannot converge, when j tends to infinity, to the Haar measure of G, because this measure has the whole set G for support.

But on the other hand the Fourier transforms $\widehat{S}_j$ of the distributions S_j defined on the discrete dual group $\widehat{G}$ of G satisfy, as is easily seen,

$$\widehat{S}_j(\chi) = \widehat{T}(\hat{h}_j(\chi)) = \widehat{T}(t_j\hat{h}(\chi)), \quad \text{for all } \chi \in \widehat{G}.$$

where $\hat{h}$ (resp. $\hat{h}_j$) is the adjoint homomorphism of h (resp. of h_j) of $\widehat{G}$ in $\mathbf{R}$ defined by the duality:

$$h: \ \mathbf{R} \longrightarrow G$$
$$\hat{h}: \ \mathbf{R} \simeq \hat{\mathbf{R}} \longleftarrow \widehat{G}.$$

Since the homomorphism h has, by hypothesis, a dense image, its adjoint $\hat{h}$ will be injective. It follows that for $\chi \neq 0$ in $\widehat{G}$, $\hat{h}(\chi) \neq 0$, so that $t_j\hat{h}(\chi) \to \infty$ and therefore $\widehat{T}(t_j\hat{h}(\chi)) \to 0$. On the other hand, for $\chi = 0$, $\widehat{S}_j(0) = \widehat{T}(0) = 1$.

From this we conclude that the sequence S_j converges to the Haar measure of G (whose Fourier transform takes the value 1 at point 0 and the value 0 elsewhere); this contradicts what has been seen before and completes the proof.

Proposition 15.4.1 can be immediately extended to groups other than $\mathbf{R}$, in particular to other locally compact fields such as $\mathbf{C}$, $\mathbf{Q}_p$, etc. It suffices to use on these groups dilations similar to the functions $x \mapsto t_jx$ and construct with them progressively faster"windings" of the group on the compact group G.

15.4.2 Adeles and associated canonical windings

Given a real Pisot number θ, there remains to apply Proposition 15.4.1 to construct a homomorphism h of $\mathbf{R}$ into a compact group. Introducing for this purpose the ring of adeles of the field $\mathbf{Q}_p$ is illuminating, as will be seen shortly.

We recall the following fundamental classical results ([29], [5]):

Proposition 15.4.2 *Let k be a field of algebraic numbers, k_A the ring of its adeles, j the canonical injection of k in k_A. Then*

1. *The image $j(k)$ is a discrete subgroup of the locally compact abelian group k_A and the quotient $k_A/j(k)$ is compact.*
2. *For each absolute value ν on k, if k_ν denotes the completion of k for ν, then the canonical composed application*

$$h_\nu: \quad k_\nu \to k_A \to k_A/j(k)$$

is a group homomorphism, with an everywhere-dense image in $k_A/j(k)$. (Strong approximation theorem).

15.4.3 Application to the Salem–Zygmund theorem

Let now θ be a real Pisot number. There is a real absolute value ν on $k = \mathbf{Q}(\theta)$ for which $|\theta|_\nu > 1$. For the other archimedean absolute values ρ distinct from ν we have $|\theta|_\rho < 1$. We identify the group $\mathbf{R}$ on which the set E_θ is defined with k:

$$E_\theta = \{\sum_{k\geq 0} \varepsilon_k \theta^{-k}; \qquad \varepsilon_k = 0 \text{ or } 1\}.$$

Let us denote by Λ_θ the set of all *finite* sums:

$$\Lambda_\theta = \{\sum_{k\geq 1} \varepsilon_k \theta^{k}; \qquad \varepsilon_k = 0 \text{ or } 1\}.$$

and by k'_ν the cofactor of k_ν in k_A (k'_ν is the ring of the adeles of k constructed with all the absolute values different from ν):

$$k_A = k_\nu \times k'_\nu.$$

Let B_1 be the "unit ball" of k'_ν, that is the set of the adeles $\{x_\rho\}$ of k'_ν satisfying $|x_\rho|_\rho \leq 1$ for every absolute value ρ, $\rho \neq \nu$, on k, and let

$$B_\tau = \underbrace{B_1 + \cdots + B_1}_{\tau \text{ terms}}.$$

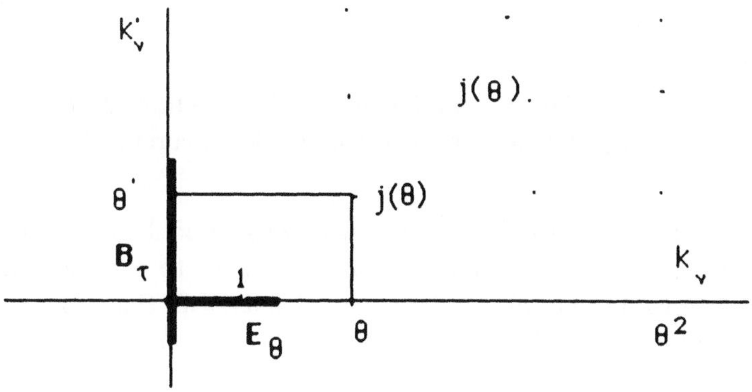

Let θ' be the canonical image of θ in k'_ν. For every finite sequence (ε_k) of numbers equal to 0 or 1 we have, in k_A:

$$\sum_{k\geq 1} \varepsilon_k \theta^k + \sum_{k\geq 1} \varepsilon_k \theta'^k \in j(k).$$

But, for τ large enough,

$$\sum_{k\geq 1} \varepsilon_k \theta'^k \in B_\tau,$$

because, for all ρ, $|\theta'_\rho| \leq 1$ if ρ is ultrametric and $|\theta'_\rho| < 1$ if ρ is archimedean.

It follows that, in k_A, Λ_θ is congruent, modulo $j(k)$, to a subset of B_τ. Let us then suppose that $|\theta|_\nu > 2$; the Lebesgue (or Haar) measure of E_θ in k_ν is null. From Fubini's theorem it follows then that the Haar measure in k_A of $E_\theta + B_\tau$ is zero. And, $j(k)$ being countable, the measure of the set $H = E_\theta + B_\tau + j(k)$ is also zero.

The canonical image in $k_A/j(k)$ of the compact set $E_\theta + B_\tau$ is therefore a compact set of measure zero and is then a proper compact subset K_θ of $k_A/j(k)$.

To conclude the proof we apply Proposition 15.4.1, taking

- as compact group: the group $k_A/j(k)$;
- as homomorphism: the canonical homomorphism h_ν of k_ν into $k_A/j(k)$;
- as proper compact subset of G: the compact K_θ;
- as sequence $(t_j)_{j\geq 1}$: the sequence $(\theta^j)_{j\geq 1}$.

By observing that, for every j,

$$\theta^j E_\theta \subset \Lambda_\theta + E_\theta \quad \text{in } k_\nu,$$

we verify that $h_j(E_\theta) \subset K_\theta$.

We have proved that if $|\theta| > 2$ and if θ is a Pisot number, E_θ is a set of uniqueness. This completes the proof of the whole Theorem 15.3.1 of Salem–Zygmund. ∎

Remark: If the aim is not to obtain an easy extension of the method to Pisot–Chabauty numbers in $\mathbf{Q}_p$ it would perhaps be simpler to consider only archimedean absolute values and obtain homomorphisms of $\mathbf{R}$ that take values in a group isomorphic to $\mathbf{T}^n$ (n being the degree of θ) as in the classical proof.

For this it suffices to replace k_A by $k_A^\infty = k_\nu \times k''_\nu$, where k''_ν is the product $k_{\nu_2} \times \cdots \times k_{\nu_m}$, $\nu_2, \ldots, \nu_m$ the archimedean absolute values on k distinct from ν. Then $j(k)$ is replaced by the image in k_A^∞ of the ring $\mathcal{A}$ of algebraic integers of k. The quotient $k_A^\infty/j(\mathcal{A})$ is then a compact group isomorphic to $\mathbf{T}^n$

The rest of the argument is unchanged.

15.4.4 Generalizations to other fields

The complex case. As in proposition 15.3.3, let θ be a complex number, non-real, algebraic over $\mathbf{Q}$, satisfying $|\theta| > 1$, and all whose conjugates over $\mathbf{Q}$ non-conjugate over $\mathbf{R}$ have absolute value less than 1. We also suppose that θ is an algebraic integer, that is to say, that $|\theta|_\rho \leq 1$ for every non-archimedean absolute value ρ on $\mathbf{Q}(\theta)$.

The preceding construction can now be repeated with $k_\nu \sim \mathbf{C}$. The remainder of the argument is unchanged and the compact set

$$E_\theta = \{\sum_{k\geq 0} \varepsilon_k \, \theta^{-k}; \ \ \varepsilon_k = 0 \ \text{ or } \ 1\}$$

is a set of uniqueness if it has measure zero in $\mathbf{C}$.

The p-adic case. Let $\theta \in S_p^0$ be a Pisot–Chabauty number in $\mathbf{Q}_p$. To show that the subset $E_\theta = \{\sum_{k\geq 0} \varepsilon_k \, \theta^{-k}; \ \ \varepsilon_k = 0 \ \text{ or } \ 1\}$ of $\mathbf{Q}_p$ is a set of uniqueness it suffices to apply the preceding construction with $k_\nu \sim \mathbf{Q}_p$ (and k'_ν equal to the ring of adeles constructed with all other absolute values, archimedean or not, on $\mathbf{Q}_p$). Here also the proof is at all points the same as above and we get that E_θ is a set of uniqueness.

Adapting this to finite extensions of $\mathbf{Q}_p$ is immediate: the field k_ν used is precisely isomorphic to the extension (see[26]).

For the ring of adeles, a proof of the Salem–Zygmund theorem had been given in [2].

15.5 A theorem by Senge and Strauss

15.5.1 Salem's problem

In §3 we saw how the behavior of the function $\Gamma_\theta(u) = \prod_{k\geq 0} \cos \pi u \theta^{-k}$ at infinity depends on θ belonging or not to S. Salem ([21, p. 62]) asked what the behavior at infinity of a product $\Gamma_\theta \Gamma_\phi$ was for two Pisot numbers θ and ϕ.

The solution of this problem is due to Senge and Strauss ([28]):

Theorem 15.5.1 *Let θ and ϕ be two Pisot numbers. Then*

$$\lim_{u\to\infty} \Gamma_\theta(u)\Gamma_\phi(u) = 0$$

if and only if $\log \theta / \log \phi$ is an irrational number.

We shall reproduce the proof of Senge and Strauss which is based on an a result of Mahler on approximation by algebraic numbers.

15.5.2 Proof of the necessary condition

If $\log\theta/\log\phi$ is a rational number there exist two integers m and n such that $\theta^m = \phi^n$. Let us denote by ω this common value, and set

$$A = \prod_{k\geq 1} \cos \pi\theta^k$$

and

$$B = \prod_{k\geq 1} \cos \pi\phi^k,$$

which are both non-zero, because θ^k and ϕ^k, being powers of Pisot numbers, are algebraic integers and therefore cannot be equal to $1/2$ modulo 1.

We have

$$\Gamma_\theta(\omega^r)\Gamma_\phi(\omega^r) = \prod_{k=1}^{mr} \cos(\pi\theta^k) \prod_{k=1}^{mr} \cos(\pi\phi^k)\Gamma_\theta(1)\Gamma_\phi(1),$$

and then

$$|\Gamma_\theta(\omega^r)\Gamma_\phi(\omega^r)| \geq |A||B||\Gamma_\theta(1)||\Gamma_\phi(1)|,$$

for all r.

Therefore the product $\Gamma_\theta\Gamma_\phi$ does not tend to zero at infinity.

15.5.3 Sequences (u_k) for which $\liminf |\Gamma_\theta(u_k)| > 0$

Lemma 15.5.1 *Let $\theta > 1$ and (u_k) be an increasing sequence tending to infinity and such that $|\Gamma_\theta(u_k)| \geq \delta > 0$ for every k. It is then possible to extract a subsequence (still denoted by (u_k)) of the type*

$$u_k = \lambda_1\theta^{m_1(k)} + \cdots + \lambda_s\theta^{m_s(k)} + \lambda_{s+1}(k)\theta^{m_{s+1}(k)},$$

where:

the integer s is fixed (independent of k),

for $1 \leq i \leq s$, $\lambda_i \in \mathbf{Q}(\theta)$ is fixed with $1 \leq |\lambda_i| \leq \theta$,

for $i = s+1, 1 \leq |\lambda_{s+1}(k)| < \theta$ for all k;

and

$$\textit{for } 1 \leq i \leq s, \qquad m_i(k) - m_{i+1}(k) \longrightarrow \infty;$$
$$\textit{for } 1 \leq i \leq s-1, \qquad m_{i+1}(k)/m_i(k) \longrightarrow 1;$$
$$\textit{for } i = s, \qquad \liminf(m_{s+1}(k)/m_s(k)) < 1\,.$$

Proof: Let
$$u_k = \lambda_1(k)\theta^{m_1(k)}$$
with $1 \leq \lambda_1(k) < \theta$.

We extract a subsequence (denoted from now on in the same way as the original sequence) for which $\lambda_1(k) \longrightarrow \lambda_1$ $(1 \leq \lambda_1 \leq \theta)$.

From
$$\prod_{j=o}^{m_1(k)} |\cos \pi\lambda_1(k)\theta^j| \geq |\Gamma_\theta(u_k)| \geq \delta > 0$$
it follows that
$$\sum_{j=0}^{m_1(k)} \sin^2(\pi\lambda_1(k)\theta^j) \leq \log\frac{1}{\delta^2}\,.$$
If we choose $k' \geq k$ we get
$$\sum_{j=0}^{m_1(k)} sin^2(\pi\lambda_1(k')\theta^j) \leq \sum_{j=0}^{m_1(k')} sin^2(\pi\lambda_1(k')\theta^j) \leq \log\frac{1}{\delta^2}\,.$$
Hence, making first k' tend to infinity, then k,
$$\sum_{j=0}^{\infty} \sin^2(\pi\lambda_1\theta^j) \leq \log\frac{1}{\delta^2}\,.$$
From this it follows that $\lambda_1 \in \mathbf{Q}(\theta)$.

We now write
$$u_k = \lambda_1\theta^{m_1(k)} + \lambda_2(k)\theta^{m_2(k)},$$
with $1 \leq |\lambda_2(k)| < \theta$.

It is clear that $m_1(k) - m_2(k) \longrightarrow \infty$. If $\liminf m_2(k)/m_1(k) < 1$ the construction is finished (and $s = 1$). If not, we extract a subsequence in such a way that $m_2(k)$ increases to infinity and $\lambda_2(k) \longrightarrow \lambda_2$, $1 \leq |\lambda_2| \leq \theta$ and take δ such that
$$\delta \leq |\Gamma_\theta(u_k)| \leq \prod_{j=0}^{m_2(k)} |\cos \pi(\lambda_1\theta^{m_1(k)} + \lambda_2(k)\theta^{m_2(k)})\theta^{-j}|.$$

Hence, after a change in the summation index

$$\sum_{j=0}^{m_2(k)} \sin^2 \pi(\lambda_1 \theta^{m_1(k)-m_2(k)+j} + \lambda_2(k)\theta^j) \leq \log \frac{1}{\delta^2}.$$

We again take $k' > k$. Then

$$\sum_{j=0}^{m_2(k)} \sin^2 \pi(\lambda_1 \theta^{m_1(k')-m_2(k')+j} + \lambda_2(k')\theta^j) \leq \log \frac{1}{\delta^2}.$$

When k' tends to infinity $\theta^{m_1(k')-m_2(k')+j}$ tends to 1 modulo 1 so we get

$$\sum_{j=0}^{\infty} \sin^2(\pi\lambda_2\theta^j) \leq \log \frac{1}{\delta^2},$$

which proves that $\lambda_2 \in \mathbf{Q}(\theta)$.

We define in the same way $\lambda_3, m_3, \ldots, \lambda_s, m_s$ until we arrive at an index s for which $m_{s+1}(k)$ satisfies

$$\liminf \frac{m_{s+1}(k)}{m_s(k)} < 1.$$

This must occur for a finite value of s. For consider the following inequality, for an integer a and for $u_k = \lambda_1\theta^{m_1(k)} + \cdots + \lambda_s\theta^{m_s(k)} + \lambda_{s+1}(k)\theta^{m_{s+1}(k)}$:

$$\begin{aligned}
\delta &\leq |\Gamma_\theta(u_k)| \\
&\leq |\cos \pi u_k \theta^{-m_1(k)-a}| \cdots |\cos \pi u_k \theta^{-m_s(k)-a}| \\
&\leq |\cos \pi(\lambda_1\theta^{-a} + \lambda_2\theta^{m_2(k)-m_1(k)-a} + \cdots)| \times \cdots \\
&\quad \cdots \times |\cos \pi(\lambda_1\theta^{m_1(k)-m_s(k)-a} + \cdots + \lambda_s\theta^{-a} + \lambda_{s+1}(k)\theta^{m_{s+1}(k)-m_s(k)-a})|.
\end{aligned}$$

If we now take a large enough so that $1/\theta^{a-1} \leq 1/2$ and then k so big that all quantities except $\lambda_1\theta^{-a}, \ldots \lambda_s\theta^{-a}$ are very small modulo 1, we see that

$$\delta \leq |\cos \pi\theta^{-a}|^s$$

or

$$s \leq \frac{\log \delta}{\log|\cos(\pi\theta^{-a})|}.$$

This completes the proof of Lemma 15.5.1. ∎

15.5.4 *An algebraic relation between θ and ϕ*

Let θ and ϕ be two Pisot numbers, satisfying $\liminf |\Gamma_\theta(u)\Gamma_\phi(u)| > 0$. Then

$$\liminf(\min(|\Gamma_\theta(u)|, |\Gamma_\phi(u)|)) \geq \delta > 0.$$

By Lemma 15.5.1 it is possible to find a sequence u_k of the form

$$\begin{aligned} u_k &= \lambda_1\theta^{m_1(k)} + \cdots + \lambda_s\theta^{m_s(k)} + \lambda_{s+1}(k)\theta^{m_{s+1}(k)} \\ &= \mu_1\phi^{n_1(k)} + \cdots + \mu_t\phi^{n_t(k)} + \mu_{t+1}(k)\phi^{n_{t+1}(k)} \quad , \end{aligned}$$

where the μ_j and n_j satisfy conditions similar to the conditions given on the λ_i and m_i in the lemma. More precisely let us suppose that we have, for $c < 1$ and every k,

$$\frac{m_{s+1}(k)}{m_s(k)} \leq c \quad \text{and} \quad \frac{n_{t+1}(k)}{m_t(k)} \leq c\,.$$

We then have the following lemma.

Lemma 15.5.2 *The equality*

$$\lambda_1\theta^{m_1(k)} + \cdots + \lambda_s\theta^{m_s(k)} = \mu_1\phi^{n_1(k)} + \cdots + \mu_t\phi^{n_t(k)}$$

is satisfied for infinitely many values of k.

Proof. For sake of simplicity in what follows, we will not write the index k. For all $\varepsilon > 0$ we have, from a certain rank on

$$m_s \leq m_1 \leq m_s(1+\varepsilon) \text{ and } n_t \leq n_1 \leq n_t(1+\varepsilon)\,.$$

As

$$\lim_{k\to\infty} \frac{\lambda_1\theta^{m_1}}{\mu_1\phi^{n_1}} = 1\,,$$

it is possible to find constants $c_1$and c_2, independent of k with

$$c_1\theta^{-\varepsilon m_s} \leq \frac{\phi^{n_t}}{\theta^{m_s}} \leq c_2\theta^{\varepsilon m_s}.$$

Therefore

$$\begin{aligned} |\lambda_1\theta^{m_1} + \cdots + \lambda_s\theta^{m_s} - \mu_1\phi^{n_1} - \cdots - \mu_t\phi^{n_t}| &= |\lambda_{s+1}\theta^{m_{s+1}} - \mu_{t+1}\phi^{n_{t+1}}| \\ &= \mathcal{O}(\theta^{cm_s}) + \mathcal{O}(\phi^{cn_t}) \\ &= \mathcal{O}(\theta^{c(1+\varepsilon)m_s}). \end{aligned}$$

This means that

$$\left|1 - \frac{\phi^{n_t}}{\theta^{m_s}} \frac{\mu_1\phi^{n_1-n_t} + \cdots + \mu_t}{\lambda_1\theta^{m_1-m_s} + \cdots + \lambda_s}\right| = \mathcal{O}(\theta^{[c(1+\varepsilon)-1]m_s}).$$

We shall prove that this bound implies that the left member is zero from a certain rank on by applying the following theorem, which is a variant of a theorem of Mahler–Lang ([9]).

Lemma 15.5.3 *If K is an algebraic number field of finite degree over $\mathbf{Q}$, ξ an algebraic number over K, $\omega_1, \ldots, \omega_r$ elements of K , then, for all $\delta > 0$ there exist at most a finite number of elements x of K and a finite number of integers $a_1, \ldots, a_r$ such that*

$$0 < |\xi - \omega_1^{a_1} \ldots \omega_r^{a_r} x| < \frac{1}{H(\omega_1)^{a_1\delta} \cdots H(\omega_r)^{a_r\delta} H(x)^{2+\delta}} ,$$

the height $H(x)$ of the elements of K being the product $\prod_\nu \max(1, |x|_\nu)$ taken on a complete system of absolute values on K satifying the product formula.

The lemma is here applied to $K = \mathbf{Q}(\theta, \phi)$ of degree d on $\mathbf{Q}$, to $\xi = 1$, to $\omega_1 = \phi$, and to $\omega_2 = \theta^{-1}$. For all $\delta > 0$ there exists a constant $D > 0$ such that the inequality

$$|1 - \phi^n \theta^{-m} x| < \frac{D}{H(\phi)^{n\delta} H(\theta)^{m\delta} H(x)^{2+\delta}}$$

implies that $\phi^n \theta^{-m} = 1$.

Let us verify that this is what happens for $n = n_t(k)$, $m = m_s(k)$ and

$$x = \frac{\mu_1\phi^{n_1-n_t} + \cdots + \mu_t}{\lambda_1\theta^{m_1-m_s} + \cdots + \lambda_s} ,$$

when k is large enough.

As θ and ϕ are Pisot numbers we obtain immediatly

$$H(\theta) \le \theta^d \quad \text{and} \quad H(\phi) \le \phi^d .$$

For evaluating $H(x)$ we may, by multiplying the coefficients λ_i and μ_j by an integer, suppose that they are algebraic integers. Then from

$$H(x) \le H(\lambda_1\theta^{m_1-m_s} + \cdots + \lambda_s)\, H(\mu_1\phi^{n_1-n_t} + \cdots + \mu_t)$$

we deduce that for a certain constant C (independent of k)

$$H(x) \leq C\theta^{(m_1-m_s)d}\phi^{(n_1-n_t)d}.$$

We have then, for ε and δ smaller than 1

$$\begin{aligned}\frac{D}{H(\theta)^{m_s\delta}H(\phi)^{n_t\delta}H(x)^{2+\delta}} &\geq \frac{D}{C}\theta^{-m_sd\delta}\phi^{-n_td\delta}\theta^{(m_s-m_1)d(2+\delta)}\phi^{(n_t-n_1)d(2+\delta)} \\ &\geq \frac{D}{C}\theta^{-m_sd\delta}\theta^{-m_s(1+\varepsilon)d\delta}\theta^{-m_sd(2+\delta)\varepsilon(1+(1+\varepsilon))} \\ &\geq \frac{D}{C}\theta^{-m_s(3\delta+9\varepsilon)d}\ .\end{aligned}$$

To conclude, given the bound

$$|1-\phi^{n_t}\theta^{-m_s}x| = \mathcal{O}(\theta^{m_s[c(1+\varepsilon)-1]})$$

obtained above, it suffices to choose ε and δ small enough to have

$$c(1+\varepsilon)-1 < -(3\delta+9\varepsilon)d.$$

15.5.5 End of the proof

We now know that the equality

$$\lambda_1\theta^{m_1(k)} + \cdots + \lambda_s\theta^{m_s(k)} = \mu_1\phi^{n_1(k)} + \cdots \mu_t\phi^{n_t(k)} \qquad (1)$$

is true for infinitely many values of k. We will deduce that $\log\theta/\log\phi$ is rational by studying separately two cases.

• **If θ and ϕ are units:** From equation (1) we see that if $|\theta|_\nu < 1$ for an archimedean absolute value ν on $\mathbf{Q}(\theta,\phi)$ then $|\phi|_\nu \leq 1$ and therefore $|\phi|_\nu < 1$ because $\phi \in S$. That is to say, for all archimedean absolute values other than the natural absolute value, the ratio

$$\frac{|\theta|_\nu^{m_s}}{|\phi|_\nu^{n_t}} = \left|\frac{\lambda_1\theta^{m_1-m_s} + \cdots + \lambda_s}{\mu_1\phi^{n_1-n_t} + \cdots + \mu_t}\right|_\nu$$

takes its values in a compact subset of $\mathbf{R}_+$; but then, as θ^{m_s}/ϕ^{n_t} is a unit, it follows from the product formula that the natural absolute value $|\theta^{m_s}/\phi^{n_t}|$ will also take its values in a compact subset of $\mathbf{R}_+$.

So the numbers θ^{m_s}/ϕ^{n_t} cannot take more than a finite number of values. Hence for distinct numbers k and k' we have

$$\frac{\theta^{m_s(k)}}{\phi^{n_t(k)}} = \frac{\theta^{m_s(k')}}{\phi^{n_t(k')}},$$

and therefore

$$\frac{\log\theta}{\log\phi} \in \mathbf{Q}.$$

• **If θ is not a unit:** There exists a non-archimedean absolute value ν on $\mathbf{Q}(\theta,\phi)$ for which $|\theta|_\nu < 1$. From equation (1) it follows that $|\phi|_\nu \le 1$. In fact, necessarily $|\phi|_\nu < 1$, because if $|\phi|_\nu = 1$, then, for all $\varepsilon > 0$, we would have

$$\begin{aligned}|\mu_1\phi^{n_1} + \cdots \mu_t\phi^{n_t}|_\nu &= |\mu_1\phi^{n_1-n_t} + \cdots + \mu_t|_\nu \\ &\ge |N(\mu_1\phi^{n_1-n_t} + \cdots + \mu_t)|^{-1} \\ &\ge C\phi^{-(n_1-n_t)d} \ge C\phi^{-\varepsilon n_t d} \\ &\ge \theta^{-2\varepsilon m_s d}.\end{aligned}$$

But the left member of equation (1) is not greater than $C|\theta|_\nu^{m_s}$. (The constants in the different lines are different.)

That would give

$$|\theta|_\nu \ge C^{1/m_s}\theta^{-2\varepsilon d},$$

which is absurd for ε small enough and $m_s(k)$ large enough. Then if $|\theta|_\nu < 1$ then, necessarily, $|\phi|_\nu < 1$.

From Equation (1) it follows that from a certain rank on,

$$|\lambda_s|_\nu|\theta|_\nu^{m_s(k)} = |\mu_t|_\nu|\phi|_\nu^{n_t(k)}.$$

Let us consider the couples of integers (m,n) of the form

$$m = m_s(k) - m_s(k_0) \quad \text{and} \quad n = n_t(k) - n_t(k_0)$$

and let (m',n') be the smallest.

From the relations

$$|\theta|_\nu^m = |\phi|_\nu^n,$$

$$|\theta|_\nu^{m'} = |\phi|_\nu^{n'},$$

it follows that

$$\frac{m}{n} = \frac{m'}{n'} .$$

Consequently there exists a coefficient α such that $(m,n) = \alpha(m',n')$. Given $\varepsilon > 0$ we have seen that we can write, from a certain rank on,

$$c_1 \theta^{-\varepsilon m_s(k)} \leq \frac{\phi^{n_t(k)}}{\theta^{m_s(k)}} \leq c_2 \theta^{\varepsilon m_s(k)}$$

for two constants c_1, c_2. Then, for constants d_1, d_2,

$$d_1 \theta^{-\varepsilon m} \leq \frac{\phi^n}{\theta^m} \leq d_2 \theta^{\varepsilon m} ,$$

and

$$d_1^{1/\alpha} \theta^{-\varepsilon m'} \leq \frac{\phi^{n'}}{\theta^{m'}} \leq d_2^{1/\alpha} \theta^{\varepsilon m'} .$$

By taking large α, and then small ε , we see that $\phi^{n'}/\theta^{m'} = 1$ and therefore that $\log\theta/\log\phi$ is rational.

This concludes the proof of Theorem 15.5.1. ■

Remark: This result may be used to obtain examples of sets of multiplicity of the type $E_\theta + E_\phi$. In [10] an application to generalizations of these sets can be found.

References

[1] N.A. BARY, *Treatise on Trigonometric Series, vol. I and II.* (translated from Russian); Pergamon Press, New York, (1964).

[2] F. BERTRANDIAS, Ensembles remarquables d'adèles algébriques. *Thèse. Bull. Soc. Math. France. Mémoire 4*, (1964).

[3] F. BRUHAT, Distributions sur un groupe localement compact, *Bull. Soc. Math. France*, 89, (1961), 43-75.

[4] G. CANTOR, Über die Ausdehnung eines Satzes aus der Theorie der trigonometrischen Reihen. *Math. An.*, 5, (1872), 123-132.

[5] J.W.S. CASSELS AND A. FRÖLICH, editors, *Algebraic Number Theory.* Academic Press, London, (1967).

[6] L.J. GOLDSTEIN, *Analytic Number Theory.* Prentice Hall, London, (1971).

[7] J.P. KAHANE, *Ensembles parfaits et séries trigonométriques.* Hermann, Paris, (1972).

[8] A.S. KECHRIS AND A. LOUVEAU, Descriptive set theory and the structure of sets of uniqueness. *London Math. Soc. lectures notes, 128,* Cambridge Univ. Press, (1987).

[9] S. LANG, On a theorem of Mahler. *Mathematika*, 7, (1960), 139-140.

[10] M. LAURENT, Exemples d'ensembles de multiplicité. *C. R. Acad. Sc. Paris*, 292, (1981), 367-370.

[11] N. LOHOUÉ, Nombres de Pisot et synthèse harmonique dans les algèbres $A_p(\mathbf{R})$. *C. R. Acad. Sc. Paris*, 270, (1971), 37-52.

[12] K. MAHLER, *Lectures on diophantine approximations.* Notre Dame, (1961).

[13] D. MENČOV, On uniform convergence of Fourier series, (in Russian).*Sbornik*, 11, (1942), 67-96.

[14] Y. MEYER, *Nombres de Pisot, nombres de Salem et analyse harmonique.* Lect. Notes 117, Springer Verlag, (1970).

[15] Y. MEYER, Nombres algébriques et analyse harmonique. *An. Scient. Ecole Norm. Sup. 3*(1970), 75-110.

[16] Y. MEYER, Nombres de Pisot et la synthèse harmonique. *An. Scient. Ecole Norm. Sup. 3* (1970), 235-246.

[17] Y. MEYER, Les nombres de Pisot et la Synthèse spectrale. *Bull. Soc. math. France, Mémoire 37*, (1974), 117-20.

[18] Y. MEYER, *Algebraic Numbers and Harmonic Analysis.* North Holland, (1972).

[19] I.I. PYATETSKI-SHAPIRO, On the problem of uniqueness of expansion of a function in a trigonometric series. (In Russian) *Učenyie zapiski Mosc.* 5, (1952), 54-72; Supplement to the work ... *ibid.* 7, (1954) 78-97.

[20] A. RAJCHMAN, Sur l'unicité du développement trigonométrique. *Fundam. Math.* 3, (1922), 286-302.

[21] R. SALEM, *Algebraic Numbers and Fourier Analysis.* Heath, Boston (1963).

[22] R. SALEM, (I) Sets of uniqueness and sets of multiplicity.*Trans.Amer. Math. Soc.* 54, (1943), 218-228; (II) *ibid.*, 56, (1944), 32-49; (III) Rectification to the papers: Sets of uniqueness and sets of multiplicity (I) and (II). *ibid.*, 63, (1948), 595-598.

[23] R. Salem and A. Zygmund, Sur un théorème de Piatecki-Shapiro. *C. R. Acad. Sc. Paris,* 240, (1955), 2040-2042.

[24] J.P. Schreiber, Sur les nombres de Chabauty-Pisot-Salem des extensions algébriques de $\mathbf{Q}_p$. *C. R. Acad. Sc. Paris,* 269, (1969), 71-73.

[25] J.P. Schreiber, Une caractérisation des nombres de Pisot-Salem des corps p-adiques. *Bull. Soc. Math. France.* Mémoire 19, (1964), 53-63.

[26] J.P. Schreiber, *Approximations diophantiennes et problèmes additifs dans les groupes abéliens localement compacts.* Thèse, Orsay, (1972).

[27] J.P. Schreiber, Nombres de Pisot et travaux d'Yves Meyer. *Séminaire Bourbaki, 22e année, Paris* n° 379, (1969–70).

[28] M.G. Senge and E.G. Strauss, P-V numbers and sets of multiplicity. *Periodica Math. Hungarica,* 3, (1973) 93-100.

[29] A. Weil, *Basic Number Theory.* Springer-Verlag, Berlin, (1967).

[30] A. Zygmund, *Trigonometric Series.* Cambridge Univ. Press, Second ed., (1968).